Structural Design for Fire Safety

Telephone (+44) 1243 779777

Email (for orders and customer service enquiries): cs-books@wiley.co.uk
Visit our Home Page on www.wileyeurope.com or www.wiley.com

Reprinted April and November 2002, August 2004, June 2005, September 2007, June 2008, March 2009
Reprinted with corrections August 2006

Other Wiley Editorial Offices

John Wiley & Sons Inc., 111 River Street, Hoboken, NJ 07030, USA

Jossey-Bass, 989 Market Street, San Francisco, CA 94103-1741, USA

Wiley-VCH Verlag GmbH, Boschstr. 12, D-69469 Weinheim, Germany

John Wiley & Sons Australia Ltd, 33 Park Road, Milton, Queensland 4064, Australia

John Wiley & Sons (Asia) Pte Ltd, 2 Clementi Loop #02-01, Jin Xing Distripark, Singapore 129809

John Wiley & Sons Canada Ltd, 22 Worcester Road, Etobicoke, Ontario, Canada M9W 1L1

British Library Cataloguing in Publication Data

A catalogue record for this book is available from the British Library

ISBN 13: 978 0 471 88993 9 (H/B) ISBN 13: 978 0 471 89060 7 (P/B)

Typeset in 10/12 Times from author's own disks by Dobbie Typesetting Ltd, Tavistock, Devon, England.

Structural Design for Fire Safet

Andrew H. Buchanan
University of Canterbury, New Zealand

JOHN WILEY & SONS, LTD
Chichester • New York • Weinheim • Brisbane • Singapore • Toronto

Contents

Preface

Fires in buildings have always been a threat to human safety. The threat increases as larger numbers of people live and work in larger buildings throughout the world. My interest in the fire resistance of structures was initiated by Brady Williamson in the 1970s when I was completing a Master's degree in structural engineering at the University of California at Berkeley. During my subsequent career as a structural engineer, briefly in California then mainly in New Zealand, I was often involved in problems of fire safety and fire resistance. Frequent projects included designing fire resisting components for many buildings, assisting manufacturers of fire protecting materials, and serving on national fire safety committees.

Soon after I joined the University of Canterbury in the late 1980s, New Zealand became one of the first countries to adopt a performance-based building code, stimulating a demand for qualified fire engineers. This led to the establishment of a Master's Degree in Fire Engineering, where one of the core courses is on structural fire resistance. The lecture notes for that course have grown into this book. Many of my Master's and PhD students have conducted research which has contributed to my knowledge of fire safety, and much of that is reported here.

Preparation of this book would not have been possible without the help of many people and institutions. The New Zealand Fire Service Commission has strongly supported the Canterbury programme since its inception. I wish to thank Charley Fleischmann and my colleagues in the Department of Civil Engineering at the University of Canterbury for continual support and encouragement. The first draft was written when I was a visiting researcher at the National Institute of Standards and Technology near Washington DC, on study leave in 1996.

Many people have provided helpful comments on the manuscript, especially Jean-Marc Franssen, Paul Clancy and Jürgen König. Other help has come from Yngve Anderberg, Charles Clifton, Mario Fontana, Hans Gerlich, Marc Janssens, Peter Moss, Len McSaveney, Mike Spearpoint, and a large number of graduate students. Val Grey and Melody Callahan produced all of the line drawings through many revisions.

This book is only a beginning. The problem of fire safety is very old and will not go away. I hope that this book helps to encourage rational improvements to fire safety in buildings throughout the world.

Andy Buchanan
University of Canterbury
2000

Notation

Greek symbols	Description	Units
α	Stress ratio	
α	Thermal diffusivity	m^2/s
α	Ratio of hot wood strength to cold wood strength	
α_h	Horizontal openings ratio	
α_v	Vertical openings ratio	
β	Reliability index	
β	Measured charring rate	mm/min
β_1	Effective charring rate if corner rounding ignored	mm/min
β_n	Nominal charring rate	mm/min
β_{par}	Charring rate for parametric fire exposure	mm/min
δ	Beam deflection	mm
Δ	Deflection	mm
Δ_L	Maximum permitted displacement	mm
Δ_0	Mid-span deflection of the reference specimen	mm
χ	Buckling factor for columns	
χ_{LT}	Buckling factor for beams	
ε	Strain	
ε_σ	Stress-related strain	
ε_{cr}	Creep strain	
ε_i	Initial strain	
ε_{th}	Thermal strain	
ε_{tr}	Transient strain	
ε	Emissivity	
φ	Configuration factor	
Φ	Strength reduction factor	
Φ_f	Strength reduction factor for fire design	
κ	Elastic curvature	1/m
γ_G	Partial safety factor for dead load	
γ_M	Partial safety factor for material	
γ_Q	Partial safety factor for live load	
Γ	Fictitious time factor	
η	Temperature ratio	
θ	Plastic hinge rotation	radians
ρ	Density	kg/m^3
ρ_i	Density of insulation	kg/m^3
ρ_s	Density of steel	kg/m^3

σ	Stefan–Boltzmann constant	$kW/m^2 K^4$
σ	Stress	MPa
ν_p	Regression rate	m/s
ξ	Reduction coefficient for charring of decks	

Alphabetic symbols

A	Depth of heat affected zone below char layer	mm
A	Depth of rectangular stress block	mm
a_f	Depth of stress block, reduced by fire	mm
a_{fi}	Thickness of wood protection to connections	mm
A	Cross section area	mm^2, m^2
A_f	Floor area of room	m^2
A_{fi}	Area of member, reduced by fire	mm^2, m^2
A_{fuel}	Exposed surface area of burning fuel	m^2
A_h	Area of horizontal ceiling opening	m^2
A_i	Cross sectional area of insulation	mm^2
A_l	Area of radiating surface	m^2
A_r	Cross section area reduced by fire	mm^2, m^2
A_s	Cross sectional area of steel	mm^2
A_t	Total internal surface area of room	m^2
A_v	Window area	m^2
b	Breadth of beam	mm
b_f	Breadth of beam reduced by fire	mm
b	$\sqrt{}$Thermal inertia $= \sqrt{k \rho c_p}$	$W s^{0.5}/m^2 K$
b_v	Vertical opening factor	
b_w	Breadth of web of T-beam	mm
B	Breadth of window opening	m
c	Thickness of char layer	mm
c_e	Effective concrete cover to center of reinforcing	mm
c_i	Specific heat of insulation	J/kg K
c_s	Specific heat of steel	J/kg K
c_p	Specific heat	J/kg K
c_v	Concrete cover to reinforcing	mm
C	Compressive force	kN
d	Depth of beam, effective depth of concrete beam	mm
d	Thickness of timber deck	mm
d	Diameter of circular column or width of square column	mm
d_f	Depth of beam reduced by fire	mm
d_i	Thickness of insulation	mm
d_T	Distance of line of thrust from top surface	mm
D	Depth of compartment	m
D	Thickness of slab of burning wood	m
D_b	Reinforcing bar diameter	mm
e	Eccentricity	mm
e_f	Fuel load energy density (per unit floor area)	MJ/m^2

e_t	Fuel load energy density (per unit area of internal surfaces)	MJ/m^2
E	Modulus of elasticity	GPa
E_k	Characteristic earthquake load	
E_f	Total energy contained in fuel	MJ
f	Factor in concrete-filled steel column equation	
f	Stress	MPa
f^*	Calculated stress in member	MPa
f_a	Allowable stress	MPa
f_b	Characteristic bending strength	MPa
$f_{b,t}$	Characteristic bending strength in fire conditions	MPa
f_c	Crushing strength	MPa
f'_c	Characteristic compressive strength of concrete	MPa
$f'_{c,T}$	Compressive strength of concrete at elevated temperature	MPa
f_t	Characteristic tensile strength	MPa
$f_{t,f}$	Characteristic tensile strength in fire conditions	MPa
f_y	Yield strength at 20°C	MPa
$f_{y,T}$	Yield strength at elevated temperature	MPa
F	Surface area of unit length of steel	m^2
F_c	Crushing load of column	kN
F_{crit}	Critical buckling load of column	kN
F_v	Ventilation factor ($A_v\sqrt{H_v}/A_t$)	$m^{0.5}$
F/V	Section factor	m^{-1}
g	Char parameter	
G	Dead load	
G_k	Characteristic dead load	
h	Slab thickness	mm
h	Height from mid-height of window to ceiling	m
h_c	Convective heat transfer coefficient	$W/m^2 K$
h_r	Radiative heat transfer coefficient	$W/m^2 K$
h_t	Total heat transfer coefficient	$W/m^2 K$
H	Height of radiating surface	m
H_p	Heated perimeter of steel cross section	m
H_r	Height of room	m
H_v	Height of window opening	m
ΔH_c	Calorific value of fuel	MJ/kg
I	Moment of inertia	mm^4
j_d	Internal lever arm in reinforced concrete beam	mm
k	Growth parameter for t^2 fire	
k	Thermal conductivity	W/m K
k_{20}	Factor to convert 5th percentile to 20th percentile	
k_i	Thermal conductivity of insulation	W/m K
k_a	Ratio of allowable strength to ultimate strength	
k_b	Compartment lining parameter	min m^2/MJ
k_c	Compartment lining parameter	min $m^{2.25}/MJ$
k_f	Strength reduction factor for heated wood	
k_{mean}	Factor to convert allowable stress to mean failure stress	
$k_{c,T}$	Reduction factor for concrete strength	

$k_{E,T}$	Reduction factor for modulus of elasticity	
$k_{y,T}$	Reduction factor for yield strength	
k_d	Duration of load factor for wood strength	
k_p	Char factor for parametric fire	
K	Effective length factor for column	
l_1, l_2	Dimensions of floor plan	m
L	Length of structural member	mm
L_f	Factored load for fire design	
L_u	Factored load for ultimate limit state	
L_w	Load for working stress design	
L_v	Heat of gasification	MJ/kg
m	Moisture content	%
m_c	Moisture content as % by weight	%
m_d	Moisture content as % of dry weight	%
$\dot{m}$	Rate of burning	kg/s
M	Mass per unit length of steel cross section	kg
M	Mass of fuel	kg
M	Bending moment	kN.m
M_f	Flexural strength in fire conditions	kN.m
M_n	Nominal moment capacity	kN.m
M_p	Moment capacity of plastic hinge	kN.m
M_y	Bending moment at first yield	kN.m
M^*_{cold}	Design bending moment in cold conditions	kN.m
M^*_{fire}	Design bending moment in fire conditions	kN.m
$M^*_{fire, red}$	Redistributed bending moment in fire conditions	kN.m
N	Axial load	kN
N_{crit}	Critical buckling load	kN
N_n	Nominal axial load capacity	kN
N_f	Axial load capacity in fire conditions	kN
N^*	Design axial force	kN
N^*_{fire}	Design axial force in fire conditions	kN
p	Perimeter of fire exposed cross section	m
P	Axial force causing instability	kN
Q	Rate of heat release	MW
Q_p	Peak heat release rate	MW
Q_{fuel}	Rate of heat release for fuel controlled fire	MW
Q_{vent}	Rate of heat release for ventilation controlled fire	MW
Q	Live load	
Q_k	Characteristic live load	
$\dot{q}''$	Heat flux	kW/m^2
r	Radius of gyration	mm
r	Radius of charred corner	mm
r	Distance from radiator to receiver	m
r_{load}	Load ratio	
R	Load capacity	
R_a	Ratio of actual to allowable load at normal temperature	
R_{cold}	Load capacity in cold conditions	

R_{fire}	Load capacity in fire conditions	
s	Heated perimeter	mm
S	Plastic section modulus	mm^3
S_k	Characteristic snow load	
t	Time	hr, min or sec
t	Thickness of steel plate	mm
t_b	Duration of burning	min
t_h	Time	hour
t_o	Initial char time in parametric fire	min
t_r	Time of fire resistance	min
t_s	Time of fire severity	min
t_v	Time delay	min
T	Axial thrust	kN
T	Flange thickness	mm
T	Temperature	°C
T_c	Concrete temperature	°C
T_f	Fire temperature	°C
T_i	Initial temperature of wood	°C
T_{lim}	Limiting temperature	°C
T_m	Maximum temperature	°C
T_p	Temperature of wood at start of charring	°C
T_0	Ambient temperature	°C
T_s	Steel temperature	°C
T_w	Surface temperature	°C
T_y	Tensile force at yield	kN
U^*	Load effect	
U^*_{fire}	Load effect in fire conditions	
V	Volume of unit length of steel member	m^3
V_c	Shear capacity in cold conditions	kN
V_f	Shear capacity in fire conditions	kN
V	Shear force	kN
V^*	Design shear force	kN
V^*_{fire}	Design shear force in fire conditions	kN
V/F	Effective thickness	mm
W	Width of compartment	m
W	Width of radiating surface	m
W_k	Characteristic wind load	
w	Ventilation factor	
w	Uniformly distributed load on beam	kN/m
w_c	Uniformly distributed load on beam, in cold conditions	kN/m
w_f	Uniformly distributed load on beam, in fire conditions	kN/m
x	Distance	
x	Height ratio	
y	Width ratio	
y_b	Distance from neutral axis to extreme bottom fibre	mm
z	Effective thickness of concrete member	mm
z	Thickness of zero strength layer	mm

z	Load factor	
z	Distance to neutral axis	mm
Z	Elastic section modulus	mm^3
Z_f	Elastic section modulus in fire conditions	mm^3

1

Introduction

1. OVERVIEW

This book is an introduction to the structural design of buildings and building elements exposed to fire. Structural fire resistance is discussed in relation to overall concepts of building fire safety.

The book brings together, from many sources, a large volume of material relating to the fire resistance of building structures. The book starts with fundamentals, giving an introduction to fires and fire safety, outlining the important contribution of structural fire resistance to overall fire safety. Methods of calculating fire severity and achieving fire resistance are described, including fire performance of the main structural materials. The most important parts of the book are the design sections, where the earlier material is synthesized and recommendations are made for rational design of building elements and structures exposed to fires.

This book refers to codes and standards as little as possible. The emphasis is on understanding structural behaviour in fire from first principles, allowing structural fire safety to be provided using rational engineering methods based on national structural design codes.

1.2 OBJECTIVE OF THIS BOOK

A structural engineer who has followed this book should be able to:

- interpret the intentions of code requirements for fire safety,
- understand the concepts of fire severity and fire resistance,
- estimate time–temperature curves for fully developed compartment fires,
- design steel, concrete or timber structures to resist fire exposure,
- assess the fire performance of existing structures.

1.3 TARGET AUDIENCE

This book is primarily written for practising structural engineers and students in structural engineering who need to assess the structural performance of steel, concrete or timber structures exposed to unwanted fires. A basic knowledge of structural mechanics and structural design is assumed. The coverage of fire science in this book is superficial, but sufficient as a starting point for structural engineers and building designers. For more detail, readers should consult recognized texts such as Drysdale (1998), Quintiere (1998) and the Handbook of the Society of Fire Protection Engineers (SFPE, 1995).

This book will help fire engineers in their discussions with structural engineers, and also be useful to architects, building inspectors, code officials, fire-fighters, students, researchers and others interested in building fire safety.

1.4 FIRE SAFETY

Unwanted fire is a destructive force that causes many thousands of deaths and billions of dollars of property loss each year. People around the world expect that their homes and workplaces will be safe from the ravages of an unwanted fire. Unfortunately, fires can occur in almost any kind of building, often when least expected. The safety of the occupants depends on many factors in the design and construction of buildings, including the expectation that certain buildings and parts of buildings will not collapse in a fire or allow the fire to spread.

Fire safety science is a rapidly expanding multi-disciplinary field of study. It requires the integration of many different fields of science and engineering, some of which are summarized in this book.

Fire deaths and property losses could be eliminated if all fires were prevented, or if all fires were extinguished at the size of a match flame. Much can be done to reduce the probability of occurrence, but it is impossible to prevent all major fires. Given that some fires will always occur, there are many strategies for reducing their impact, and some combination of these will generally be used. The best proven fire safety technology is the provision of automatic fire sprinklers because they have been shown to have a very high probability of controlling or extinguishing any fire. It is also necessary to provide facilities for the detection and notification of fires, safe travel paths for the movement of occupants and fire-fighters, barriers to control the spread of fire and smoke, and structures which will not collapse prematurely when exposed to fire. The proper selection, design and use of building materials is very important, hence this book.

1.5 PERFORMANCE-BASED BUILDING CODES

Until recently, most design for fire has been based on *prescriptive* building codes, with little or no opportunity for designers to take a rational engineering approach to the provision of fire safety. Many countries have recently adopted *performance-based* building codes which allow designers to use any fire safety strategy they wish, provided that adequate safety can be demonstrated (Bukowski, 1997). In general terms, a prescriptive code states how a building is to be constructed whereas a performance based code states how a building is to perform under a wide range of conditions (Custer and Meacham, 1997).

Performance-based design is not totally new. Even within some prescriptive codes, there has been the opportunity for performance-based selection of structural assemblies. For example, if a code specifies a floor with a fire resistance rating of two hours, the designer has had the freedom to select from a wide range of listed systems which have sufficient fire resistance. This book provides tools for assessing the fire performance of systems which have not been tested, systems without listed approvals, or systems with different geometry, loads or fire exposure from those tested.

In the development of new codes, many countries have adopted a multi-level code format as shown in Figure 1.1. At the highest levels, there is legislation specifying the overall goals, functional objectives and required performance which must be achieved in all buildings. At a lower level, there is a selection of alternative means of achieving those goals. The three most common options are to comply with a prescriptive 'Acceptable Solution', to comply with an approved standard calculation method, or to perform a performance-based fire engineering design from first principles.

Standard calculation methods have not yet been developed for widespread use, so compliance with performance-based codes in most countries is usually achieved by simply meeting the requirements of the Acceptable Solution ('deemed-to-satisfy' solution), or alternatively carrying out a performance-based 'alternative design' based on fire engineering principles. Alternative designs can often be used to justify variations from the Acceptable Solution in order to provide cost savings or other benefits.

The code environment in New Zealand (described by Buchanan, 1994(a), 1999), is similar to that in England, Australia and some Scandinavian countries. Moves towards performance-based codes are being taken in the United States (IFCI, 2000, SFPE, 2000). Codes are different around the world, but the objectives are similar; that is to protect life and property from the effects of fire (SFPE, 1996, 1998). It is not easy to produce or use performance-based fire codes for many reasons: fire safety is part of a complex system of many interacting variables, there are so many possible strategies that it is not simple to assess performance in quantitative terms, and there

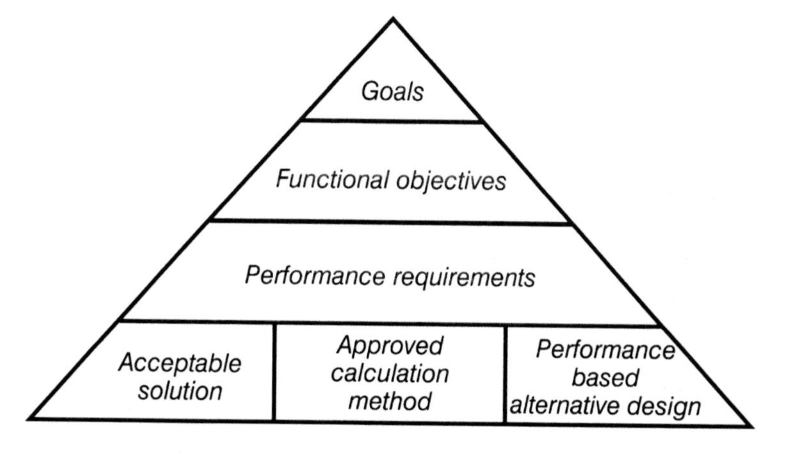

Figure 1.1 Hierarchical relationship for performance-based design

is a lack of information on the behaviour of fires and the performance of people and buildings exposed to fires.

A number of useful documents have been produced to assist users of performance-based codes, including BSI (1997), Buchanan (2001), Custer and Meacham (1997), FCRC (1996) and ISO (1998). This book provides useful additional information, addressing the design of structures for fire safety, which is a small but important segment of the overall provision of fire safety.

1.6 DOCUMENTATION AND QUALITY CONTROL

As the provision of fire safety in buildings moves away from blind adherence to prescriptive codes towards rational engineering which meets specified performance goals, the need for comprehensive documentation and quality control becomes increasingly important. It is recommended (FCRC, 1996, ISO, 1998) that quantitative calculations should be put in context with a 'qualitative design review' which defines the objectives and acceptance criteria for the design, identifies potential hazards and fire scenarios, and reviews the overall design and fire safety features. The review and accompanying calculations should be included in a comprehensive report which describes the building and the complete fire design process (Caldwell *et al.*, 1999). The report should address the installation and maintenance of the fire protection features, and the management of the building to ensure fire safety, with reference to drawings and documentation from other consultants. It is important to consider quality control throughout the design process, construction and eventual use of the building. Changes to the design often occur during construction, and these may affect fire safety if the significance of the original details is not well documented and well understood on the job site. The approving or checking authorities should also prepare a comprehensive report describing the design and the basis on which it is accepted or rejected. Those taking responsibility for design, approval and site inspection must be suitably qualified. The reliability of active and passive fire protection will depend on the quality of the construction, including workmanship and supervision.

1.7 FIRE RESISTANCE

This book concentrates on fire resistance, which is an important part of any design for fire safety. In most buildings, selected structural members and non-structural barriers are provided with fire resistance in order to prevent the spread of fire and smoke, or to prevent structural collapse during an uncontrolled fire.

The provision of fire resistance is just one part of the overall fire design strategy for protecting the lives of occupants and fire-fighters, and for limiting property losses. Fire resistance is often described as *passive* fire protection, which is always ready and waiting for a fire, as opposed to *active* fire protection such as automatic sprinklers which are required to activate after a fire is detected. Design strategies often incorporate a combination of active and passive fire protection measures.

Fire resistance is of little significance in the very early stages of a fire, but becomes increasingly important as a fire gets out of control and grows beyond flashover to full room involvement. The importance of fire resistance depends on the size of the building and the fire safety objectives. To provide life safety, fire resistance is essential in all buildings where a fire could grow large before all the occupants have time to escape. This includes large and tall buildings, and those where the occupants have difficulty in moving. Fire resistance is also important for Fire Service access and rescue, because fire-fighters may need to be inside a building well after all occupants have escaped. Fire resistance is also most important for property protection in buildings of any size, especially if the fire is not controlled with a fire suppression system while it is still small.

1.8 RISK ASSESSMENT

Fire safety is all about risk. The probability of a serious fire in any building is low, but the possible consequences of such a fire are enormous. The objectives of design for fire safety are to provide an environment with an acceptably low probability of loss of life or property due to fire. Tools for quantitative risk assessment in fire are still in their infancy, so most fire engineering design is deterministic. The design methods in this book are deterministic, and must be applied with appropriate safety factors to ensure that they produce an acceptable level of safety.

1.9 ENGINEERING JUDGEMENT

Fire safety engineering is not a precise discipline, because any assessment of safety requires judgement as to how fire and smoke will behave in the event of an unplanned ignition, and how fire protection systems and the occupants of the building will respond. Design to provide fire safety is based on scenario analysis. For any scenario it is possible to calculate some responses, but the level of accuracy can only be as good as the design assumptions, the input data and the analytical methods available. Fire safety engineering is a very new discipline. The precision of calculation methods will improve as the discipline matures, but it will always be necessary to exercise engineering judgement based on experience and logical thinking, using all the information that is available. Analysis of past fire disasters and visits to actual fires and fire-damaged buildings are excellent ways of gaining experience.

1.10 UNITS

This book uses metric units throughout. These are generally SI (Système International) units. The basic SI unit for length is the *metre* (m), for time the *second* (s), and for mass the *kilogram* (kg). Weight is expressed using the *Newton* (N) where one Newton is the force that gives a mass of one kilogram an acceleration of one metre per second per second. On the surface of the Earth, one kilogram weighs approximately 9.81 N because the acceleration due to gravity is 9.81 m/s^2. The basic unit of stress or pressure is the *Pascal* (Pa) which is one Newton per square metre (N/m^2). It is more common to express stress using the megapascal (MPa) which is one meganewton per square metre (MN/m^2) or identically one Newton per square millimetre (N/mm^2).

The basic unit of heat or energy or work is the *Joule* (J) defined as the work done when the point of application of one Newton is displaced one metre. Heat or energy is more often expressed in thousands of Joules (kilojoules (kJ)) or millions of Joules (megajoules (MJ)). The basic unit for rate of power or heat release rate is the *Watt* (W). One Watt is one Joule per second, hence a kilowatt (kW) is a thousand Joules per second and a megawatt (MW) is a megajoule per second.

Temperature is most often measured in degrees *Celsius* (°C), but for some calculations the temperature must be the *absolute* temperature in *Kelvin* (K). Zero degrees Celsius is 273.15 Kelvin, with the same intervals in each system. A list of units and conversion factors is included in Appendix A. A more extensive list of units and conversion factors can be found in the SFPE Handbook (SFPE, 1995).

1.11 ORGANIZATION OF THIS BOOK

This book is organized in a form suitable for teaching a fire safety design course to structural engineering students. Chapter 2 is a discussion of fire safety in buildings, looking at overall strategies and the importance of preventing the spread of fire or structural collapse within the whole context of fire safety. Chapter 3 is an elemental review of combustion and heat transfer for those with no background in those subjects, and Chapter 4 describes fire behaviour in rooms in order to give an indication of the impact of an uncontrolled fire on the building structure. Chapter 5 describes fire severity by comparing post flashover fires with standard test fires, and Chapter 6 describes methods of achieving fire resistance, including standard test results and calculation methods. The structural engineering section of the book starts in Chapter 7 where structural design for fire conditions is contrasted with structural design at normal temperatures, and important concepts such as flexural continuity, moment redistribution and axial restraint are introduced. The next four chapters address the fire behaviour and design of structural materials and assemblies. Chapters 8, 9 and 10 describe steel, concrete and timber structures respectively, with light frame structures of mixed materials covered in Chapter 11. Chapter 12 gives a summary of the recommended fire design methods for structures of different materials.

2

Fire Safety in Buildings

2.1 OVERVIEW

This chapter gives an introduction to the overall strategy for providing fire safety in buildings, and identifies the role of fire resistance and structural performance as important parts of that strategy.

2.2 FIRE SAFETY OBJECTIVES

The primary goal of fire protection is to limit, to acceptable levels, the probability of death, injury, and property loss in an unwanted fire.

The balance between life safety and property protection varies in different countries, depending on the type of building and its occupancy. The earliest fire brigades and fire codes were promoted by insurance companies who were more interested in property protection than life safety. This was certainly the case at the time of the great fire of London in 1666 (Figure 2.1).

A recent trend has been for national codes to give more emphasis to life safety than to property protection. Many codes consider that fire damage to a building is the problem of the building owner or insurer, with the code provisions only intended to provide life safety and protection to the property of other people. Many fire protection features such as automatic sprinkler systems provide both life safety and property protection. The distinction between life safety and property protection becomes important if the owner is unaware of the likely extent of fire damage to the building and contents, even if the building complies with minimum code requirements.

2.2.1 Life Safety

The most common objective in providing life safety is to ensure safe escape. To do this it is necessary to alert people to the fire, provide suitable escape paths, and ensure that they are not affected by fire or smoke while escaping through those paths to a safe place. In some buildings it is necessary to provide safety to people unable to escape, such as those under restraint, in a hospital, or in a place of refuge within the building. People in adjacent buildings must also be protected, and provision must be made for the safety of fire-fighters who enter the building for rescue or fire control purposes.

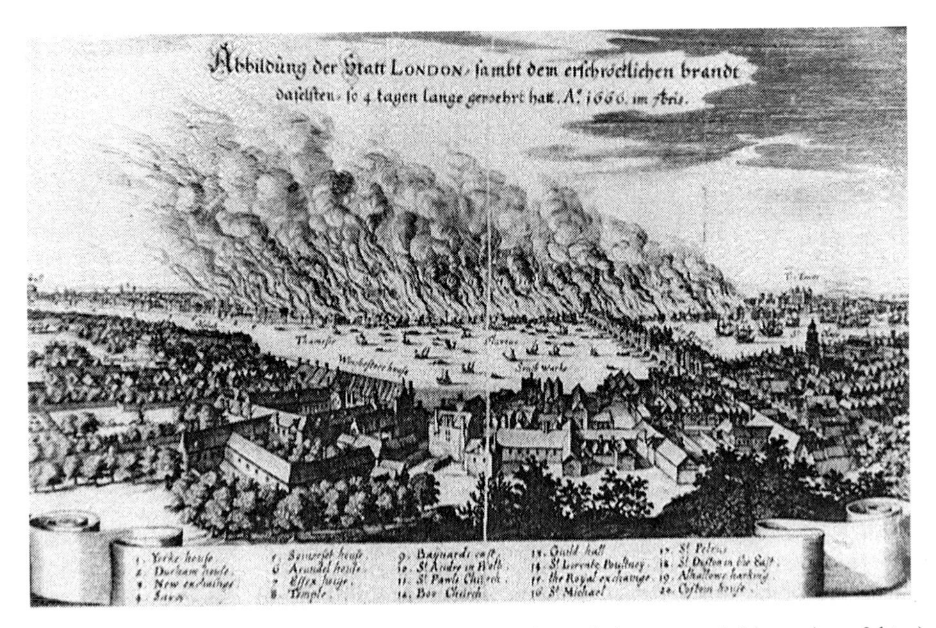

Figure 2.1 The Great Fire of London, 1666 (www.jmccall.demon.co.uk/history/page2.htm)

2.2.2 Property Protection

The objective of protecting property includes protecting the structure and fabric of the building, and the moveable contents. This protection must also apply to neighbouring buildings. An extra level of protection may be necessary if rapid repair and re-use after a fire are important. In many cases an important objective may be to protect intangible items such as possible loss of business or irreplaceable loss of heritage values. A loss disproportionate to the size of the original fire can occur if there is major damage to 'lifelines' such as energy distribution or telecommunications facilities.

2.2.3 Environmental Protection

In many countries an additional objective is to limit environmental damage in the event of a major fire. The primary concerns are emissions of gaseous pollutants in smoke, and liquid pollution in fire-fighting run-off water, both of which can have major environmental impacts. The best way to prevent these emissions is to extinguish any fire while it is small.

All of the above objectives can be met if any fire is extinguished before growing large, which can be accomplished most easily with an automatic sprinkler system.

2.3 PROCESS OF FIRE DEVELOPMENT

Fire safety objectives are usually met with a combination of active and passive fire protection systems. *Active systems* control the fire or fire effects by some action taken by a person or an automatic device. *Passive systems* control the fire or fire effects by

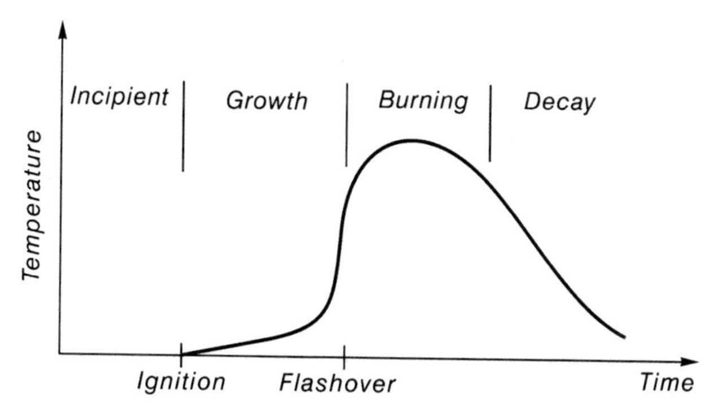

Figure 2.2 Time–temperature curve for full process of fire development

systems that are built into the structure or fabric of the building, not requiring specific operation at the time of a fire. The typical development of a fire in a room is described below as an introduction to the need for fire protection systems.

Figure 2.2 shows a typical time–temperature curve for the complete process of fire development inside a typical room, assuming no fire suppression by sprinklers or fire-fighters. Not all fires follow this development because some fires go out prematurely and others do not reach flashover, especially if the fuel item is small and isolated or if there is not enough air to support continued combustion. If a room has very large window openings, too much heat may flow out the windows for flashover to occur.

Table 2.1 is a summary of the main periods of fire behaviour shown in Figure 2.2. The following brief discussion relates to the figure and the table, as an introduction

Table 2.1 Summary of periods of typical fire development

	Incipient period	Growth period	Burning period	Decay period
Fire behaviour	Heating of fuel	Fuel controlled burning	Ventilation controlled burning	Fuel controlled burning
Human behaviour	Prevent ignition	Extinguish by hand, escape	Death	
Detection	Smoke detectors	Smoke detectors, heat detectors	External smoke and flame	
Active control	Prevent ignition	Extinguish by sprinklers or fire fighters; control smoke	Control by fire-fighters	
Passive control		Select materials with resistance to flame spread	Provide fire resistance; contain fire, prevent collapse	

to the discussion of fire safety strategies later in this chapter and the description of fire behaviour in the following chapter.

2.3.1 Fire Behaviour

In the *incipient* period of fire development, heating of potential fuel is taking place. *Ignition* is the start of flaming combustion, marking the transition to the *growth* period. In the growth period, most fires spread slowly at first on combustible surfaces, then more rapidly as the fire grows, providing radiant feedback from flames and hot gases to other fuel items. If upper layer temperatures reach about 600°C, the burning rate increases rapidly, leading to *flashover* which is the transition to the *burning* period (often referred to as 'full room involvement' or 'fully developed fire'). The rate of burning in the growth period is generally controlled by the nature of the burning fuel surfaces, whereas in the burning period the temperatures and radiant heat flux within the room are so great that all exposed surfaces are burning and the rate of heat release is usually governed by the available ventilation. It is the burning period of the fire which impacts on structural elements and compartment boundaries. If the fire is left to burn, eventually the fuel burns out and temperatures drop in the *decay* period, where the rate of burning again becomes a function of the fuel itself rather than the ventilation.

2.3.2 Human Behaviour

People in the room of origin may see or smell signs of the potential fire during the incipient period when the fuel is being heated by some heat source. Many fires are averted by occupants who prevent ignition by removing the fuel or eliminating the ignition source in the incipient period. After ignition the fire will be more obvious, giving occupants the opportunity to extinguish it while it is very small if they are awake and mobile. Once the fire grows to involve a whole item of furniture or more, it cannot be extinguished by hand, but active occupants may have time for escape, provided that smoke has not blocked the escape routes. Conditions in a room fire become life-threatening during the growth period. After flashover, survival is not possible because of the extreme conditions of heat, temperature and toxic gases. People elsewhere in the building may not know about the fire until it is large, leading to hazardous conditions (Figure 2.3). In order to ensure life safety in the event of a fire, it is essential that the fire be detected, and the occupants be alerted with sufficient information to make a decision to move, and with sufficient time to reach a safe place before conditions become untenable. The SFPE Handbook (SFPE, 1995) gives more information on human behaviour and tenability limits.

2.3.3 Fire Detection

In the incipient period of a fire, human detection is possible by sight or smell. Automatic detection before ignition is possible if a very sensitive aspirating smoke

Figure 2.3 Hotel fire where spread of smoke remote from the fire killed 84 people [MGM Grand Hotel, Las Vegas, 1980] (Coakley *et al.*, 1982)

detector has been installed, which is only likely in special buildings containing very valuable items or equipment. After ignition, a growing fire can be detected by the occupants, or by a smoke detector or heat detector usually located on the ceiling. Smoke detectors are more sensitive than heat detectors, especially for smouldering fires (e.g. mattress fires) where there may be life threatening smoke but little heat. Automatic sprinkler systems are activated by heat detecting devices. After flashover, neighbours will detect smoke and flames coming out of windows or other openings.

2.3.4 Active Control

Active control refers to control of the fire by some action taken by a person or an automatic device. The best form of active fire protection is an automatic sprinkler system which sprays water over a local area under the activated sprinkler head. A sprinkler system will extinguish most fires, and prevent growth of others. A sprinkler system must operate early in a fire to be useful because the water supply system is designed to extinguish only a certain size of fire.

Active control of smoke requires the operation of fans or other devices to remove smoke from certain areas or to pressurise stairwells. Active control of smoke may require sophisticated control systems to ensure that smoke and toxic products are removed from the building and not circulated to otherwise safe areas.

Occupants can prevent ignition or extinguish very small fires. Fire-fighters can actively control or extinguish a fire, only if they arrive before it gets too large. Time is critical because it takes time for detection, time for notification of the fire-fighters, time for travel to the fire and time for locating the fire in the building and setting up water supplies. Fire-fighters usually have insufficient water to extinguish a large post-flashover fire, in which case they can provide control by preventing further spread and extinguishing it during the decay period.

2.3.5 Passive Control

Passive control refers to fire control by systems that are built into the structure or fabric of the building, not requiring operation by people or automatic controls. For pre-flashover fires, passive control includes selection of suitable materials for building contents and interior linings that do not support rapid flame spread in the growth period. In post-flashover fires, passive control is provided by structures and assemblies which have sufficient fire resistance to prevent both spread of fire and structural collapse.

2.4 CONCEPTUAL FRAMEWORK FOR FIRE SAFETY

Building codes are different in every country. In a prescriptive code environment, designers have little choice but to follow a book of rules. With more modern performance-based codes, designers have unlimited freedom to design innovative solutions to fire safety problems, provided that the required levels of safety and performance can be demonstrated to the satisfaction of the approving authorities. Whatever type of code is used, design for fire safety will include a combination of reducing the probability of ignition, controlling the spread of fire and smoke, allowing for occupant escape and fire-fighter access, and preventing structural collapse. It is difficult to visualize or demonstrate safety without a conceptual framework because of the large number of interacting variables. Several related frameworks are described briefly below. Even a simple ‘checklist’ of fire safety and fire protection items can be of considerable assistance in seeing the big picture (Buchanan, 2001, ISO, 1998).

2.4.1 Scenario Analysis

One framework for demonstrating fire safety is scenario analysis. In this method a number of ‘worst case’ scenarios are analysed. In each scenario the likely growth and spread of fire and smoke is compared with detection and occupant movement, taking into account all the active and passive fire protection features and structural behaviour, to establish whether the performance requirements have been satisfied. An overview of scenario analysis is shown in Figure 2.4. This type of scenario analysis is the most often used basis of fire engineering design (Buchanan, 2001, FCRC, 1996).

Within the selected scenarios it is possible to ask a large number of ‘what if?’ questions to find the worst cases and optimize the design. Another more systematic approach is to carry out a *Failure Modes and Effects Analysis* (FEMA), either on the whole fire safety system or on particular components (Custer and Meacham, 1997).

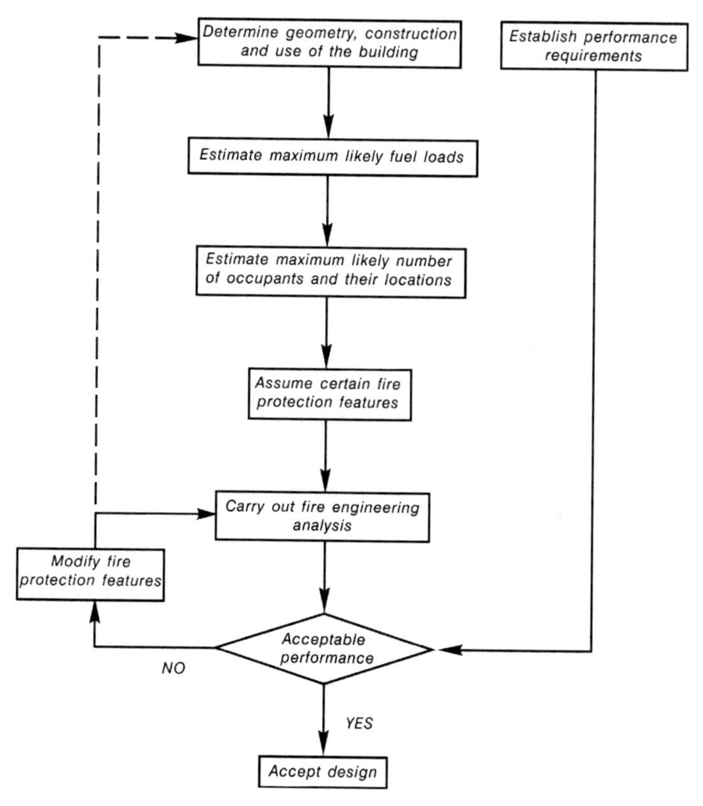

Figure 2.4 Overview of scenario analysis (Reproduced from Buchanan (2001) by permission of Centre for Advanced Engineering)

2.4.2 Quantitative Risk Assessment

In any study of safety there is a need for quantification, in order to answer the question 'how safe?'. Quantitative risk assessment is a rapidly growing discipline which is increasingly being applied to fire safety, although most current performance-based design does not quantify the level of safety.

A risk analysis can be based on existing historical data for the type of building under consideration, but such data are extremely limited. Safety can be quantified using fault tree analysis or event tree analysis if sufficient input data can be derived or estimated. A summary of risk assessment methods for fire safety is given by Watts (1997).

There are a number of computational fire risk assessment programs under development which are able to carry out probabilistic calculations of the scenario analyses described above, in order to quantify the overall expected fire loss and expected risk-to-life (e.g. Beck and Yung, 1994). Such programs are more useful for research and code-writing than for design.

In the absence of simple probabilistic design methods, most design calculations will be made deterministically, with appropriate safety factors applied to provide the required level of safety. Structural designers are very familiar with this process, where design codes provide partial safety factors for applied loads and material strength, calibrated so that the deterministic design process provides sufficient safety. The determination of suitable safety factors for fire design is in its infancy, so there will be many occasions when a large degree of professional judgement may be required by the fire designer and consequently by the approving authority.

2.4.3 Fire Safety Concepts Tree

One of the more durable frameworks for fire safety assessment is the Fire Safety Concepts Tree developed by the National Fire Protection Association (NFPA, 1997). Figure 2.5 shows an edited summary of the Fire Safety Concepts Tree. The following paragraphs give a brief explanation of the tree, as a guide to establishing the relative importance of the various components of a fire safety strategy.

Prevention versus management

Line 2 of the tree states the obvious: fire management is unnecessary if ignition can be prevented, but if not, the impact of the fire must be managed. In reality there will always be unplanned ignitions, but the probability of these can be reduced with fire prevention programmes. Arson is a growing cause of fires which cannot easily be controlled by building designers. Unless stated otherwise, this tree shows alternative strategies, whereby the objectives on one line can be met by any one of the items on the following line.

Manage fire impact

Line 3 shows that managing the impact of a fire can be achieved either by managing the fire itself or by management of exposed persons and property.

Management of exposed persons and property

Line 4 shows that exposed persons and property can be managed by moving them from the building or by defending them in place. The usual strategy is to move people from a building, unless they are incapacitated or under restraint. An intermediate position for very large buildings is to move people to a place of refuge within the building. Most exposed property must be defended in place.

In order for people to move, the fire must be detected, the people must be notified, and there must be a suitable safe path for movement (Line 5). The 'AND' symbol indicates that success in both boxes is required to meet the objective. Human behaviour and escape route design is beyond the scope of this book. Refer to the SFPE Handbook (SFPE, 1995) for more information.

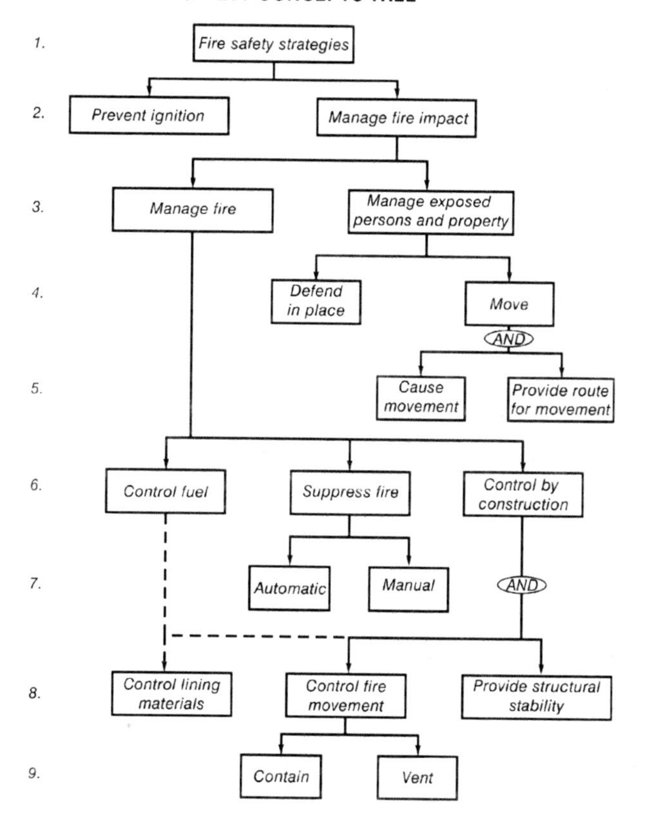

Figure 2.5 Fire safety concepts tree (adapted from NFPA, 1997)

Control of the fire

Line 6 shows three options for managing a fire. In the first case the fuel source can be controlled, by limiting the geometry or amount of fuel. For example this could be a limit on the amount of combustible material stored in a space. The second option is to suppress the fire and the third is to control the fire by construction. Fire suppression is a huge topic beyond the scope of this book, but as shown in Line 7 suppression can either be automatic or manual. In either case suppression depends on early detection of the fire and application of sufficient quantities of appropriate suppressant, usually water.

Control by construction

Control of fire by construction is the subject of this book. Line 8 of the concepts tree shows that in order to control fire by construction it is necessary to both control the movement of the fire **and** provide structural stability. The left-hand box on Line 8 indicates that fire growth and severity can be controlled by limiting the fuel in combustible room linings. This box is connected by dotted lines because, strictly speaking, it should be a sub-set of 'control fuel' from Line 6, but it has been placed on Line 8 because selection and installation of the linings is part of the construction process, rather than a building management issue.

Provide structural stability

The provision of structural stability is essential if a building is to remain standing during a fire, and be easily repaired for subsequent use. Structural stability is also essential to protect people or property elsewhere in the building at the time of the fire. Some elements such as walls and floors have a separating function as well as a load-bearing function. Building elements such as beams and columns only have a load-bearing function. Structural stability in fire is covered in later chapters.

Control fire movement

The two strategies for controlling fire movement are either to contain the fire or vent it to the outside (Line 9). Fire venting is a useful strategy for reducing the impact of fires, especially in single storey buildings (or the top storey of taller buildings). Venting can be by an active system of mechanically operated vents, or a passive system that relies on the melting of plastic skylights. In either case, the increased ventilation may increase the local severity of the fire, but fire spread within the building and the overall thermal impact on the structure will be reduced.

Containment of a fire to prevent spread is the principal tool of passive fire protection. The walls and floors of most buildings are provided with fire resistance in order to contain any fire to the room of origin. Preventing fires growing to a large size is one of the most important components of a fire safety strategy. Radiant spread of fire to neighbouring buildings must also be prevented, by limiting the size of openings in exterior walls. Fire resistance of walls and floors is covered in detail elsewhere in this book.

Smoke movement can also be controlled by venting or containment. Smoke removal is an important strategy in fires whose size has been limited by automatic sprinkler systems. Pressurization and smoke barriers can both be used to contain the spread of smoke in a building (Buchanan, 2001, Klote and Milke, 1992).

2.5 FIRE RESISTANCE

As stated earlier, fire safety objectives are usually met with a combination of active and passive fire protection systems. *Active systems* control the fire or fire effects by some action taken by a person or an automatic device. *Passive systems* control the fire or fire effects by systems that are built into the structure or fabric of the building, not requiring specific operation at the time of a fire. The most important component of passive fire protection is *fire resistance*, which is designed to prevent spread of fire and structural collapse in fire.

2.5.1 Objectives for Fire Resistance

The objectives for providing fire resistance need to be established before making any design, recognizing that fire resistance is only one component of the overall fire safety strategy. Structural elements can be provided with fire resistance for controlling the spread of fire or to preventing structural collapse, or both, depending on their function (Figure 2.6). A review of the requirements in several current or draft performance-based codes (BCA, 1996, BIA, 1992, NKB, 1994) shows a similar approach to the requirements for fire resistance. The objectives for providing fire resistance are outlined below.

- To prevent internal spread of fire, a building can be divided into 'fire compartments' or 'firecells' with barriers which prevent fire spread for the *fire design time*. The many reasons for providing compartmentation include increasing the time available for escape, limiting the area of possible loss, reducing the fire impact on the structure, separating different occupancies, isolating hazards, and protecting escape routes. The separating barriers are usually floors or walls.

- To reduce the probability of fire spread to other buildings, boundary walls must have sufficient fire resistance to remain standing and to contain a fire for the *fire design time*.

- To prevent structural collapse, structural elements must be provided with sufficient fire resistance to maintain stability for the *fire design time*. Prevention of collapse is essential for load-bearing structural members and for load-bearing barriers which also provide containment. Structural fire resistance must be provided to the main load-bearing structural elements, and to secondary elements which support or provide stability to barriers or main members.

- Prevention of collapse is also essential if there are people or property to be protected elsewhere in the building, and for a building which is to be repaired after a fire.

2.5.2 Fire Design Time

The term *fire design time* is not precisely defined. Depending on the importance of the building, the requirements of the owner, and the consequences of a structural collapse or spread of fire, the *fire design time* will be selected by the designer as one of the following:

(1) the time required for occupants to escape from the building,

(2) the time for fire-fighters to carry out rescue activities,

(3) the time for fire-fighters to surround and contain the fire,

(4) the duration of a burnout of the fire compartment with no intervention.

Figure 2.6(a) Severe fire in a warehouse storing foamed plastic materials. Note that the fire has not spread into the office area at the left end of the building

Figure 2.6(b) Collapsed steel beams and damaged concrete masonry wall after the fire. All of the fuel and the timber roof purlins have completely burned away

Figure 2.6(c) View of the light timber frame wall separating the warehouse from the office area. The gypsum board on the fire side has been removed by fire fighters and the gypsum board on the cold side is undamaged. The fire did not spread into the offices

Codes in various countries use these times in different ways for different occupancies. Many small single-storey buildings may be designed to protect the escape routes and to remain standing only long enough for the occupants to escape (time (1)) after which the fire will destroy the building. Alternatively, all very tall buildings, or buildings where people cannot escape, should be designed to prevent major spread of fire and structural collapse for a complete burnout of one or more fire compartments (time (4)). Times (2) and (3) are intermediate times which may be applied to medium sized buildings, to provide life safety or property protection, respectively.

It can be seen that the provision of structural fire resistance may be essential, or unimportant, or somewhere between these two extremes (Almand, 1989). On one hand there may be a major role for the structure so that collapse is unacceptable even in the largest foreseeable fire. This may occur where evacuation is likely to be slow or impossible, or where great value is placed on the building or its contents. On the other hand, there may be virtually no role for the structure so that structural collapse after some time of fire exposure is acceptable where the building can be readily evacuated, there is little value placed on the building and there is no fire threat to adjoining properties.

Design for burnout of a fire compartment is a conservative approach which is likely to be used in many situations. Many modern codes require design of certain buildings for burnout. This book gives design methods which can be used for calculating structural fire resistance for complete burnout, or partial burnout with intervention from fire-fighters or suppression systems.

The *fire design time* must be assessed carefully, because it is not the same as the *fire resistance* time specified by a building code or measured in a fire resistance test. The *fire design time* includes time for ignition, growth and fire spread before flashover. The *fire design time* should include a safety factor to allow for the number of people in the building, the size of the building and the consequences of failure of the building. Schleich (1996) proposes safety factors ranging from 1.0 for small single-storey buildings to 2.5 for large multi-storey buildings.

2.5.3 Trade-offs

One of the difficulties in assessment of fire safety is the extent to which some fire protection measures can be 'traded off' against others. For example, some prescriptive codes allow fire resistance ratings or fire compartment areas to be reduced if an automatic sprinkler system is installed, or they allow travel distances to be increased when smoke or heat detectors or sprinklers are installed. Trade-offs do not apply in a totally performance-based environment, because the designer will produce a total package of fire protection features contributing to the required level of safety. However, in practice, most designs are partly based on prescriptive codes, so it is often necessary to make trade-offs.

It can be difficult to justify trade-offs, especially reductions of fire resistance if automatic sprinkler systems are installed. It is generally accepted that if an automatic suppression system could be relied on to control a fire with total certainty, no fire resistance or passive fire protection would be necessary. However, no system is 100%

effective, so the question is how much fire resistance should be provided for the remote probability that the suppression system fails to operate or fails to control the fire. As an example, it could be argued that if the suppression system fails when street water supplies are destroyed by an earthquake, the resulting fire will have the same severity as if there had been no suppression system at all, so there should be no trade-off for sprinklers.

No codes allow a total trade-off for sprinklers, but many national codes allow a partial trade-off, assuming that in a sprinklered building, the probability of an uncontrolled fire is much less likely than the probability of a sprinkler-controlled fire. Quantitative justification for partial trade-offs is not easy, but two possible probabilistic arguments are as follows.

(1) Many national codes allow a 50% reduction in fire resistance of structural members if the building is sprinklered. A possible justification for this approach is based on safety factors. If, for example, the fire resistance normally specified for a burnout of a fire compartment in an unsprinklered building has an inherent safety factor of 2.0, then in the unlikely event of a fire and a sprinkler failure, that safety factor could be reduced to as low as 1.0, hence the 50% reduction. Such an argument can only be used if the method of specifying fire resistance for unsprinklered buildings is sufficiently conservative in the first instance.

(2) The Eurocodes (EC1, 1994) suggest that for calculating fire resistance, the fuel load in a sprinklered building be taken as 60% of the design fuel load. This approach can be justified by considering sprinkler failure to be such an unlikely event that the fuel load should be the most likely fuel load rather than the 90 percentile fuel load used for design of unsprinklered buildings. Suggestions for additional trade-offs are given by Schleich (1996, 1999).

2.5.4 Repairability and Reserviceability

Repair and reserviceability may be important for some building owners. A building designed to resist a complete burnout will be severely damaged after such a fire, even if the fire is contained and the structure is intact. Most performance-based codes do not require that the structure remains undamaged following a fire. For example, Eurocode 1 (EC1, 1994) states that when designing for a required fire resistance period, the performance of the structure beyond that time need not be considered. A requirement for little or no damage to the building structure may be requested by some codes or some building owners, but this will require a greater level of passive fire protection than required to only prevent collapse.

A reserviceability requirement would limit damage so that the building could be re-occupied with no (or very little) time for repairs. Such a requirement might be imposed on buildings of social, cultural or economic importance. This is only possible with the use of active fire suppression systems such as sprinklers to prevent the fire from becoming large and destructive.

An example of the importance of repairability is the 38 storey Meridian Plaza office tower in Philadelphia which had a serious fire in 1991. Severe damage occurred between levels 22 and 30. The building remained unoccupied for more than five years after the fire because of a dispute about the structural integrity of the structural elements which were damaged and distorted by the fire (Gilvary and Dexter, 1997).

2.6 CONTROLLING FIRE SPREAD

The larger a fire, the greater its destructive potential. Many facets of fire protection are aimed at preventing small fires from becoming large ones. The control of fire movement, or fire spread, is discussed here in four categories; within the room of origin, to other rooms on the same level, to other storeys of the same building, and to other buildings.

2.6.1 Fire Spread Within the Room of Origin

Fire spread within the room of origin depends largely on the heat release rate of the initially burning object. Initial fire spread can be by flame impingement or radiant heat transfer from one burning item to another. As the fire grows, the movement of buoyant hot gases under the ceiling can cause the fire to spread to other parts of the room. Vertical and horizontal fire spread will be greatly increased if the room is lined with combustible materials susceptible to rapid flame spread on the walls and especially on the ceilings. Most countries have prescriptive codes which place limits on the combustibility or flame spread characteristics of linings in particular buildings.

The properties of interest are ignitability, heat release, flame spread and the amount of smoke produced. These are often called the 'early fire hazard' properties or 'reaction to fire' properties. There are many different test methods for assessing the early fire hazard properties of materials in different countries, which makes international comparisons very difficult. In North America the principal test is the ASTM E-84 Steiner Tunnel test using a 7.6 m long tunnel. Most other countries have a variety of tests which expose materials in various sizes to heating by a radiant panel. Recent international developments include the cone calorimeter test of small specimens (100 × 100 mm) and the full-scale room fire test for evaluating the fire performance of room lining materials. The most recent international test method is the Single Burning Item test which will become the test procedure for classifying building products in a harmonized European system. This is an intermediate scale test where two test samples are mounted in a corner configuration where they are subjected to a gas flame ignition source. The rates of heat release and smoke production are measured. All of the above tests have been the subject of much recent research and international standardization.

With regard to ignition and fire spread, unprotected wood-based materials are safer than many plastic or synthetic materials, but are less safe than materials such as paper-faced gypsum plaster or completely non-combustible materials such as concrete. The early fire hazard properties of wood materials can be improved with the use of special paints or pressure treatment.

2.6.2 Fire Spread to Adjacent Rooms

Spread of fire and smoke to adjacent rooms is a major contributor to fire deaths. Fire and smoke movement depends very much on the geometry of the building. If doors are open, they can provide a path for smoke and toxic combustion products to travel from the hot upper layer of the fire room into the next room or corridor. These hot gases can pre-heat the next area leading to subsequent rapid spread of fire.

Keeping doors closed is essential to preventing fire spread from room to room. Doors through fire barriers must be able to maintain the containment function of the barrier through which they pass, whether for smoke control or fire resistance. Door closing devices which operate automatically when a fire is detected are very effective for greatly increasing fire safety. Other recent innovations to improve door performance include smoke control strips to reduce the spread of smoke around the door, and strips of intumescent material that swell when heated to prevent fire spreading through gaps around the door.

Concealed spaces are one of the most dangerous paths for the spread of fire and smoke. A hazardous situation occurs if there are concealed spaces which allow the spread of fire and smoke to adjacent rooms, or even to rooms some distance from the fire. Figure 2.7 shows the spread of smoke through a concealed ceiling cavity. Concealed cavities are a particular problem in old buildings, especially if a number of new ceilings or partitions have been added over the years.

Fire can also spread to adjacent rooms by penetrating the surrounding walls (Figure 2.14). Walls can be designed with sufficient fire resistance to prevent the spread of fully developed fires, but they must be constructed with attention to detail if fire performance is to be ensured. Fire resisting walls must extend through suspended ceilings to the floor or roof above so that the fire does not spread by travelling through a concealed space above the wall. In order to prevent fire spreading over the top of a fire-resisting wall at roof level, the wall can be extended above the roof line to form a parapet, or the roof can be fire-rated for some distance either side of the top of the wall.

A severe fire will find any weakness in a separating barrier, and many such weaknesses are not visible during normal operation of the building. Care must be

Figure 2.7 Spread of smoke and fire through a ceiling cavity

taken to ensure that poor quality workmanship or penetrations for services and fittings do not compromise the performance of fire-resisting walls. The term 'fire-stopping' refers to the sealing of penetrations and cavities through which fire might spread (O'Hara 1994). There are many techniques for fire stopping of penetrations, construction joints and seismic gaps (Abrams and Gustaferro, 1971). Materials for fire stopping include mineral wool, wood blocks, gypsum board, metal brackets and a wide array of excellent proprietary products such as fire-resisting putty, board materials and intumescent pillows and collars (Figure 2.9).

Air-handling ducts which pass through fire resistant walls and floors can create paths for the spread of fire. This can be prevented by the use of fire-resistant insulating duct materials and internal 'fire dampers' which are designed to close off the opening in the event of a fire (see Section 6.7.9).

2.6.3 Fire Spread to Other Storeys

Fire can spread to other storeys by a variety of paths, inside and outside the building. Internal routes for fire spread include failure of the floor/ceiling assembly, and fire spread through vertical concealed spaces, service ducts, shafts or stairways. Vertical services must either be enclosed in a protected duct or have fire-resistant penetration closers at each floor level, as shown in Figures 2.8 and 2.9. Vertical shafts and stairways must be fire-stopped or separated from the occupied space at each level to avoid producing a path for spread of fire and smoke from floor to floor. A particularly dangerous situation can arise if there are interconnected horizontal and vertical concealed spaces, within the building or on the façade.

Another potential path for vertical fire spread is through gaps at the junction of the floor and the exterior wall, just inside the façade. This is particularly important for 'curtain-wall' construction where the exterior panels are not part of the structure.

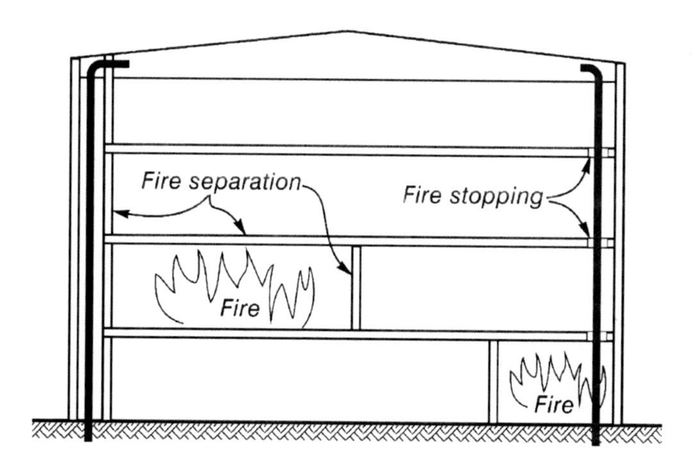

Figure 2.8 Fire separation of vertical services (Reproduced from Buchanan (2001) by permission of Centre for Advanced Engineering)

Figure 2.9 Fire protection to service penetrations through a fire resisting floor. Many proprietary products are available for preventing fire spread through service penetrations

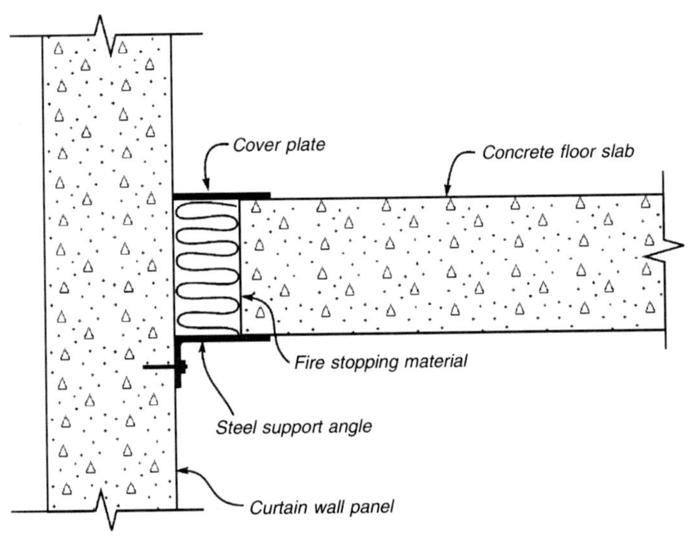

Figure 2.10 Fire stopping between slab and curtain wall

A possible detail is shown in Figure 2.10. Careful detailing and installation is necessary to ensure that the entire gap is sealed, especially at corners and junctions, to eliminate any possible path for fire spread (Gustaferro and Martin, 1988). Gaps such as these between structural and non-structural elements are often filled with non-rigid fire-stopping materials to allow for seismic or thermal movement. The filling material must be able to provide the necessary fire resistance both before and after the anticipated movement (including earthquake movement in seismic areas). Filling material may be mineral or ceramic fibre batts or blankets, which must be adequately held in place. Glass fibre materials are not suitable for fire stopping because they shrink and melt at temperatures over about 300°C. Metal brackets or

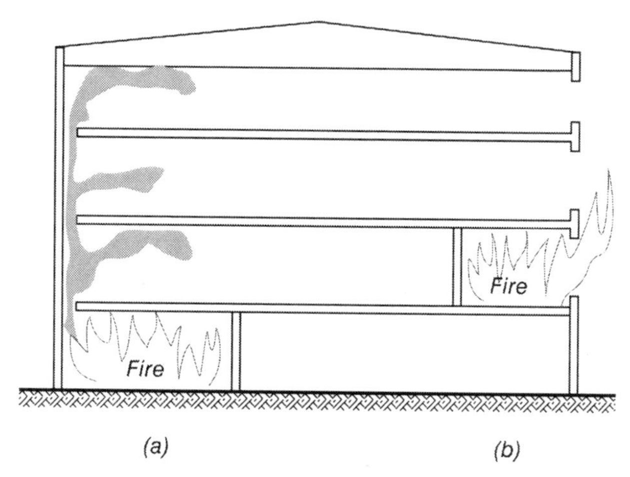

Figure 2.11 Fire spread from storey to storey (Reproduced from Buchanan (2001) by permission of Centre for Advanced Engineering)

angles supporting the filling material must not be made from aluminium alloys because these melt at temperatures over 500°C.

Vertical fire spread can also occur outside the building envelope, via combustible cladding materials or exterior windows as shown in Figure 2.11. Combustible cladding susceptible to rapid flame spread should not be used on the exterior of tall buildings.

Vertical spread of fire from window to window is a major hazard in multi-storey buildings (Figure 2.12). This hazard can be partly controlled by keeping windows small and well separated, and by using horizontal aprons which project above window openings (Oleszkiewicz, 1991). Flames from small narrow windows tend to project further away from the wall of the building than flames from long wide windows, leading to lower probability of storey to storey fire spread (Drysdale, 1998).

2.6.4 Fire Spread to Other Buildings

Fire can spread from a burning building to adjacent buildings by flame contact, by radiation from windows, or by flaming brands. Fire spread can be prevented by providing a fire-resisting barrier or by providing sufficient separation distances (Figure 2.13). If there are openings in the external wall, the probability of fire spread depends greatly on the distances between the buildings and the size of the openings. Exterior fire-resisting walls must have sufficient structural fire resistance to remain in place for the duration of the fire. This becomes a particular problem if the structure which normally provides lateral support to the walls is damaged or destroyed in the fire. Collapse of exterior walls can be a major hazard for fire-fighters and bystanders, and can lead to further spread of fire to adjacent buildings (Figure 2.14).

Fire spread by flame contact is only possible if the buildings are quite close together, whereas fire spread by radiation can occur over many metres. Radiant heat flux from the window of a building fire can ignite combustible cladding on a nearby

Figure 2.12 Fire on the 12th floor of a 62 storey building, illustrating the importance of providing both containment and structural stability for fire safety in tall buildings [First Interstate Bank building, Los Angeles, 1988] (Reproduced by permission of Boris Yaro)

building, or combustible products inside the windows. The calculation of radiant heat flux from one building to another is described in Chapter 3. Fire can also travel large distances between buildings if combustible vegetation is present.

Flaming brands carried by the wind can cause fire spread between buildings with combustible roofing materials as shown in Figure 2.15. This can be controlled by restricting the use of combustible roofing materials. Fire spread between adjacent buildings also depends on the relative heights of the buildings. A fire burning through the roof of a low building can spread into windows of an adjacent tall building as shown in Figure 2.16 unless adequate fire resistance is provided.

2.7 BUILDING CONSTRUCTION FOR FIRE SAFETY

2.7.1 Fire Following Earthquake

The possibility of fire following an earthquake is a major threat in seismic regions of the world. There are many examples of serious fires following earthquakes (Figures 2.17 and 2.18), including San Francisco in 1906 and 1989, Tokyo, 1923, and Napier, New Zealand in 1931 (Botting, 1998, Steinbrugge, 1982).

Figure 2.13 Severe fire in a department store. The entire building is fully involved in the fire. The roof framing is about to collapse. The fire was prevented from spreading to adjacent properties by fire resisting walls

There are three main factors increasing the problem of fire after an earthquake. The probability of ignition is high because of toppled furniture, electrical malfunction and movement of hot equipment. Active and passive fire protection systems may be damaged by the earthquake, and the probability of prompt Fire Service attention is much lower than in normal conditions (Scawthorn, 1992).

Active suppression systems such as automatic sprinklers are of particular concern because of heavy reliance on such systems for fire safety in large modern buildings. There is a high probability of earthquake damage to pipework in the building or the complete loss of city water supplies. For these reasons greater attention should be given to passive containment and structural fire resistance in seismic regions than in non-seismic regions, especially in tall or large buildings where there could be significant danger to life or property.

2.7.2 Fire During Construction and Alterations

The possibility of a fire occurring during the construction process, or during alterations, is often overlooked, despite many serious fire losses. The fire hazard will

Figure 2.14 The masonry walls of a large department store after a severe fire. This fire was similar to that in the previous photo, except that the fire spread through unprotected openings in the walls to engulf the whole complex. All the combustible material was burned and 43 people were killed. [Ballantynes department store, New Zealand, 1947] (Reproduced by permission of The New Zealand Herald)

Figure 2.15 Fire spread by flaming brands

usually be greater during construction than at any other time in the life of the building, because of the increased number of ignition sources and incomplete fire protection features.

There are many recorded cases of ignition from cutting or welding during construction, some leading to massive fire losses. Many fire protection systems are not commissioned and activated until a building is complete, and much passive fire protection such as fire-stopping and structural fire-cladding is not completed until late in the building process.

Figure 2.16 Fire spread from low building to taller building

Figure 2.17 Fire damage following the San Francisco earthquake, 1906. Conflagrations such as this have occurred following several major earthquakes (Reproduced from Walker (1982) by permission of the Bancroft Library)

The probability of fire losses during construction can be reduced by the implementation of a safety plan which recognizes the hazards and the condition of the building at each stage in the construction process, and brings active and passive fire protection systems on line as soon as possible.

Figure 2.18 One of the many severe fires which destroyed buildings after the Hanshin-Awaji earthquake, Kobe, Japan 1995

Poorly executed alterations may reduce the fire safety of a building. There are many documented cases of fires spreading through unprotected concealed spaces which were created during alterations. The persons carrying out repairs may damage or remove passive fire safety features because they are not aware of their importance. For example, new building services installed above suspended ceilings may penetrate important fire walls without fire-stopping, creating paths for unseen spread of fire and smoke. Weaknesses such as these are not obvious without careful inspection.

2.7.3 Assessment and Repair of Fire Damage

Structural engineers are sometimes engaged to report on options for re-use or repair of fire-damaged buildings. If there is a danger of local collapse, immediate concerns about the stability of free-standing residual parts of a fire damaged building will have to be addressed very quickly. More often the owner will want to know if the damaged building can be rehabilitated, in which case there will be time for a more complete investigation.

Table 2.2 Approximate melting temperature of materials (from Gustaferro and Martin, 1988)

Material	Approximate melting temperature (°C)
Polyethylene	110–120
Lead	330
Zinc	420
Aluminium alloys	500–650
Aluminium	650
Glass	600–750
Silver	950
Brass and bronze	850–1000
Copper	1100
Cast iron	1150–1300
Steel	>1400

Inspection

It is valuable to visit the fire scene as soon as possible after the fire while all the fire debris and non-structural damage is visible. This visit can provide essential information on the extent of the fire, the location of the most severe burning and the maximum temperatures reached during the fire. It is also important to re-visit the fire scene after debris and non-structural items have been removed, when it becomes possible to inspect structural members in more detail. It is very important to inspect the details of connections between structural members for cracking of concrete, damage to welded connections or distortion of bolts.

Maximum local temperatures reached in a fire can be estimated from an inspection of materials which have melted. Approximate melting temperature of several materials are given in Table 2.2. The duration of the fully-developed period of the fire can be roughly estimated from the residual size of heavy timber members which will have charred at approximately 0.6 mm per minute as described in Chapter 10.

Most of the significant fire damage which is likely to occur in a structure will be readily visible. With the exception of temperature-related loss of material strength, significant damage will usually be visible as large deflections, local deformations, spalling of concrete or charring of timber. Most members which have deformed during the fire will have to be replaced, unless the deformations do not affect the future use of the building. Deflections can be estimated by sighting along straight members, or by using surveying equipment. Allowance must be made for deflections which may have existed before the fire. If a large number of members have significant distortion the entire structure may have to be demolished.

Steel

Unprotected steel members often suffer large deformations in fully-developed fires, whereas well-protected members usually exhibit no damage. In most cases no further assessment is necessary of fire-exposed steel members which remain straight after cooling (Tide, 1998). The most common grades of structural steel do not suffer significant loss of strength when cooled after heating to temperatures up to about 600°C. Heating to higher temperatures can result in a strength reduction of up to 10%. The reduction in strength is greater for high-strength steels containing alloys such as vanadium and niobium. If necessary, hardness tests or small tensile test specimens can be used to determine whether there has been a reduction in strength. Many types of high-strength bolts have been heat treated during manufacture which makes them susceptible to loss of strength after heating, in which case they can often be replaced without difficulty. Extensive guidance is given by Kirby *et al.* (1993).

Concrete and masonry

Concrete structures generally behave well in fires. Concrete slabs or beams which have excessive deflections will have to be replaced. Cover concrete which has spalled off, or which is badly cracked, can be replaced with poured or sprayed concrete, incorporating additional reinforcing if necessary. Concrete members exhibiting no visible damage may have reduced strength due to elevated temperatures of the concrete or the reinforcing. Typical mild steel reinforcing regains any lost strength when it cools. High strength steels, especially cold-drawn prestressing tendons, are susceptible to high temperatures if heated to temperatures above 400°C. Prestressing steels cooled after heating to 500°C can have a 30% loss of strength and heating to 600°C can result in a 50% loss of strength (Gustaferro and Martin, 1988).

Loss of strength of the concrete is usually of less concern than loss of strength of the reinforcing. The heat affected region is often not very deep because of the low thermal conductivity of concrete. In simply supported flexural members, the compression zone on the top of the slab or beam is often not exposed to high temperatures. Loss of strength of concrete near the surface can be estimated with an impact rebound hammer or other tests (Purkiss, 1996). Some types of concrete change colour after heating to elevated temperatures, depending on the aggregate. Marchant (1972) describes a design procedure for reinstatement of fire damaged reinforced concrete buildings, and reports that typical concrete heated to less than 300°C will have no colour change, concrete heated to 300°C–600°C may be pink, concrete heated to 600°C–950°C may be whitish-grey and concrete heated over 950°C may be a buff colour. Fire-exposed concrete suffers no significant loss of residual strength when heated to 300°C, whereas for higher temperatures the strength loss will be similar to that shown in Figure 9.14. When the concrete cools after heating, it regains strength slowly but never reaches the original strength (Lie, 1992).

Ceramic clay bricks lose very little strength after heating to temperatures as high as 1000°C, but the mortar may suffer some damage. Concrete masonry will need to be assessed in the same way as reinforced concrete.

Timber

Because wood burns, fire damage to exposed timber surfaces is immediately visible. Heavy timber structural members such as beams, columns, or solid wood floors will be charred on the surface, with undamaged wood in the centre, as described in Chapter 10. The residual wood under the charred layer can be assumed to have full strength. The size of the residual cross section can be determined by scraping away the charred layer and any wood which is significantly discoloured. Fire-exposed timber members tend to deform less than equivalent steel members.

Fire-damaged timber members do not need to be replaced if the residual cross section has sufficient strength to carry the design loads. For future fire resistance it may be necessary to apply additional protection such as gypsum board because of loss of the sacrificial wood intended to provide fire resistance. Severely damaged members will need to be replaced, but it may be possible to glue extra laminations on to laminated beams or columns which have only suffered moderate damage.

Light timber frame structures are protected from fire by linings of material such as gypsum board, as described in Chapter 11. After a severe fire, the linings on the underside of ceilings and the fire side of walls will be damaged. Some linings may have fallen off due to the effects of the fire or fire-fighting activities. All damaged linings should be removed to inspect damage to the studs or joists. Any charred timber will have reduced load capacity. Calculations will be necessary to assess the strength of the residual members.

Inspection of gypsum board can give an indication of the duration of fully-developed burning. When gypsum board is exposed to fire it dehydrates steadily from the hot surface. The depth of dehydration can be observed by breaking open a small piece of board to locate the transition between the soft dehydrated plaster and the solid gypsum of the original board. Typical gypsum board dehydrates at approximately 0.5 mm per minute.

3

Fire and Heat

3.1 OVERVIEW

This chapter is a basic summary of heat transfer and the processes involved in the combustion of fuels in typical building fires. Simple descriptions of pre-flashover fires are provided. For more information on these topics, refer to Drysdale (1998), or Quintiere (1998).

3.2 FUELS

3.2.1 Materials

Most of the potential fuel in building fires is organic material originally derived from plants, animals or petrochemicals. The material available as fuel may be part of the building structure, lining materials, or the permanent or temporary contents of the building.

Plant-based materials include products from plants like cotton, jute, straw, food crops or trees. Wood-based materials include solid wood, plywood and other panel products, paper and cardboard, cork, and many others. Animal-based materials include wool, meat and many food products. Petrochemicals include many liquid and gaseous fuels, and almost all plastic materials, also known as polymers. All of these materials are hydrocarbons, their molecules consisting mainly of carbon and hydrogen atoms, with the addition of oxygen, nitrogen and others in some cases.

3.2.2 Calorific Value

The rate of heat release from a combustion reaction depends on the nature of the burning material, the size of the fire, and the amount of air available. The calorific value or heat of combustion is the amount of heat released during complete combustion of a unit mass of fuel. Most solid, liquid and gaseous fuels have a calorific value between 15 and 50 MJ/kg.

Net calorific values ΔH_c (MJ/kg) for a range of common fuels are shown in Table 3.1. For materials such as wood which contain moisture under normal conditions, the effective calorific value $\Delta H_{c,n}$ (MJ/kg) can be calculated from

$$\Delta H_{c,n} = H_c(1 - 0.01\, m_c) - 0.025\, m_c \tag{3.1}$$

where m_c is the moisture content as a percentage by weight, given by

$$m_c = \frac{100\, m_d}{100 + m_d}$$

Table 3.1 Net calorific values of combustible materials (EC1, 1994)

Solids	MJ/kg	Plastics	MJ/kg
Anthracite	34	ABS	36
Asphalt	41	Acrylic	28
Bitumen	42	Celluloid	19
Cellulose	17	Epoxy	34
Charcoal	35	Melamine resin	18
Clothes	19	Phenolformaldehyde	29
Coal, Coke	31	Polyester	31
Cork	29	Polyester, fibre-reinforced	21
Cotton	18	Polyethylene	44
Grain	17	Polystyrene	40
Grease	41	Petroleum	41
Kitchen refuse	18	Polyisocyanurate foam	24
Leather	19	Polycarbonate	29
Linoleum	20	Polypropylene	43
Paper, Cardboard	17	Polyurethane	23
Paraffin wax	47	Polyurethane foam	26
Foam rubber	37	Polyvinylchloride	17
Rubber isoprene	45	Urea formaldehyde	15
Rubber tire	32	Urea formaldehyde foam	14
Silk	19		
Straw	16		
Wood	19		
Wool	23		
Particle board	18		
Liquids	**MJ/kg**	**Gases**	**MJ/kg**
Gasoline	44	Acetylene	48
Diesel oil	41	Butane	46
Linseed oil	39	Carbon monoxide	10
Methanol	20	Hydrogen	120
Paraffin oil	41	Propane	46
Spirits	29	Methane	50
Tar	38	Ethanol	27
Benzene	40		
Benzyl alcohol	33		
Ethyl alcohol	27		
Isopropyl alcohol	31		

where m_d is the moisture content as a percentage of the dry weight, as usually used for wood products.

For typical dry wood fuel with $\Delta H_c = 19\,\text{MJ/kg}$, and a moisture content of $m_d = 12\%$, the above equations give $m_c = 10.7\%$ hence $\Delta H_{c,n} = 16.7\,\text{MJ/kg}$ at 12% moisture content. For wood fuel, Parker and Tran (1992) give average typical heat of combustion of oven dry wood around 20 MJ/kg while noting that the effective heat of combustion in the OSU Apparatus was in the range 13 to 15 MJ/kg. The NFPA Handbook gives the typical range of heat of combustion as 18 to 21 MJ/kg. Babrauskas (1995) recommends a value of only 12 MJ/kg for heat of combustion of wood, which includes an allowance for incomplete combustion.

The maximum possible energy that can be released when fuel burns is the energy contained in the fuel, E (MJ) given by

$$E = M\Delta H_c \tag{3.2}$$

for dry fuel, or

$$E = M\Delta H_{c,n}$$

for fuels containing moisture, where M is the mass of the fuel (kg).

3.2.3 Fire Load

Fire load in buildings is most often expressed as *Fire Load Energy Density* (FLED) per square metre of floor area. For each room, the fire load energy density e_f (MJ/m^2 floor area) is given by

$$e_f = E/A_f \tag{3.3}$$

where A_f is the floor area of the room

Many European references express fire load as energy density per square metre of total bounding surfaces of the room. This energy density e_t (MJ/m^2 total room surface area) is given by

$$e_t = E/A_t \tag{3.4}$$

where A_t is the total area of the bounding surfaces of the room (floor, ceiling and walls, including window openings).

It can be seen that for a given room, e_f is larger than e_t by the ratio A_t/A_f. It is essential to know which fuel load density is being used in any given situation. Major errors can be caused if the distinction is not clear.

The results of several extensive fuel load surveys are quoted in CIB (1986), with an extract included as Appendix B to this book. Typical values of fire load energy density range from 100 to 10 000 MJ/m^2 floor area. The Acceptable Solution to the New Zealand Building Code (BIA 1992) gives design values of 400, 800 and 1200 MJ/m^2 floor area for residential, office and retail occupancies respectively, requiring storage areas to be assessed separately. These values appear to be rather too low for safe design. The Eurocode (EC1, 1994) gives five classes (I to V) with fire load ranging from 250 to 2000 MJ/m^2 floor area.

Design fire loads should be determined in a similar way to design loads for other extreme events such as wind or earthquake, so that the design fire represents an extreme value of the likely fire scenarios. The design fire load should be that which has less than 10% probability of being exceeded in the 50 year life of the building. The design fire load will therefore be close to the maximum fire load expected in the life of the building. Both fixed and moveable fire loads should be included. When the fire load is determined from representative surveys, the design load should be the 90 percentile value of surveyed loads, which will be much larger than the typical average fire load at a random point in time. For a coefficient of variation of 50% to 80% of the average value, the 90 percentile value will be 1.65 to 2.0 times the average value (CIB, 1986). The recommended design fire load is therefore two times the average values given in Appendix B.

3.2.4 Heat Release Rate

For any fire, the heat release rate in MW can be calculated if the amount of heat released as MJ in a certain time (in seconds) is known. Hence the average heat release rate Q (MW) is given by

$$Q = E/t \tag{3.5}$$

where E is the total energy contained in the fuel (MJ), and t is the duration of the burning (s).

3.3 COMBUSTION

3.3.1 Chemistry

In its most simple form, the combustion of organic material is an exothermic chemical reaction involving the oxidation of hydrocarbons to produce water vapour and carbon dioxide. For example, the chemical reaction for the complete (stoichiometric) combustion of propane is given by

$$C_3H_8 + 5O_2 \rightarrow 3CO_2 + 4H_2O \tag{3.6}$$

This is a simplification of the chemistry. There are many chemical processes involved, depending on temperatures, pressures and the availability of the materials. Intermediate reactions involve a large number of atoms and free radicals. In many fire situations there will be incomplete combustion, leading to the production of carbon monoxide gas (CO) or solid carbon (C) as soot particles in the flames or smoke. The chemistry is changing throughout the combustion process.

3.3.2 Phase Change and Decomposition

At room temperatures, some fuels are gases but most are solids or liquids. Gases can mix with air to burn directly without any phase change, but all solid and liquid fuels must be converted to the gaseous phase before they can burn. For most liquids, the transition from liquid to the gaseous phase under the application of heat is *evaporation*. For some polymers the process is thermal decomposition into new volatile products.

Many solid fuels *melt* when heated, producing a liquid which can then evaporate or thermally decompose into a gas. Some other fuels, including most wood products, thermally decompose with a transition directly from solid to gaseous phase. This thermal decomposition of wood is known as *pyrolysis*.

3.3.3 Premixed Flames

The combustion process for any material requires the availability of oxygen for the oxidation reaction to occur. The most efficient combustion is premixed burning where the gaseous fuel is mixed with oxygen or air containing oxygen before the ignition source is introduced (as in a Bunsen burner). Combustion will be very rapid if the gases are mixed in the right proportions (for example in an internal combustion engine). Combustion will not occur if the mixture has too much or too little oxygen for the given conditions of temperature and pressure. The limiting conditions are called the limits of flammability.

3.3.4 Diffusion Flames

In most building fires there is no premixed burning, and the rate of combustion depends on the rate of mixing of air with the gaseous fuels as they become available. The combustion takes place in the region where the gases mix. The mixing is usually driven by buoyancy and turbulence resulting from the convective movement of the flame and combustion products in the plume above the fire.

3.3.5 Flame Temperature

The maximum temperature that can be reached in a flame is known as the *adiabatic flame temperature*. This is the theoretical maximum temperature that can be reached when the combustion products are heated from their initial temperature by the heat released in the combustion reaction, with no losses. In flames from a typical burning object, the adiabatic flame temperature may be reached in a small region in the centre of the flame, but the average temperature of the flame will be considerably less.

3.3.6 Established Burning

For an object to be first ignited, it is necessary for there to be an external source of heat to raise the temperature of the object to its ignition temperature. If the fire grows after ignition, it may reach *established burning* after which the flames are large enough to sustain the combustion reaction with no assistance from any external source of heat. The burning is driven by heat from the flames which heats the remaining fuel to a sufficient temperature for the production of volatile combustible gases which burn in a dynamic process, producing more volatiles and more flames.

3.3.7 Smouldering Combustion

Smouldering is the term given to flameless combustion such as in a cigarette. Smouldering combustion is much slower than flaming combustion, and temperatures are also lower (Ohlemiller, 1995). Smouldering combustion is a particular hazard in building fires, because insufficient heat or noise is generated to wake sleeping occupants who can be overcome by the smoke and toxic combustion products. The smoke from smouldering combustion will activate smoke detectors, but usually has insufficient temperature to activate heat detectors or automatic sprinkler systems. Smouldering combustion does not produce temperatures sufficient to affect structures, so is not considered further in this book, although smouldering can develop into flaming combustion.

3.3.8 Calorimetry

A useful similarity between all hydrocarbon-based fuels is that the heat release rate per unit of oxygen consumption is almost constant. This forms the basis of oxygen consumption calorimetry experiments. For almost all fuels, approximately 13 MJ of heat is released for each kilogram of oxygen consumed, which is approximately 3 MJ of heat for each kilogram of air involved in the burning. This relationship allows the rate of heat release in experimental fires to be obtained by sampling the oxygen concentration in the flue gases. This procedure is becoming widely used in fire testing: on a small scale in cone calorimeters, at medium scale in furniture calorimeters and at a larger scale in room calorimeters.

3.4 FIRE INITIATION

3.4.1 Sources and Mechanisms

Ignition occurs when a combustible mixture of gases is heated to temperatures that will trigger the exothermic oxidation reaction of combustion. Ignition almost always requires the input of heat from an external source. The few cases where self heating within solid materials can cause *spontaneous combustion* is a special subject which is not covered in this book.

There are numerous possible heat sources that cause building fires to ignite. These include flaming sources (matches, candles, gas heaters, open fires), smouldering sources (cigarettes), electrical sources (arcing, overheating), radiant sources (sunlight, hot items, heaters, fires), also hot surfaces, friction, lightning and others. Arson is a significant problem in many countries. War and terrorism have also been the cause of many fires in buildings. Many sources can be reduced or controlled by fire prevention strategies, but some unwanted fires will always occur.

The amount of heat and the temperature required to cause ignition depend on the material properties of the fuel, the size and shape of the ignited object, and the time of exposure to heat. A competent ignition source is one which has sufficient heat and temperature to cause ignition in the expected time of exposure.

The time to ignition of materials depends on the *thermal inertia* of the material itself. Thermal inertia is defined later in this chapter as the product of thermal conductivity, density and specific heat. When exposed to the same heat source, the surface of materials with low thermal inertia (e.g. polystyrene foam) will heat more rapidly than materials with higher thermal inertia (e.g. wood) leading to much more rapid ignition.

3.4.2 Pilot Ignition and Auto-ignition

It is useful to distinguish between *pilot ignition*, which occurs in the presence of a flame or spark, and *auto-ignition*, which causes the spontaneous ignition of volatile gases from a fuel source in the absence of any flame or spark. Auto-ignition requires the gases to be at a higher temperature than for pilot ignition. For surfaces exposed to radiant heat flux, the heat flux intensity required to cause auto-ignition is higher than that required for pilot ignition.

3.4.3 Flame Spread

After ignition has occurred somewhere in a building, fire safety depends greatly on the rate of fire spread. Initial fire spread is caused by the spread of flames on the burning object or adjacent combustible materials.

The main factor affecting flame spread is the rate of heating of the fuel ahead of the flame. This in turn depends on the size and location of the flame (causing radiant heating), the air flow direction (causing convective heating), the thermal properties of the fuel (affecting the rate of temperature rise), and the flammability of the fuel (Drysdale, 1998).

Heating ahead of the flame will be more rapid if there are heat sources in addition to the flame itself, such as radiation from a layer of hot gases under the ceiling. Air movement is very important. Flame spread will be much more rapid with air flow in the direction of spread ('wind aided') than air flow in the other direction ('wind opposed'). Upward flame spread is always rapid because the flame can rapidly preheat the material ahead of the burning region.

Flames tend to spread most rapidly on surfaces which have a high rate of temperature increase on exposure to heat flux. These are materials with low thermal inertia which are also more susceptible to ignition. Materials such as low density plastic foams experience rapid flame spread and fire growth for this reason.

3.5 BURNING OBJECTS

There has been a large amount of research into the open-air burning of flammable liquids and solid objects such as items of furniture. The rate of heat release from a pool or solid item burning in the open depends on the rate at which heat from the flames can evaporate or pyrolyse the remaining fuel, and the rate at which oxygen can mix with the unburned fuel vapour to form diffusion flames. A *plume* of smoke and hot gases rises directly above the fire, cooling as it rises because of the large amount of surrounding air *entrained* into the plume.

If an object such as a furniture item is ignited and allowed to burn freely in the open, the heat release rate tends to increase exponentially as the flames get larger and they radiate more heat back to the fuel. A *peak heat release rate* is usually reached, followed by steady-state burning and eventual decay. The peak heat release rate for open air burning depends on the geometry and nature of the fuel within the object. In a room fire this peak heat release rate may not be reached because of lack of available ventilation.

There is a large amount of information available on the heat release rate of burning items. Many objects such as furniture items and appliances have been burned in furniture calorimeters (Figure 3.1), producing much valuable information

Figure 3.1 Burning sofa in a furniture calorimeter test. If this sofa was burning in a room, the room would be full of hot toxic smoke

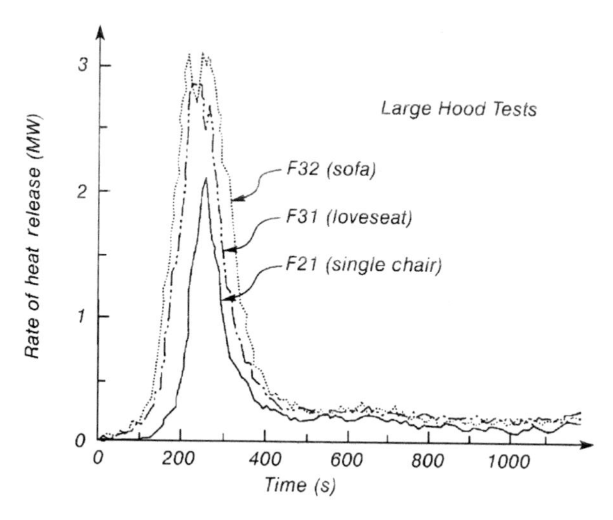

Figure 3.2 Heat release rate for furniture items (Reproduced from Babrauskas (1995) by permission of National Fire Protection Association)

including the rate of heat release, production of smoke and combustion gases. The furniture calorimeter simulates free burning in the open air. The burning rate may be very different under a ceiling, or inside a room, as described in the next chapter. Burning rates of many materials are described by Babrauskas and Grayson (1992) and Babrauskas (1995). For example, heat release rates for some typical items of furniture are given in Figure 3.2.

Many studies have attempted to predict the burning of an item of furniture from information on the burning characteristics of the individual components, most often measured in a cone calorimeter (ISO, 1993). These studies have only had limited success because of difficulty scaling the complex phenomena from bench scale to complete items or rooms. For fire engineering design it is preferable to use the results of tests on realistic full size objects to develop fire design curves.

Burning rates for some liquid and solid fuels are given in Table 3.2 (derived from Babrauskas, 1995). These are based on a constant rate of burning per square metre of exposed surface area. The figures for liquids and plastics are measurements from open-air burning experiments where the rate of burning depends on the radiation back to the fuel surface from flames above the burning object. The figures for liquids are for pool fires over 2 m in diameter. The figures for wood are based on a constant regression rate of 40 mm/h which is typical for the burning of wood in a fully developed room fire. The values for wood furniture will be similar to those given for wood cribs. Table 3.2 shows how the surface regression rate, the mass loss rate and the burning rate are all related, and they can be combined with the calorific value of the fuel to give the heat release rate. These figures can be used to estimate the rate of heat release in a large open air fire, such as may occur in an industrial building after the roof has collapsed.

Table 3.2 Burning rates for some liquid and solid fuels

	Density (kg/m^3)	Regression rate (mm/h)	Mass loss rate (kg/h)	Surface burning rate ($kg/s/m^2$) (surface)	Ratio of surface area to floor area	Total burning rate ($kg/s/m^2$) (floor)	Net calorific value (MJ/kg)	Heat release rate (MW/m^2) (floor)
Liquids								
LPG (C_3H_8)	585	609	356	0.099	1.0	0.099	46.0	4.55
Petrol	740	268	198	0.055	1.0	0.055	43.7	2.40
Kerosene	820	171	140	0.039	1.0	0.039	43.2	1.68
Ethanol	794	68	54	0.015	1.0	0.015	26.8	0.40
Plastics								
PMMA	–	–	–	0.054	1.0	0.054	24.0	1.34
Polyethylene	–	–	–	0.031	1.0	0.031	44.0	1.36
Polystyrene	–	–	–	0.035	1.0	0.035	40.0	1.40
Wood								
Flat wood	500	40	20	0.056	1.0	0.0056	16	0.09
1 m cube	500	40	20	0.056	6.0	0.033	16	0.53
100 mm in crib	500	40	20	0.056	20	0.11	16	1.8
25 mm in crib	500	40	20	0.056	47	0.26	16	4.2
Softboard	300	108	32	0.009	1.0	0.009	16	0.14

3.6 *t*-SQUARED FIRES

The growth rate of a design fire is often characterized by a parabolic curve known as a *t-squared fire* such that the heat release rate is proportional to the time squared. The *t*-squared fire can be thought of in terms of a burning object with a constant heat release rate per unit area, in which the fire is spreading in a circular pattern with a constant radial flame speed.

The *t*-squared heat release rate is given by:

$$Q = (t/k)^2 \tag{3.7}$$

where Q is the heat release rate (MW), t is the time (s), and k is a growth constant ($s/\sqrt{MW}$).

Values of k are given in Table 3.3 for slow, medium, fast and ultrafast fire growth, producing the heat release rates shown in Figure 3.3. The numerical value of k is the time for the fire to reach a size of 1.055 MW. The choice of growth constant depends on the type and geometry of the fuel. Values of k and peak heat release rate for many different burning objects are given by Babrauskas (1995). An alternative formulation which gives identical results is to describe the heat release rate Q (MW) for a t^2 fire by:

Table 3.3 Fire growth rates for t^2 fires

Fire growth rate	Value of k	Value of α	Typical real fire
Slow	600	0.00293	Densely packed wood products
Medium	300	0.0117	Solid wood furniture such as desks Individual furniture items with small amounts of plastic
Fast	150	0.0466	Some upholstered furniture High stacked wood pallets Cartons on pallets
Ultrafast	75	0.1874	Most upholstered furniture High stacked plastic materials Thin wood furniture such as wardrobes

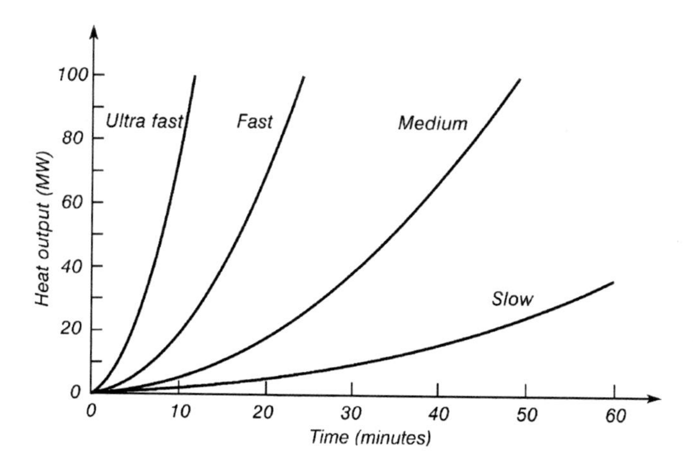

Figure 3.3 Heat release rate for t^2 fires

$$Q = \alpha\, t^2 \tag{3.8}$$

where α is the fire intensity coefficient (MW/s^2). Values of α are also given in Table 3.3.

3.6.1 Calculations

Calculations for a t^2 fire are given below with reference to Figure 3.4. Using Equation (3.7), the time t_1 (s) for the fire to reach the peak heat release rate Q_p (MW) is given by

$$t_1 = k\sqrt{Q_p} \tag{3.9}$$

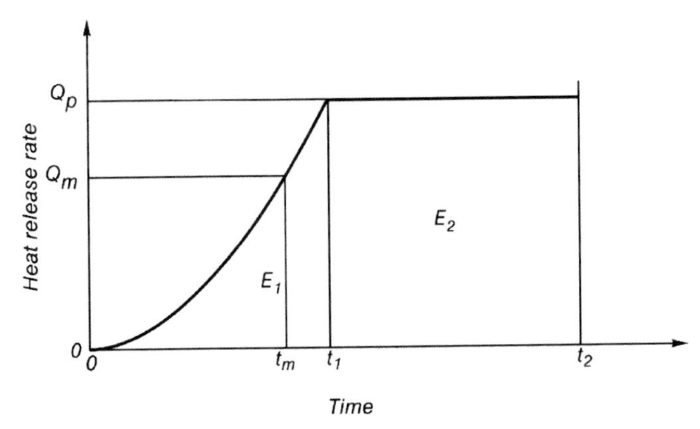

Figure 3.4 Calculation of heat release rates (Reproduced from Buchanan (2001) by permission of Centre for Advanced Engineering)

The energy released is the area under the curve of heat release rate vs.. time. Because the area under a parabola is one third of the enclosing rectangle, the energy E_1 (MJ) released to time t_1 is given by

$$E_1 = t_1 Q_p / 3 \tag{3.10}$$

If the total energy E (MJ) in the fuel has not been released at time t_1, the energy released in the steady burning phase E_2 (MJ) is given by

$$E_2 = E - E_1 \tag{3.11}$$

and the duration t_b (s) of the steady burning phase is given by

$$t_b = t_2 - t_1 = E_2 / Q_p \tag{3.12}$$

If the fuel has insufficient time to reach its peak heat release rate, all of the fuel will be consumed in time t_m (s) where

$$t_m = (3\ E\ k^2)^{1/3} \tag{3.13}$$

and the burning rate Q_m at time t_m is given by

$$Q_m = (t_m / k)^2 \tag{3.14}$$

Figure 3.5 shows the resulting heat release rates for a fire in office furniture with slow, medium and fast fire growth rates. The peak heat release rate has been taken as 9 MW. The furniture item weighs 160 kg with a calorific value of 20 MJ/kg, giving a total energy load of 3200 MJ, which is the area under each of the curves shown in Figure 3.5. A spreadsheet program or the MAKEFIRE routine in the FPEtool computer package can be used to construct t-squared design fires such as these (Deal, 1993).

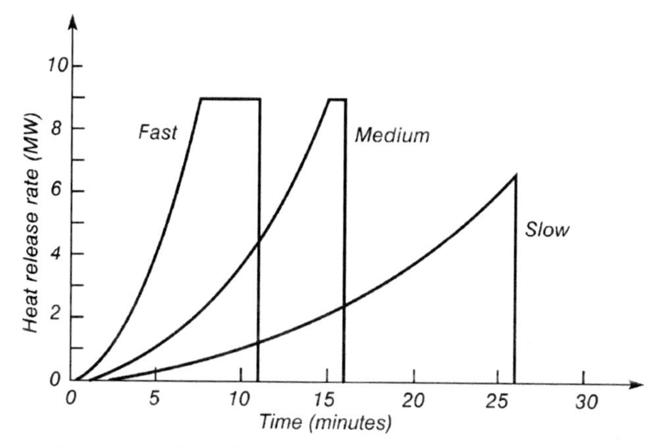

Figure 3.5 Heat release rates for a fire load of 3200 MJ (Reproduced from Buchanan (2001) by permission of Centre for Advanced Engineering)

3.7 PRE-FLASHOVER DESIGN FIRES

The *t*-squared fires described above can be used to construct pre-flashover design fires, as input for calculating fire growth in rooms.

3.7.1 Fire Spread to Other Items

The fires described above are generally used to describe the heat release rate for burning of a single object. Fire can spread from the first burning object to a second object by flame contact if it is very close, or by radiant heat transfer if it is further away. The time to ignition of a second object depends on the intensity of radiation from the flame and the distance between the objects. When the time to ignition of the second object has been calculated, the combined heat release rate can be added at any point in time to give the total heat release rate for these two, and subsequent objects. This combined curve then becomes the input design fire for the room under consideration. There may be many more items involved, and the resulting combination may itself be approximated by a *t*-squared fire for simplicity.

For example, Figure 3.6(a) shows the *t*-squared heat release rate separately for two objects. The first burns with medium growth rate for 10 minutes, followed by 1 minute of steady burning at its peak heat release rate of 4.0 MW. The second object ignites after 3 minutes, burning with fast growth rate for 4 minutes followed by steady burning at 2.5 MW for 2 minutes. Figure 3.6(b) shows the combined heat release rate curve for the two objects. The FREEBURN routine in the FPEtool computer package can be used to calculate the time of ignition of the second object, and to construct design fires for two up to five fuel objects (Deal, 1993).

3.7.2 Room Fires

The design fires described above all assume open-air burning with unlimited ventilation, and no suppression by sprinklers or fire-fighters. When these burning

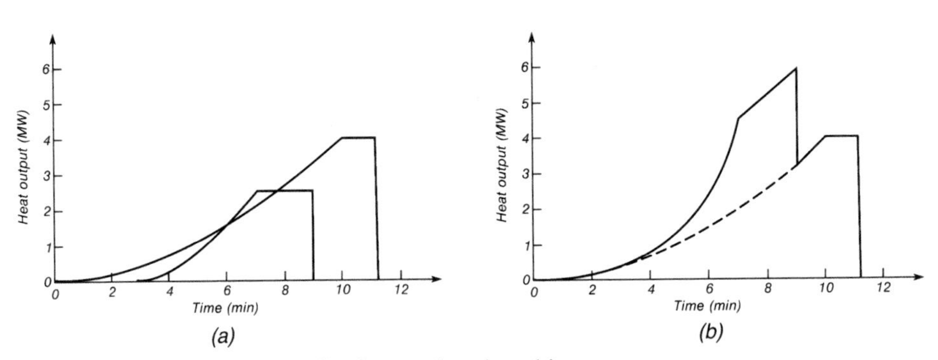

Figure 3.6 Combined design fire for two burning objects

objects are in a room they burn differently, with the burning rate enhanced by radiation, but limited by the available ventilation, as described in the next chapter. The above design fires are used as input to room fire computer models which calculate the actual process of fire development.

Hand calculations can be made for room fires using Equations (3.9) to (3.12), where the value of Q_p should be the lesser of the total peak heat release rate for all the fuel items, or the ventilation-controlled heat release rate from Equation (4.4).

3.8 HEAT TRANSFER

Some knowledge of heat transfer is essential to the understanding of fire behaviour. Heat transfer occurs by the three processes of conduction, convection and radiation, which can occur separately or together depending on the circumstances.

3.8.1 Conduction

Conduction is the mechanism for heat transfer in solid materials. In materials which are good conductors of heat, the heat is transferred by interactions involving free electrons, hence materials which are good electrical conductors are usually also good conductors of heat. In other materials which are poor conductors, heat is conducted by mechanical vibrations of the molecular lattice. Conduction of heat is an important factor in the ignition of solid surfaces, and in the fire resistance of barriers and structural members.

Several material properties are needed for heat transfer calculations in solid materials. These are the *density*, *specific heat* and *thermal conductivity*. Density, ρ, is the mass of the material per unit volume in kg/m^3. Specific heat c_p is the amount of heat required to heat a unit mass of the material by one degree, with units of J/kgK. Thermal conductivity, k, represents the rate of heat transferred through a unit thickness of the material per unit temperature difference, with units W/mK.

There are two derived properties which are often needed. These are the *thermal diffusivity* given by $\alpha = k/\rho c_p$ with units m^2/s, and the *thermal inertia* $k\rho c_p$ with units W^2 s/m^{4}K^2.

Table 3.4 Thermal properties of some common materials[a] (Drysdale, 1998)

Material	Thermal conductivity k (W/mK)	Specific heat c_p (J/kgK)	Density ρ (kg/m^3)	Thermal diffusivity α (m^2/s)	Thermal inertia $k\rho c_p$ (W^2 s/m^4 K^2)
Copper	387	380	8940	1.14×10^{-4}	1.3×10^{9}
Steel (mild)	45.8	460	7850	1.26×10^{-5}	1.6×10^{8}
Brick (common)	0.69	840	1600	5.2×10^{-7}	9.3×10^{5}
Concrete	0.8–1.4	880	1900–2300	5.7×10^{-7}	2×10^{6}
Glass (plate)	0.76	840	2700	3.3×10^{-7}	1.7×10^{6}
Gypsum plaster	0.48	840	1440	4.1×10^{-7}	5.8×10^{5}
PMMA[b]	0.19	1420	1190	1.1×10^{-7}	3.2×10^{5}
Oak[c]	0.17	2380	800	8.9×10^{-8}	3.2×10^{5}
Yellow pine[c]	0.14	2850	640	8.3×10^{-8}	2.5×10^{5}
Asbestos	0.15	1050	577	2.5×10^{-7}	9.1×10^{4}
Fibre insulating board	0.041	2090	229	8.6×10^{-8}	2.0×10^{4}
Polyurethane foam[d]	0.034	1400	20	1.2×10^{-6}	9.5×10^{2}
Air	0.026	1040	1.1	2.2×10^{-5}	–

[a]From Pitts and Sissom (1977) and others. Most values for 0 or 20°C. Figures have been rounded off.
[b]Polymethylmethacrylate. Values of k, c_p and for ρ other plastics are given in Drysdale (1998), Table 1.2.
[c]Properties measured perpendicular to the grain.
[d]Typical values only.

When materials with low thermal inertia are exposed to heating, surface temperatures increase rapidly, so that these materials ignite more readily. For a given fire load, rooms lined with materials of low thermal inertia will experience higher temperatures than rooms lined with materials of higher thermal inertia. Thermal properties for common materials are given in Table 3.4 from Drysdale (1998). A more extensive list is given in Appendix A of the SFPE Handbook (1995), including temperature dependant thermal properties for metals.

In the steady-state situation, the transfer of heat by conduction is directly proportional to the temperature gradient between two points, with a constant of proportionality known as the thermal conductivity, k, so that

$$\dot{q}'' = k\,\mathrm{d}T/\mathrm{d}x \tag{3.15}$$

where $\dot{q}''$ is the heat flow per unit area (W/m^2), k is the thermal conductivity (W/mK), T is temperature (°C or K), and x is distance in the direction of heat flow (m). The steady-state calculation does not require consideration of the heat required to change the temperature of material that is being heated or cooled.

For transient heat flow when temperatures are changing with time, the amount of heat required to change the temperature of the material must be included. For one-dimensional heat transfer by conduction in a material with no internal heat being released, the governing equation is

$$\frac{\delta^2 T}{\delta x^2} = \frac{1}{\alpha} \frac{\delta T}{\delta t} \tag{3.16}$$

where t is time (s), and α is thermal diffusivity (m^2/s).

It can be seen that materials with low thermal diffusivity will conduct more heat than materials with high thermal diffusivity, when exposed to increasing surface temperatures in unsteady-state conductive heat transfer.

This can be extended to two or three dimensions as necessary. There are many methods of solving the heat conduction equation using analytical, graphical or numerical methods. Some methods are given by Drysdale (1998) and there are many standard textbooks on heat transfer. Calculation of conductive heat transfer can be by simple formulae, the use of design charts or by numerical analysis.

Simple formulae

Standard textbooks on heat transfer contain simple formulae for conductive heat transfer in various materials and geometries. Some of these are available for common materials exposed to fire. A lumped heat capacity formula can be used for a protected or unprotected steel element on the assumption that the internal steel temperatures are constant. This type of formula is most accurate where it is used repeatedly with sequential time steps. Some formulae can take account of the heat required to heat up heavy insulating materials, or the time delay resulting from driving off moisture. Typical formulae are given by ECCS (1985), Milke (1995) and Pettersson *et al.* (1976). A simple spreadsheet formulation is described by Gamble (1989). Examples of these methods are given in Chapter 8.

A semi-infinite slab analysis can be used in situations where the heat transfer is essentially one-dimensional, such as with large flat surfaces. This analysis assumes that the heat is absorbed before reaching the unexposed side, so the material has to be relatively thick. Schaffer (1977) has applied this type of analysis to wood slabs.

Design charts

Design charts are available as graphical solutions for many situations where structural materials are exposed to fire temperatures. Most graphical solutions are for exposure to the standard test fire which is described in Chapter 5. Available charts include those by Milke (1995) and Fleischmann (1995) for standard fire exposure of steel and concrete elements respectively. Figure 8.6 is a chart for estimation of temperatures in steel members exposed to the standard fire. Lie (1972) provides design charts for walls or slabs exposed on one or two sides to the standard fire, which can be used for any inert material such as steel, concrete or wood before charring occurs. Design charts for steel members exposed to realistic fires are provided by Pettersson *et al.* (1976).

Numerical analysis

The most powerful tools for calculating conductive heat transfer are computer-based numerical methods such as finite-element or finite-difference formulations. These techniques are well established, but there are not many user-friendly commercial computer packages customized for fire applications. Special characteristics needed for structural fire applications include internal voids and temperature dependant thermal properties. The most widely used finite-element programs for fire design include FIRES-T3 (Iding *et al.*, 1977(a)), SAFIR (Franssen *et al.*, 2000), TASEF (Sterner and Wickström, 1990), and TEMPCALC (Anderberg, 1988). A review of some of these programs is given by Sullivan *et al.* (1994).

Many generic finite-element stress-analysis programs can calculate heat transfer by conduction. Some widely used commercial programs include ABAQUS, ANSYS and NASTRAN. These are versatile programs which can analyse any three-dimensional mesh which is input by the user.

Both two- and three-dimensional heat transfer analysis is possible, but two-dimensional analysis is adequate for almost all fire engineering applications. This is because structural elements are mostly planar or linear, and there will be no temperature gradient along an element if it can be assumed that temperatures are uniform within a post-flashover fire compartment. This assumption holds for fires in small rooms but is less accurate for fires in large spaces.

A deficiency of all the above programs is their inability to model mass transfer such as the transport of water or water vapour through permeable materials. Moisture movement has an influence on fire performance of materials such as gypsum plaster and wood as described by Fredlund (1993). Most programs do not easily model shrinkage of the material or ablation of material from fire-exposed surfaces, but effects such as these can usually be simulated by varying the temperature-dependent thermal properties in a conventional heat transfer program.

3.8.2 Convection

Convection is heat transfer by the movement of fluids, either gases or liquids. Convective heat transfer is an important factor in flame spread and in the upward transport of smoke and hot gases to the ceiling or out of the window from a room fire.

Convective heat transfer calculations usually involve heat transfer between the surface of a solid and a surrounding fluid which heats or cools the solid material. The rate of heating or cooling depends on several factors, especially the velocity of the fluid at the surface. For given conditions the heat transfer is usually taken to be directly proportional to the temperature difference between the two materials, so that the heat flow per unit area $\dot{q}''$ (W/m^2) is given by

$$\dot{q}'' = h\Delta T \tag{3.17}$$

where h is the convective heat transfer coefficient (W/m^{2}K), and ΔT is the temperature difference between the surface of the solid and the fluid (°C or K).

The value of the heat transfer coefficient h can vary depending on factors such as the geometry of the surface, the nature of the flow, and the thickness of the boundary layer. A typical value for fire-exposed structural elements is 25 $W/m^2 K$. Other values are available in many heat transfer textbooks.

3.8.3 Radiation

Radiation is the transfer of energy by electromagnetic waves which can travel through a vacuum, or through a transparent solid or liquid. Radiation is extremely important in fires because it is the main mechanism for heat transfer from flames to fuel surfaces, from hot smoke to building objects and from a burning building to an adjacent building.

The radiant heat flux $\dot{q}''$ (W/m^2) at a point on a receiving surface is given by

$$\dot{q}'' = \varphi \varepsilon_e \sigma T_e^4 \tag{3.18}$$

where φ is the configuration factor, ε_e is the emissivity of the emitting surface, σ is the Stefan–Boltzmann constant (5.67×10^{-8} $W/m^2 K^4$), and T_e is the absolute temperature of the emitting surface (K)

The resulting heat flow $\dot{q}''$ (W/m^2) from the emitting surface to the receiving surface is given by

$$\dot{q}'' = \varphi \varepsilon \sigma (T_e^4 - T_r^4) \tag{3.19}$$

where T_r is the absolute temperature of the receiving surface (K), and ε is the resultant emissivity of the two surfaces, given by

$$\varepsilon = \frac{1}{1/\varepsilon_e + 1/\varepsilon_r - 1} \tag{3.20}$$

where ε_r is the emissivity of the receiving surface.

The emissivity ε indicates the efficiency of the emitting surface as a radiator, with a value in the range from zero to 1.0. A so-called 'black-body' radiator has an emissivity of 1.0. In fire situations, most hot surfaces, smoke particles or luminous flames have an emissivity between 0.7 and 1.0. The emissivity can change during a fire; for example zinc-coated steel (galvanized steel) has a very low emissivity until the temperature reaches about 400°C when the zinc melts and the bare steel is exposed to the fire.

The configuration factor φ (sometimes called the 'view factor') is a measure of how much of the emitter is 'seen' by the receiving surface. In the general situation shown in Figure 3.7 (Drysdale, 1998) the configuration factor for incident radiation at point 2, a distance r from a radiating surface of area A_1 is

$$\varphi = \int^{A_1} \frac{\cos\theta_1 \cos\theta_2}{\pi r^2} \, dA_1 \tag{3.21}$$

where the terms are shown in Figure 3.7.

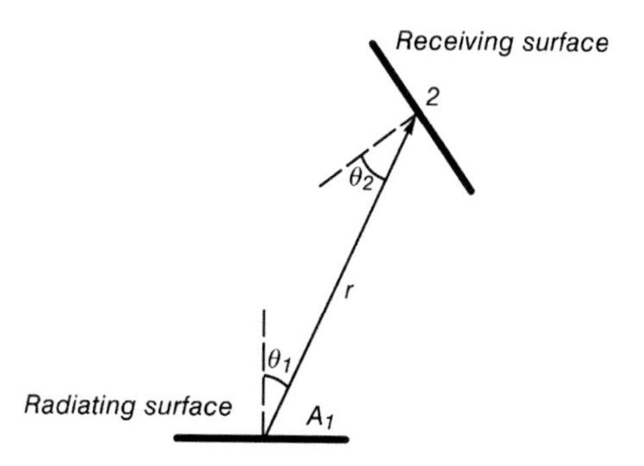

Figure 3.7 Radiation from one surface to another

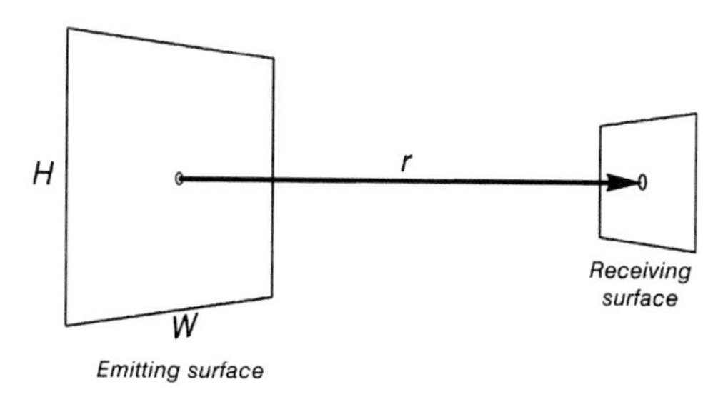

Figure 3.8 Emitting and receiving surfaces

For the particular case of two parallel surfaces as shown in Figure 3.7, the configuration factor φ at a point on the receiving surface at distance r from the centre of the rectangular radiator is

$$\varphi = \frac{1}{90}\left[\frac{x}{\sqrt{1+x^2}}\tan^{-1}\left(\frac{y}{\sqrt{1+x^2}}\right) + \frac{y}{\sqrt{1+y^2}}\tan^{-1}\left(\frac{x}{\sqrt{1+y^2}}\right)\right] \tag{3.22}$$

where $x = H/2r$, $y = W/2r$, H is the height of the rectangular source, W is the width of the rectangular source, and $\tan^{-1}$ is the inverse tangent in degree mode. If the distance r is large relative to the size of the emitting surface, the configuration factor φ is given approximately by

$$\varphi = A_1/\pi\, r^2 \tag{3.23}$$

where A_1 is the area of the emitting surface ($A_1 = HW$ in Figure 3.8).

Configuration factors for many other situations and values of emissivity are given by Drysdale (1998) and many heat transfer text books. A useful computer program for calculating the radiant heat flux from a number of radiating surfaces at any angle is available in FIREWIND (Shestopal, 1998) formerly available as FIRECALC (CSIRO, 1993).

WORKED EXAMPLE 3.1

Calculate the average heat release rate when 200 kg of paraffin wax burns in half an hour.

Mass of fuel	$M = 200$ kg
Calorific value	$\Delta H_c = 46$ MJ/kg
Energy contained in the fuel	$E = M\Delta H_c = 200 \times 46 = 9200$ MJ
Time of burning	$t = 1800$ s
Heat release rate	$Q = E/t = 9200/1800 = 5.11$ MW

Calculate the fuel load energy density in an office 5 m × 3 m containing 150 kg of dry wood and paper and 75 kg of plastic materials. Assume calorific values of 16 MJ/kg and 30 MJ/kg respectively.

Mass of wood	$M_{wood} = 150$ kg
Calorific value	$\Delta H_{c,wood} = 16$ MJ/kg
Energy contained in the wood	$E_{wood} = M\Delta H_c = 150 \times 16 = 2400$ MJ
Mass of plastic	$M_{plastic} = 75$ kg
Calorific value	$\Delta H_{c,plastic} = 30$ MJ/kg
Energy contained in plastic	$E_{plastic} = M\Delta H_c = 75 \times 30 = 2250$ MJ
Total energy in fuel	$E = E_{wood} + E_{plastic}$
	$= 2400 + 2250 = 4650$ MJ
Floor area	$A_f = 5 \times 3 = 15$ m^2
Fuel load energy density	$e_f = E/A_f = 4650/15 = 310$ MJ/m^2

WORKED EXAMPLE 3.2

A room in a storage building has 2000 kg of polyethylene covering the floor. Calculate the heat release rate and duration of burning after the roof collapses in a fire. The room is 6.0 m by 10.0 m. Use the open-air burning rates from Table 3.2.

Mass of polyethylene	$M = 2000$ kg
Calorific value	$\Delta H_c = 43.8$ MJ/kg
Energy content of fuel	$E = M\Delta H_c = 2000 \times 43.8 = 87\,600$ MJ
Surface burning rate	$q = 0.031$ kg/s/m^2
Floor area	$A_f = 6.0 \times 10.0 = 60.0$ m^2

Specific heat release rate	$Q_s = q\Delta H_c = 0.031 \times 43.8 = 1.36\,\text{MW/m}^2$
Total heat release rate	$Q = Q_s A_f = 1.36 \times 60 = 81.6\,\text{MW}$
Duration of burning	$t = E/Q = 87\,600/81.6 = 1074\,\text{sec} = 18\,\text{minutes}$

WORKED EXAMPLE 3.3

Calculate the heat release rate for 160 kg of office furniture with an average calorific value of 20 MJ/kg, if it burns as a 'fast' t^2 fire with a peak heat release rate of 9.0 MW.

Mass of fuel	$M = 160\,\text{kg}$
Calorific value	$\Delta H_c = 20\,\text{MJ/kg}$
Energy contained in fuel	$E = M\Delta H_c = 160 \times 20 = 3200\,\text{MJ}$
Growth factor for fast fire	$k = 150\,\text{s}/\sqrt{\text{MW}}$
Peak heat release rate	$Q_p = 9.0\,\text{MW}$
Time to reach peak heat release	$t_1 = k\sqrt{Q_p} = 150 \times \sqrt{9} = 450\,\text{s}$
Energy released in time t_1	$E_1 = t_1 Q_p/3 = 450 \times 9/3 = 1350\,\text{MJ}$

$E_1 < E$ so there is steady burning

Energy released in steady burning	$E_2 = E - E_1 = 3200 - 1350 = 1850\,\text{MJ}$
Duration of steady burning	$t_b = E_2/Q_p = 1850/9 = 206\,\text{s}$

The heat release rate curve for this fire is shown as the 'fast' fire in Figure 3.5.

Repeat the calculation for a 'slow' t^2 fire growth rate.

Growth factor for slow fire	$k = 600\,\text{s}/\sqrt{\text{MW}}$
Time to reach peak heat release	$t_1 = k\sqrt{Q_p} = 600 \times \sqrt{9} = 1800\,\text{s}$
Energy released in time t_1	$E_1 = t_1\, Q_p/3 = 1800 \times 9/3 = 5400\,\text{MJ}$

$E_1 > E$ so the fire does not reach steady-burning stage

Time for all fuel to burn	$t_m = (3E\,k^2)^{1/3} = (3 \times 3200 \times 600^2)^{1/3} = 1512\,\text{s}$
Heat release at time t_1	$Q_m = (t_m/k)^2 = (1512/600)^2 = 6.3\,\text{MW}$

The heat release rate curve is shown as the 'slow' fire in Figure 3.5.

WORKED EXAMPLE 3.4

Calculate the steady-state heat transfer through a 150 mm thick concrete wall if the temperature on the fire side is 800°C and the temperature on the cooler side is 200°C.

Wall thickness	$x = 0.15\,\text{m}$
Temperature difference	$\Delta T = 800 - 200 = 600°\text{C} = 600\,\text{K}$
Temperature gradient	$\text{d}T/\text{d}x = 600/0.150 = 4000\,\text{K/m}$
Thermal conductivity	$k = 1.0\,\text{W/mK}$ (from Table 3.4)
Heat transfer	$\dot{q}'' = k\,\text{d}T/\text{d}x = 1.0 \times 4000 = 4000\,\text{W/m}^2$
	$= 4\,\text{MW/m}^2$

Calculate the convective heat transfer coefficient on the cool side of the wall if the ambient temperature is 20°C and all the heat passing through the wall is carried away by convection.

Temperature of wall	$T_w = 200°\text{C}$
Ambient temperature	$T_a = 20°\text{C}$
Temperature difference	$\Delta T = T_w - T_a = 200 - 20 = 180°\text{C} = 180\,\text{K}$
Heat transfer	$\dot{q}'' = 4000\,\text{W/m}^2$
Convective heat transfer coefficient	$h = \dot{q}''/\Delta T = 4000/180 = 22.2\,\text{W/m}^2\,\text{K}$

WORKED EXAMPLE 3.5

Calculate the radiant heat flux from a window in a burning building to the surface of an adjacent building 5.0 m away. The window is 2.0 m high by 3.0 m wide and the fire temperature is 800°C. Assume an emissivity of 0.9.

Emitter height	$H = 2.0\,\text{m}$
Emitter width	$W = 3.0\,\text{m}$
Distance from emitter	$r = 5.0\,\text{m}$
Height ratio	$x = H/2r = 2/(2 \times 5) = 0.20$
Width ratio	$y = W/2r = 3/(2 \times 5) = 0.30$

Configuration factor

$$\varphi = \frac{1}{90}\left[\frac{x}{\sqrt{1+x^2}}\tan^{-1}\left(\frac{y}{\sqrt{1+x^2}}\right) + \frac{y}{\sqrt{1+y^2}}\tan^{-1}\left(\frac{x}{\sqrt{1+y^2}}\right)\right]$$

$$= 0.0703$$

Emitter temperature	$T = 800°\text{C} = 1073\,\text{K}$
Emissivity	$\varepsilon = 0.9$
Stefan–Boltzmann constant	$\sigma = 5.67 \times 10^{-8}\,\text{W/m}^2\text{K}^2$
Radiant heat flux	$\dot{q}'' = \varphi\varepsilon\sigma T^4$
	$= 0.070 \times 0.9 \times 5.6 \times 10^{-8} \times 1073^4/1000$
	$= 4.69\,\text{kW/m}^2$

4
Room Fires

4.1 OVERVIEW

This chapter reviews the behaviour of fires in rooms. Emphasis is given to post-flashover fires which provide the greatest threat to structural members and other fire resistant building elements.

4.2 PRE-FLASHOVER FIRES

Fires in rooms are described separately for pre- and post-flashover fires, beginning with pre-flashover fires. An understanding of pre-flashover fire behaviour is essential when designing for life safety in fires. A brief summary is given below. For more information on room fires, consult Drysdale (1998) and Karlsson and Quintiere (2000).

4.2.1 Burning Items in Rooms

The heat release rate of burning items of furniture or other fuel in the open has been discussed in Chapter 3. Burning objects can behave differently when they burn inside a room rather than in the open. The convective plume of hot gases above the burning object will hit the ceiling and spread horizontally to form a hot upper layer. In the early stages of the fire, the rate of burning may be significantly enhanced by radiant feedback from this hot upper layer. Later, the rate of burning may be severely reduced because of limited ventilation, which can restrict the transport of incoming air and outgoing combustion products through the openings, and reduce the oxygen concentration in the lower layer.

4.2.2 Room Fires

Figure 4.1 shows a fire in a room, at an early stage when only a single item of furniture is burning, before any spread of flame to linings or other items. The room has one open door. The combustion reaction requires the input of oxygen, initially

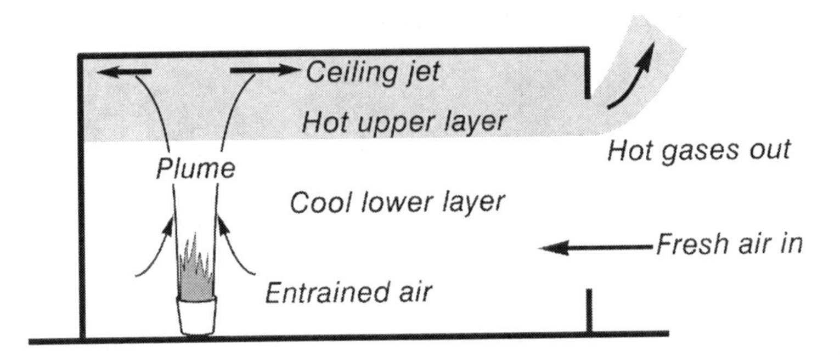

Figure 4.1 Early stages of fire in a room

obtained from the air in the room, but later from air coming in through the door opening. The energy released by the fire acts like a pump, pulling cool air into the room, entraining it into the fire plume and pushing combustion products out through the top of the door opening.

The fire *plume* provides buoyant convective transport of combustion products up to the ceiling. The plume entrains a large amount of cold air which cools and dilutes the combustion products. The diluted combustion products form a hot upper layer within the room. The thickness and temperature of the hot layer increase as the fire grows. The lower layer consists of cooler incoming air which is heated slightly by mixing and radiation from the upper layer (Figure 4.2).

Where the plume reaches the ceiling, there is a flow of hot gases radially outwards along the underside of the ceiling, called the *ceiling jet*. The direction of the ceiling jet will be influenced by the shape of the ceiling. For a smooth horizontal ceiling the flow will be the same in each radial direction. The hot gases in the ceiling jet will activate the heat detectors or fire sprinkler heads located near the ceiling.

As the fire continues to burn, the volume of smoke and hot gases in the hot upper layer increases, reducing the height of the interface between the two layers. The combustion products will start to flow out the door opening when the interface drops below the door soffit as shown in Figure 4.1. The hot layer thickness depends on the size and duration of the fire and the size of door or window openings. If there are insufficient ventilation openings, the fire will die down and may self-extinguish because of lack of oxygen.

The nature of wall, floor and ceiling linings can have a significant influence on fire growth and development in a room. Combustible lining materials can drastically increase the rate of initial fire growth due to rapid flame spread up walls and across ceilings. Temperatures will be higher and fire growth more rapid in a well-insulated room where less heat can be absorbed by the bounding elements. Computer models predicting fire growth including ignition and flame spread on combustible lining materials are under development (e.g. Wade and Barnett, 1997).

If an automatic sprinkler system is installed, it will operate early in the pre-flashover fire period, and either extinguish the fire or prevent it from growing any larger after that time.

Figure 4.2 Smoke damage following a pre-flashover fire in a room, indicating the thickness of the hot upper layer during the fire

4.2.3 Pre-flashover Fire Calculations

In the fire engineering design process, much effort is expended in calculating the effects of pre-flashover fires, because this stage of the fire has the most influence on life safety. In order to ensure safe egress of building occupants it is necessary for the designer to know the expected rate of fire growth, and the resulting depth and temperature of the hot upper layer in the fire room and adjacent corridors as the fire develops. It is also essential to know the activation time and resulting effects of automatic detection and suppression systems. These calculations are most often made using computer models such as those described below. For hand calculations of upper layer temperatures, Walton and Thomas (1995) describe equations derived by McCaffrey *et al.* (1981).

Zone models

Zone models are relatively simple computer programs which can model the behaviour of a pre-flashover fire such as that shown in Figure 4.1. Most are *two-zone* models because they consider the room fire in terms of two homogeneous layers, or zones, and the connecting plume (Quintiere, 1995). Conservation equations for mass, momentum and energy are applied to each zone in a dynamic process that calculates the size, temperature, and species concentration of each zone as the fire progresses, together with the flow of smoke and toxic products through openings in the walls and ceiling. Zone models do not calculate the growth of fire on objects or surfaces, so they require a design fire such as described in Chapter 3 to be provided as input data. Typical output includes the layer height, temperatures and concentrations of gas species in both layers, floor and wall temperatures, and the heat flux at the floor level.

Among the most simple zone models are ASET (for rooms with no significant openings) and FIRE SIMULATOR (for rooms with openings) from the FPEtool suite of programs (Nelson, 1986, Deal, 1993). One of the most versatile and widely used zone models is CFAST (Peacock *et al.*, 1993) which can calculate the movement of smoke and hot gases in up to 15 interconnected rooms. FASTLite (Portier *et al.*, 1996) is a package of fire-modelling programs which allows fire modelling of up to three interconnected rooms using CFAST. This is available free of charge on CD-ROM and on the World Wide Web. These models are produced by the Building and Fire Research Laboratory of the National Institute of Standards and Technology (NIST).

Field models

The assumption of two distinct layers of gases is a convenient way of describing and calculating fire behaviour in rooms, but in reality there is a gradual three-dimensional transition of temperature, density and smoke between the layers, and this transition limits the accuracy of zone models. *Field models* are much more sophisticated computer programs which use computational fluid dynamics (CFD) to model fires using a large number of discrete zones in a three-dimensional grid. Field models are much more difficult to run and to interpret than zone models, so they are generally used as research tools rather than design tools. Neither zone models nor field models are suitable for modelling post-flashover fires.

4.3 FLASHOVER

If the fire shown in Figure 4.1 is allowed to grow without intervention, assuming sufficient fuel in the burning item, temperatures in the hot upper layer will increase, with increasing radiant heat flux to all objects in the room. At a critical level of heat flux, all exposed combustible items in the room will begin to burn, leading to a rapid increase in both heat release rate and temperatures. This transition is *flashover*. After flashover the fire is often referred to as a 'post-flashover fire', 'fully-developed fire' or 'full-room involvement'. It is not possible for flashover to occur in an open unenclosed space because, by definition, flashover can only occur in an enclosed compartment.

The definition of *flashover* is the transition from a localized fire to combustion of all exposed combustible surfaces in a room. Drysdale (1998) points out that flashover must be considered as a transition between two states rather than a precise event.

4.3.1 Conditions Necessary for Flashover

There are certain pre-conditions necessary for flashover to occur. There must be sufficient fuel and ventilation for a growing fire to develop to a significant size. The ceiling must be able to trap hot gases, and the geometry of the room must allow the radiant heat flux from the hot layer to reach critical ignition levels at the level of the fuel items. Drysdale (1998) gives a detailed discussion of these factors, with summaries of a number of compartment tests. In a typical room flashover occurs when the hot layer temperature is about 600°C resulting in a radiant heat flux of about 20 kW/m^2 at floor level.

From an analysis of a large number of experimental fires, it has been observed that flashover will only occur if the heat output from the fire reaches a certain critical value, related to the size of the ventilation openings. For a room with one window, the critical value of heat release Q_{fo} (MW) is given by 'Thomas's flashover criterion' (Walton and Thomas, 1995):

$$Q_{fo} = 0.0078\ A_t + 0.378\ A_v\ \sqrt{H_v} \tag{4.1}$$

where A_t is the total internal surface area of the room (m^2), A_v is the area of the window opening (m), and H_v is the height of the window opening (m).

Drysdale (1998) derives a similar expression, and points out that this type of correlation is very approximate, depending on the size, shape and lining materials of the room, and even the location of the fire within the room. Walton and Thomas (1995) show a comparison of this and similar expressions. If all the burning objects in a room can be characterized by a *t*-squared design fire as described in Chapter 3, a rough method of calculating the time to flashover is to use the critical value of heat release Q_{fo} from Equation (4.1) in Equation (3.9).

4.4 POST-FLASHOVER FIRES

The behaviour of the fire changes dramatically after flashover. The flows of air and combustion gases become very turbulent. The very high temperatures and radiant heat fluxes throughout the room cause all exposed combustible surfaces to pyrolyse, producing large quantities of combustible gases, which burn where there is sufficient oxygen. The most important information for structural design is the temperature in the room during the post-flashover fire. Sometimes the burning rates are also useful, so they are described first.

4.4.1 Ventilation Controlled Burning

In typical rooms, post-flashover fires are ventilation controlled, so the rate of combustion depends on the size and shape of ventilation openings. It is usually conservatively assumed that all window glass (other than wired glass or fire resistant glass) will break and fall out at the time of flashover, as a result of the rapid rise in temperature. If the glass does not fall out the fire will burn for a longer time at a lower rate of heat release.

In a ventilation controlled fire, there is insufficient air in the room to allow all the combustible gases to burn inside the room, so the flames extend out of the windows and additional combustion takes place where the hot unburned gaseous fuels mix with outside air (Figure 4.4).

Rate of burning

When the fire is ventilation controlled, the rate of combustion is limited by the volume of cold air that can enter and the volume of hot gases that can leave the room. For a room with a single opening, Kawagoe (1958) used many experiments to show that the rate of burning of wood fuel $\dot{m}$ (kg/s) can be approximated by

$$\dot{m} = 0.092\ A_v\ \sqrt{H_v} \tag{4.2}$$

where A_v is the area of the window opening (m^2) and H_v is the height of the window opening (m). In many references the burning rate $\dot{m}$ is given as $5.5A_v\sqrt{H_v}$ kg/minute or $330A_v\sqrt{H_v}$ kg/hour, which are the same as Equation (4.2) in different units of time. Note that $A_v\sqrt{H_v}$ can be re-written as $BH_v^{1.5}$ where B is the breadth of the window opening. This shows that the burning rate is largely dependent on the area of the window opening, but more so on its height.

If the total mass of fuel in the room is known, the duration of the burning period t_b (s) can be calculated using

$$t_b = M_f/\dot{m} \tag{4.3}$$

where M_f is the total mass of fuel available for combustion (kg).

The corresponding ventilation controlled heat release rate Q_{vent} (MW) for steady-state burning is

$$Q_{vent} = \dot{m}\Delta H_c \tag{4.4}$$

where ΔH_c is the heat of combustion of the fuel (MJ/kg).

If the total amount of fuel is known in energy units (MJ), the duration of the burning period t_b (s) can be calculated from

$$t_b = E/Q_{vent} \tag{4.5}$$

where E is the energy content of fuel available for combustion (MJ).

These calculations all depend on the approximate relationship for burning rate given by Equation (4.2) which is widely used, but not always accurate. Even if the burning rate is known precisely, the calculation of heat release rate is not accurate because an unknown proportion of the pyrolysis products burn as flames outside the window rather than inside the compartment. Other sources of uncertainty arise because some proportion of the fuel may not be available for combustion, and the fire may change to fuel control after some time.

Drysdale (1998) shows how Equation (4.2) can be derived by considering the flows of air and combustion products through an opening as shown in Figure 4.3. He points out that this derivation is not appropriate for fuel other than wood cribs, where greatly different behaviour can occur.

In a ventilation controlled fire there are very complex interactions between the radiant heat flux on the fuel, the rate of pyrolysis (or evaporation) of the fuel, the rate of burning of the gaseous products, the inflow of air to support the combustion, and the outflow of combustion gases and unburned fuel through openings. The

Figure 4.3 Window flows for ventilation controlled fire

interactions depend on the shape of the fuel (cribs or lining materials), the fuel itself (wood or plastic or liquid fuel) and the ventilation openings.

The empirical dependence of the ventilation controlled burning rate on the term $A_v\sqrt{H_v}$ has been observed in many studies, but some tests have shown departures from Kawagoe's equation. Following a large number of small-scale compartment fires with wood cribs reported by Thomas and Heselden (1972), Law (1983) proposed a slightly more refined equation for burning rate, finding that the burning rate is not directly proportional to $A_v\sqrt{H_v}$ but also depends on the floor shape of the compartment. Law's equation is

$$\dot{m} = 0.18\, A_v \sqrt{\frac{H_v W}{D}}\,(1 - e^{-0.036\Omega}) \tag{4.6}$$

where

$$\Omega = \frac{A_t - A_v}{A_v\sqrt{H_v}}$$

W is the compartment width (m), D is the compartment depth (m), and A_t is the total area of the internal surfaces of the compartment (m^2).

Equation (4.6) gives approximately the same burning rate as Equation (4.2) for square compartments with a ventilation factor of $A_v\sqrt{H_v}/A_t = 0.05$ ($\Omega = 20$). The burning rate is greater than Equation (4.2) for smaller openings and wider shallower compartments. Equation (4.6) only applies directly to compartments with windows in one wall because it is not easy to differentiate the terms W and D if there are windows in two or more walls. Equation (4.6) is used in the Eurocode (EC1, 1994) for calculating the rate of burning when assessing the flame height from compartment windows.

Thomas and Bennetts (1999) have cast considerable doubt on the applicability of Kawagoe's equation, showing that the burning rate also depends heavily on the shape of the room and the width of the window in proportion to the wall in which it is located. If the width of the window is less than the full width of the wall, the burning rate is seen to be much higher than predicted by Equation (4.2) because of increased turbulent flow at the edges of the window. Despite these recent findings, Kawagoe's equation will form the basis of most post-flashover fire calculations until further research is conducted.

Figure 4.4 Post-flashover fire on the top floor of a multi-storey office building. The flames coming out the windows indicate that this fire is ventilation controlled

Ventilation factor

The amount of ventilation in a fire compartment is often described by the *ventilation factor* (or *opening factor*) F_v ($m^{0.5}$) given by

$$F_v = A_v\sqrt{H_v}/A_t \tag{4.7}$$

where A_v is the area of the window opening (m^2), H_v is the height of the window opening (m), and A_t is the total internal area of the bounding surfaces (including openings) (m^2).

The ventilation factor F_v has units of $m^{0.5}$ which has little intuitive meaning. However, if the acceleration of gravity g is introduced, the term $A_v\sqrt{gH_v}/A_t$ has units of m/s, which is related to the velocity of gas flow through the openings.

It can be seen that the rate of burning in many of the above equations is a function of $A_v\sqrt{H_v}$ ($m^{2.5}$) These units also have little intuitive meaning. If the acceleration of gravity is introduced again, the term $A_v\sqrt{gH_v}$ has units of m^3/s, which is related to the volumetric flow of gases through the opening.

Multiple openings

Equations (4.1), (4.2) and (4.6) have been written for a single window opening in one wall of the compartment. If there is more than one opening, the same equations can

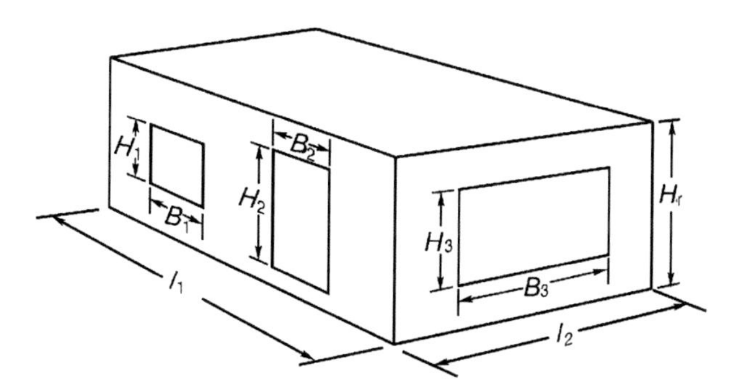

Figure 4.5 Calculation of ventilation factor for more than one window

be used, with A_v being the total area of all the openings and H_v being the weighted average height of all the window and door openings. Openings can be on several walls, which implies an assumption that the air flow is similar in all openings and there is no strong wind blowing which would create a cross flow through the room.

Referring to Figure 4.5, the weighted average height of the openings H_v and the area of the internal surfaces of the compartment A_t can be calculated using

$$H_v = (A_1 H_1 + A_2 H_2 + \dots)/A_v$$
$$A_v = A_1 + A_2 + \dots = B_1 H_1 + B_2 H_2 + \dots \qquad (4.8)$$
$$A_t = 2(l_1 l_2 + l_1 H_r + l_2 H_r)$$

The terms B_i and H_i are the breadth and height of the windows, l_1 and l_2 are the floor plan dimensions, and H_r is the room height.

4.4.2 Fuel Controlled Burning

Not all post-flashover fires are ventilation controlled. The rate of burning may sometimes be controlled by the surface area of the fuel, especially in large well-ventilated rooms containing fuel items which have a limited area of combustible surfaces. In this case the rate of burning will be similar to that which would occur for the fuel item burning in the open air, with enhancement from radiant feedback from the hot upper layer of gases or hot wall and ceiling surfaces. Most fires become fuel controlled in the decay period.

The average heat release rate from a fuel controlled fire can be calculated if the total fuel load and the duration of burning are both known. For example, Law (1983) has observed many experimental fires from which she concludes that typical domestic furniture fires have a free burning fire duration t_b of about 20 minutes (1200 s), so that the heat release rate Q_{fuel} (MW) is given crudely by

$$Q_{fuel} = E/1200 \qquad (4.9)$$

where E is the total fuel load (MJ).

More often the duration is not known, and the heat release rate needs to be estimated from information about the fuel and the temperatures in the fire compartment. When the rate of burning of a fuel item is controlled by the available surface area, Drysdale (1998) shows how the heat release rate Q_{fuel} (MW) can be estimated using

$$Q_{\text{fuel}} = \dot{q}_{\text{i}}'' A_{\text{fuel}} \, \Delta H_{\text{c}} / L_{\text{v}} \tag{4.10}$$

where $\dot{q}_{\text{i}}''$ is the incident radiation reaching the fuel surface (the total heat flux less any losses expressed as a heat flux through the fuel surface) (MW/m^2), A_{fuel} is the exposed surface area of the fuel (m^2), ΔH_{c} is the heat of combustion of the volatiles (MJ/kg), and L_{v} is the heat of gasification (MJ/kg). Some of these terms are not clearly defined for typical building fires. The incident radiation in a post-flashover fire is generally assumed to be about 70 kW/m^2.

For example, the incident radiation used in the Fire Simulator module of the FPEtool program (Deal, 1993) varies from 60 to 80 kW/m^2 which (with an emissivity of 0.8) corresponds to a radiating surface temperature of 800°C to 880°C. The FASTLite program uses Equation (4.10) with $\dot{q}_{\text{i}}'' = 70$ kW/m^2 (Buchanan, 1997).

The heat of gasification is the amount of energy required to pyrolize a unit mass of fuel. Drysdale (1998) reports heat of gasification values for wood ranging from 1.7 to 5.9 MJ/kg and for plastics ranging from 1.2 to 3.7 MJ/kg. The highest figures in these ranges should be used to get realistic results (J. Quintiere, personal communication). The rate of heat release calculated using Equation (4.10) is higher than the rate measured in some fire tests using wood fuel, which may be due to the reduction of the burning rate after a layer of char has formed on the wood surface.

Equation (4.10) is based on the assumption that the rate of heat release is proportional to the imposed heat flux, with no influence of the thickness or shape of the fuel. This is appropriate for liquid or plastic fuels, but for wood, the burning rate also depends on the thickness of the slab or the size of the sticks. Babrauskas (1981) reports that for burning of thick slabs of wood, the surface regression rate v_{p} (m/s) is approximately

$$v_{\text{p}} = 8.5 - 10.0 \times 10^{-6} \text{ m/s} \tag{4.11}$$

(0.5–0.6 mm/min) This is similar to the charring rate observed in many fire resistance tests of heavy timber construction, as described in Chapter 10. For thinner slabs the regression rate increases as a function of the thickness of the wood, according to

$$v_{\text{p}} = 2.2 \times 10^{-6} D^{-0.6} \text{ m/s} \tag{4.12}$$

where D is the thickness of the slab of wood (m).

The corresponding heat release rate Q_{fuel} (MW) is given by

$$Q_{\text{fuel}} = v_{\text{p}} \rho A_{\text{fuel}} \Delta H_{\text{c}} \tag{4.13}$$

where ρ is the density of the fuel (kg/m^3). For a room with wood fuel of known size and shape, Equation (4.13) is easier to use than Equation (4.10) because the input variables are more easily defined.

In either case, if the total amount of fuel E (MJ) is known and the heat release rate is assumed to be constant, the duration of the burning period t_b (s) can be calculated from

$$t_b = E/Q_{fuel} \qquad (4.14)$$

In all the above calculations, for both ventilation and fuel controlled burning, not all of the combustible material in the room may be available for immediate combustion. For this reason, many researchers introduce a *fuel fraction* which is an efficiency factor by which the heat of combustion or available fuel is reduced. Babrauskas (1981) suggests a value in the range 0.5 to 0.9, so 0.7 may be a suitable value for general design purposes.

As shown by the worked examples, the various equations for fuel controlled heat release rate can give very different answers, so more research in this area is necessary. This is not of major importance because most post-flashover fire calculations assume ventilation controlled burning.

4.4.3 Temperatures

Estimation of temperatures in post-flashover fires is an essential part of structural design for fire safety. Unfortunately this cannot be done precisely. This chapter describes measured and predicted temperatures from various studies, and a range of methods for estimating temperatures for design purposes.

Temperatures in post-flashover fires are usually at least 1000 °C. The temperature at any time depends on the balance between the heat released within the room and all the heat losses; through openings by radiation and convection, and by conduction into the walls, floor and ceiling.

Measured temperatures

Several experimental studies have measured temperatures in post-flashover fires. There is considerable scatter between the results of different studies. Figure 4.6 shows the shapes of typical time–temperature curves, starting at flashover, measured by Butcher *et al.* (1966) in real rooms with door or window openings and well-distributed fuel load. Figure 4.6 also shows the ISO 834 standard curve used for fire-resistance testing, as described in Chapter 5.

Figure 4.7 shows the maximum recorded temperature during the steady burning period for a large number of wood crib fires in small-scale compartments reported by Thomas and Heselden (1972). The recorded temperature was the average of a number of thermocouple readings within each compartment. An empirical equation for the line in Figure 4.7 has been developed by Law (1983), summarized by Walton and Thomas (1995). The maximum temperature T_{max} (°C) is given by

$$T_{max} = 6000(1 - e^{-0.1\Omega})/\sqrt{\Omega} \qquad (4.15)$$

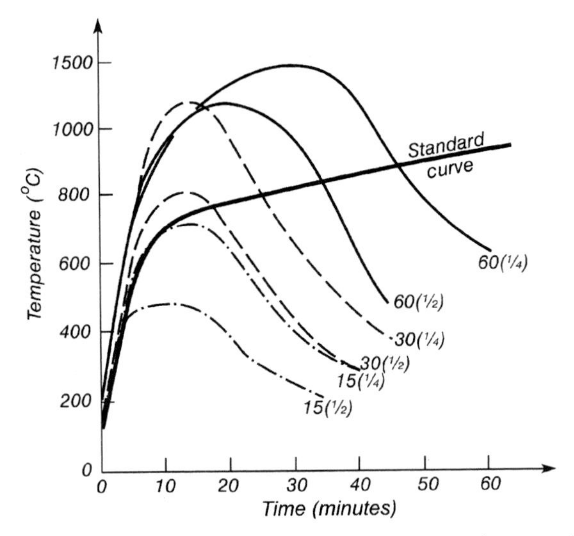

Figure 4.6 Experimental time temperature curves (From Butcher *et al.* (1966). Crown copyright is reproduced with the permission of the Controller of Her Majesty's Stationery Office)

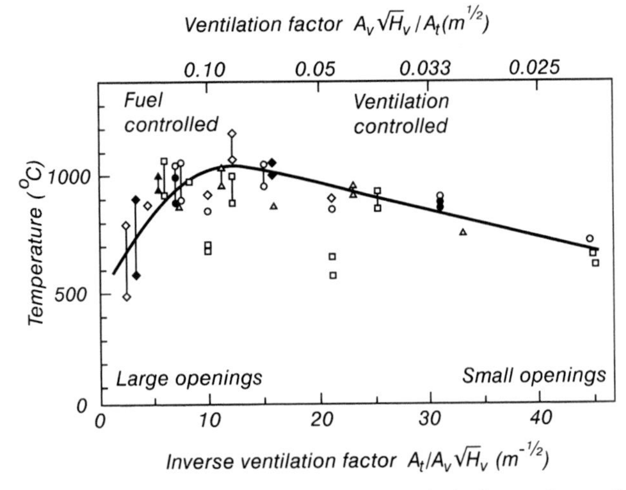

Figure 4.7 Maximum temperature in the burning period of experimental fires (Reproduced from Thomas and Heselden (1972) by permission of Building Research Establishment Ltd)

where

$$\Omega = \frac{A_t - A_v}{A_v\sqrt{H_v}}$$

The maximum temperature in Equation (4.15) may not be reached if the fuel load is small. For low fuel loads it can be reduced according to

$$T = T_{max}(1 - e^{-0.05\psi}) \tag{4.16}$$

where

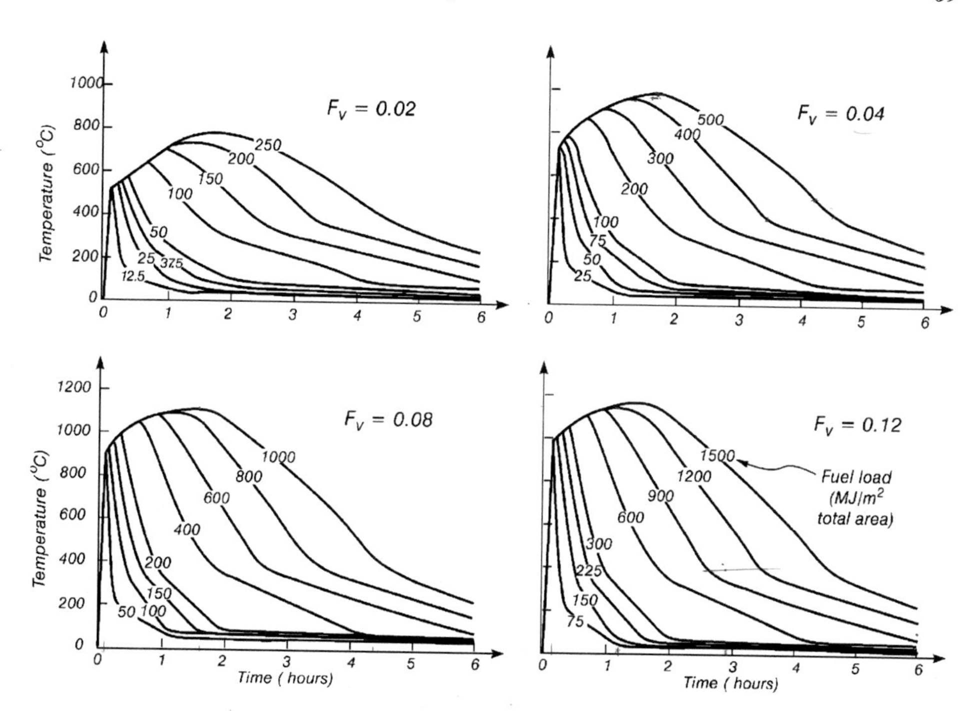

Figure 4.8 Time–temperature curves for different ventilation factors and fuel loads (MJ/m^2 of total surface area)

$$\psi = \frac{L}{\sqrt{A_v(A_t - A_v)}}$$

and L is the fire load (kg, wood equivalent).

Swedish curves

The most widely referenced time–temperature curves for real fire exposure are those of Magnusson and Thelandersson (1970), shown in Figure 4.8. These are often referred to as the 'Swedish' fire curves. They are derived from heat balance calculations, using Kawagoe's equation (Equation (4.2)) for the burning rate of ventilation controlled fires. Each group of curves is for a different ventilation factor, with fuel load as marked. Note that the units of fuel load are MJ per m^2 of total surface area (not MJ per m^2 of floor area which is more often used in design calculations). The rising branch of the curve for ventilation factor of 0.04 is very similar to the standard time–temperature curve described in Chapter 5.

To show the effects of changing fuel load and ventilation more clearly, some of the curves in Figure 4.8 have been re-drawn in the following two figures. Figure 4.9 shows the effect of varying the size of the ventilation openings, for a constant fuel load. Well-ventilated fires burn faster than poorly ventilated fires, so they burn at higher temperatures, but for a shorter duration.

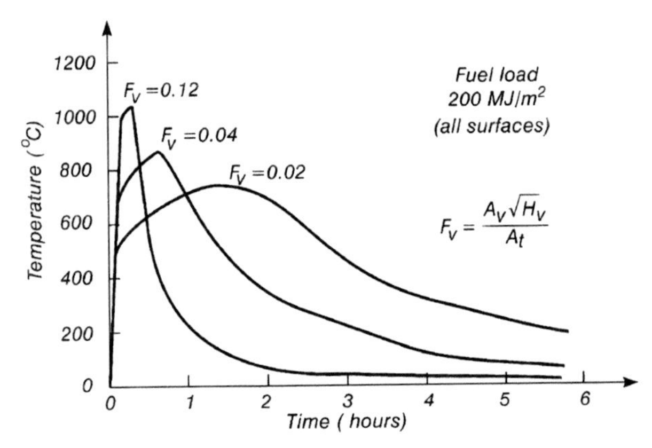

Figure 4.9 Time–temperature curves for varying ventilation and constant fuel load (MJ/m^2 total surface area)

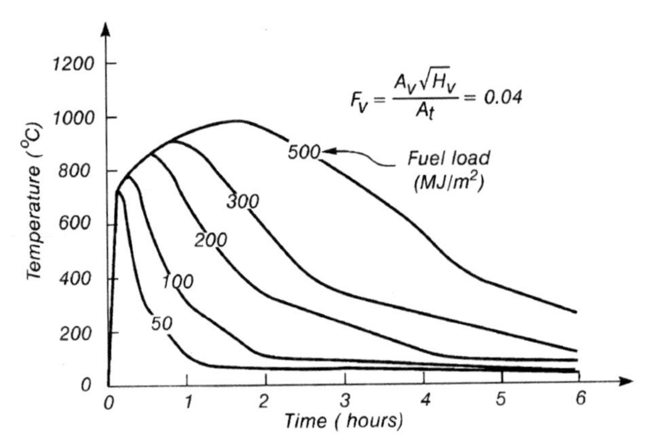

Figure 4.10 Time–temperature curves for varying fuel load (MJ/m^2 total surface area) and constant ventilation

Figure 4.10 shows the effect of varying the fuel load for a constant size of ventilation opening. The rate of burning is the same in all cases because it is controlled by the window size, but increasing the fuel load leads to longer and hotter fires before the decay period begins.

Rate of temperature decay

The rate of temperature decay in a post-flashover fire is not easy to predict. The decay rate depends mainly on the shape and material of the fuel, the size of the ventilation openings and the thermal properties of the lining materials.

If all the fuel is liquid or molten material in a pool, the burning period will end suddenly when all the fuel has been consumed. On the other hand, solid materials like wood will burn at a predictable rate, leading to long decay periods depending on the thickness of the fuel items. The burning rate will be controlled by limited

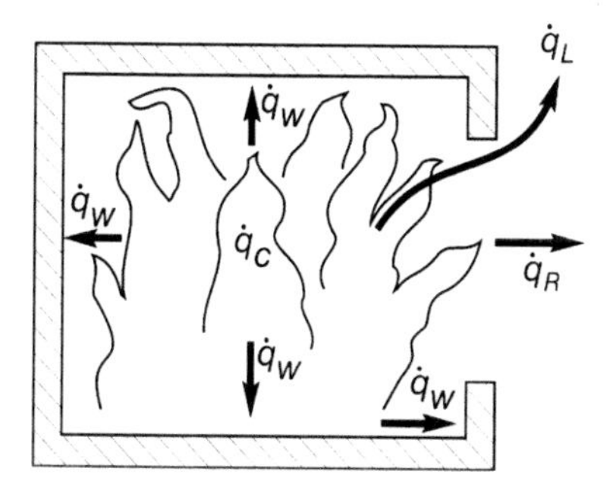

Figure 4.11 Heat balance for a post-flashover room fire

ventilation as long as the area of burning surfaces remains large. After the burning surface area reduces to a certain level, the fire will become fuel controlled and the decay rate will depend on the volume and thickness of the remaining items of fuel. If the fuel has a small ratio of surface area to volume, the fuel controlled burning in the later stages of the fire will lead to a long slow decay rate.

The rate of temperature decay depends on the ventilation openings because large openings will allow rapid heat loss from the compartment by both convection and radiation, whereas small openings will allow the heat to be trapped for much longer.

The effect of the thermal properties of the construction materials is not easy to quantify. On one hand, materials of low thermal inertia will store less heat and hence transfer less heat back into the compartment after the fire is out, leading to a rapid rate of decay. On the other hand, such materials (which also have low thermal conductivity) will insulate the compartment and result in higher temperatures if any residual burning occurs in the decay period. Of these two contradictory effects, the first is likely to predominate, so that materials of low thermal inertia will lead to more rapid decay rates.

4.4.4 Computer Models

A number of computer models have been used for calculating temperatures in post-flashover room fires. Most of these are single-zone models which consider the room to be a well-mixed reactor. It is possible to use two-zone models for post-flashover fires, but these are not generally considered appropriate because many of the pre-flashover assumptions are no longer valid. Field models are not easily applied to post-flashover fires because of excessive turbulence.

All computer models for post-flashover fires are based on heat balance. Figure 4.11 shows the main components of heat flow in a simple compartment fire. The heat produced by combustion of the fuel $\dot{q}_C$ is balanced by the heat losses, the main components being heat conducted into the surrounding structure $\dot{q}_W$, heat radiated through the opening $\dot{q}_R$, and heat carried out of the opening by convection of hot gases and smoke $\dot{q}_L$. The computer models consider this heat balance and solve the conservation equations to predict the temperature of the gases within the compartment. The single zone models assume that all combustion takes place within the compartment, temperatures are uniform within the compartment, and

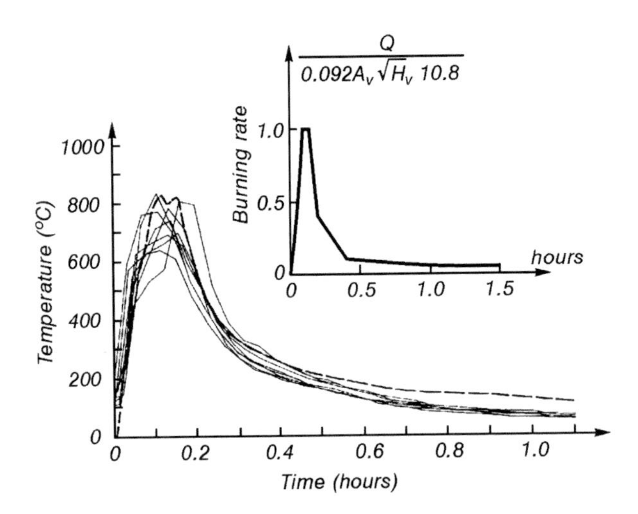

Figure 4.12 Burning rate and resulting temperatures used for calculating Swedish curves (Reproduced from (Magnusson and Thelandersson, 1970) by permission of Fire Safety Engineering Department, Lund University)

heat flow into the surrounding structure is identical on all walls and the ceiling. A description of many different models is given by Feasey (1999).

A difficulty for all models is the calculation of the burning rate. Most models simply assume ventilation controlled burning as given by Equation (4.2), but some also include fuel controlled burning. None of the available models is able to accurately include the effects of horizontal openings in the ceiling.

Swedish method

The Swedish curves have been shown in Figure 4.8. When calibrating their model, Magnusson and Thelandersson (1970) manipulated the heat release rate to produce temperatures similar to those observed in short duration test fires. The resulting shape of the heat release rate curve used in those calculations is shown in Figure 4.12. The peak rate of heat release is the theoretical rate for ventilation controlled burning given by Equation (4.2), assuming a calorific value of 10.8 MJ/kg which is lower than the value used by most other authors. Temperatures in the decay phase are calculated in the computer model using assumed burning rates such as those shown in Figure 4.12. Magnusson and Thelandersson extrapolated their computer model to much higher fuel loads and longer durations than the available test data in order to produce curves such as those shown in Figure 4.8.

Lie method

In a similar approach to the Swedish method, Lie (1995) used values from Kawagoe's original work to perform heat balance calculations for post-flashover fires with a range of ventilation factors and different wall lining materials, proposing a set of approximate equations for design purposes, including the duration of burning and arbitrary decay rates. Lie's curves are unrealistic for rooms with small windows because the proposed temperatures are not sufficient for flashover to have occurred.

COMPF2

A more comprehensive computer program for calculating temperatures in post-flashover room fires is COMPF2 (Babrauskas, 1979). This program is also a single-zone model which solves the heat balance equations to generate gas temperatures. There are several options for calculating the heat release rate, based on ventilation control, fuel control (Equations (4.11)–(4.12)), or the porosity of wood crib fuels. Feasey (1999) gives a detailed description of COMPF2 showing how he has calibrated it to a large number of European fires.

Babrauskas (1981) has developed a closed form approximation to the results of his COMPF2 computer program. The procedure, which is a little cumbersome for design purposes, involves modifying a reference temperature with five non-dimensionalized multiplicative factors to allow for the combustion chemistry, transient and steady-state wall losses, window opening height and combustion efficiency, accounting for both fuel rich and fuel lean fires. A summary is provided by Walton and Thomas (1995).

OZONE

Ozone is a one zone model recently developed in Europe as part of a major collaboration investigating the 'natural fire safety concept' for competitive design of steel buildings (Franssen *et al.*, 1999). Ozone is based on similar principles to COMPF2, with improved calculation procedures.

FASTLite

FASTLite is the only two-zone model which seriously attempts to model post-flashover fires. The post-flashover calculations use the C-FAST model with new equations for the heat release rate adapted from the FIRE SIMULATOR section of FPEtool. These have been improved to make the duration of burning directly proportional to the amount of available fuel. FASTLite calculates fuel controlled and ventilation controlled heat release rates using Equations (4.4) and (4.10), and uses the lower of these in the simulation. The fuel area and thickness can be manipulated by the user to enforce either fuel or ventilation controlled burning (Buchanan, 1997). The output temperatures from FASTLite are rather higher than predicted by COMPF2 or the Swedish curves and there is some doubt about the applicability of the two-zone equations to a post-flashover fire that is essentially a single zone of well-mixed burning.

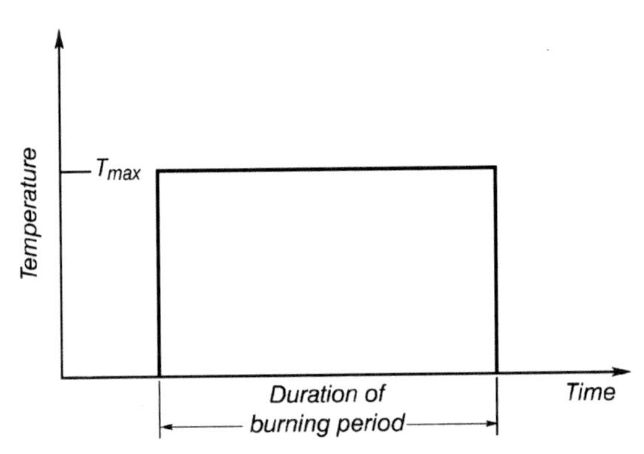

Figure 4.13 Design fire with constant temperature

Rate of temperature decay

Of the programs described above, only COMPF2 can calculate the rate of temperature reduction in the decay period for varying types and shapes of fuel. If the fuel has a small surface area to volume ratio, burning becomes fuel controlled in the later stages of the fire, leading to a long slow decay rate as thick items of fuel burn away. If all the fuel is assumed to be consumed in the burning period, the temperatures calculated by COMPF2 drop very rapidly after the fuel has burned away. FASTLite produces similar rapid decay rates, very unlike the more realistic fire behaviour shown in the Swedish curves.

4.5 DESIGN FIRES

When designing a structure to resist exposure to fire, it is often necessary to select a design fire. Alternative methods of obtaining design fires include hand calculations, published curves or parametric fire equations. Each of these are discussed in this section.

4.5.1 Hand Methods

A very simple, but crude, method is to assume that the fire has a constant temperature throughout the burning period, giving a time–temperature curve as shown in Figure 4.13. Such a time–temperature curve will be sufficiently accurate for simple designs. The maximum temperature can be estimated using Equations (4.15) and (4.16). The duration of the burning period can be calculated from Equation (4.5), assuming ventilation control.

4.5.2 Published Curves

For many applications it is possible to scale temperatures off published curves which have been derived from computer calculations. This can be done using the Swedish

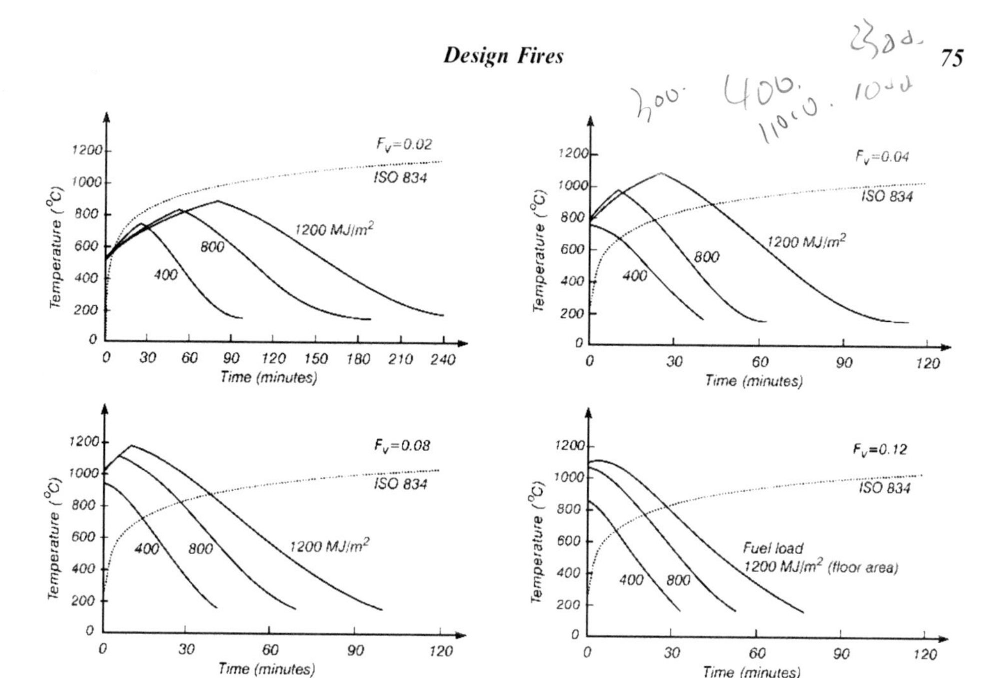

Figure 4.14 Time–temperature curves from COMPF-2

curves shown in Figure 4.8, using interpolation where necessary. Data points on the curves are also published (Magnusson and Thelandersson, 1970, Drysdale, 1998) but no simple formulae are available.

An alternative set of post-flashover design fire curves is shown in Figure 4.14. These curves have been derived by Feasey and Buchanan (2000) using the COMPF2 computer program. The fires have been calculated for a room with similar characteristics to that used in the Swedish curves after calibrating COMPF2 to a large number of experimental fires. Note that the fuel load is MJ per m² floor area. Figure 4.14 also includes the standard fire curve for comparison.

4.5.3 Eurocode Parametric Fires

The Eurocode (EC1, 1994) gives an equation for 'parametric' fires, allowing a time–temperature relationship to be produced for any combination of fuel load, ventilation openings and wall lining materials. The Eurocode parametric fire curves have been derived to give a good approximation to the burning period of the Swedish curves shown in Figure 4.8.

Equation for burning period

The Eurocode equation for temperature T (°C) is

$$T = 1325(1 - 0.324e^{-0.2t^*} - 0.204e^{-1.7t^*} - 0.472e^{-19t^*}) \tag{4.17}$$

where t^* is a fictitious time (hours) given by

$$t^* = \Gamma t \tag{4.18}$$

where t is the time (hours) and

$$\Gamma = \frac{(F_v/F_{ref})^2}{(b/b_{ref})^2}$$

where b is $\sqrt{\text{thermal inertia}} = \sqrt{k\rho c_p}$ ($Ws^{0.5}/m^2K$), F_v is the ventilation factor ($\sqrt{m}$) given by

$$F_v = A_v\sqrt{H_v}/A_t$$

F_{ref} is the reference value of the ventilation factor, $k\rho c_p$ is the thermal inertia (W^2s/m^4K^2) (see Chapter 3), and b_{ref} is the reference value of $\sqrt{k\rho c_p}$.

Equation (4.17) is a good approximation to the ISO 834 standard fire curve for temperatures up to about 1300°C. Hence the Eurocode parametric fire curve is close to the ISO 834 curve for the special case where $F_v = F_{ref}$ and $b = b_{ref}$. Larger ventilation openings or more highly insulated compartments will result in higher room temperatures. Smaller openings and poorly insulated compartments will result in lower temperatures.

In the Eurocode, the value of F_{ref} is 0.04 and the value of b_{ref} is 1160 such that

$$\Gamma_{EC} = \frac{(F_v/0.04)^2}{(b/1160)^2} \tag{4.19}$$

Recent research using the COMPF2 program calibrated to many test fires (Feasey and Buchanan, 2000) has shown that the temperatures in the Eurocode formula are often too low, and it is more accurate to use a value of $b_{ref} = 1900$. The recommended post-flashover design fire is therefore obtained from Equation (4.17), using

$$\Gamma_{DESIGN} = \frac{(F_v/0.04)^2}{(b/1900)^2} \tag{4.20}$$

Note that the recommended value of $b_{ref} = 1900\ Ws^{0.5}/m^2K$ comes from the Eurocode values of thermal properties of normal weight concrete (EC2, 1993), with thermal conductivity $k = 1.6$ W/mK, density $\rho = 2300\ kg/m^3$ and specific heat $c_p = 980$ J/kgK. These values are a little different from the figures in Table 3.4 which are from different sources.

Multiple layers of materials

The above formulation of the parametric fire curve assumes that the walls and ceiling of the fire compartment are made from one layer of material. If there are two or more layers of different materials, the Eurocode gives a formula for calculating an effective value of the b term. Franssen (1999(a)) has shown that the Eurocode formula is wrong, and has proposed an alternative approach for a wall or ceiling assembly material made up of two layers, with material 1 on the fire side and material 2 protected by material 1. The thicknesses of the two layers are s_1 and s_2 respectively and the thermal properties $b=\sqrt{k\rho c_p}$ are called b_1 and b_2 respectively.

If a heavy material is insulated by a lighter material such that $b_1 < b_2$ the value of the lighter material in layer 1 should be used in the calculations, so that $b = b_1$. If a light material is covered by a heavier material (as in sandwich panel construction) such that $b_1 > b_2$ then the b value depends on the thickness of the heavier material and the time of the heating period of the fire.

The limiting thickness $s_{lim,1}$ of the fire-exposed material is calculated from

$$s_{lim,1} = \sqrt{\frac{tk}{\rho c_p}} \tag{4.21}$$

where t is the time of the heating period of the fire (s) and the thermal properties are for material 1. If $s_1 > s_{lim,1}$ then $b = b_1$, and if $s_1 < s_{lim,1}$ then $b = (s_1/s_{lim,1})\, b_1 + (1 - s_1/s_{lim,1})\, b_2$.

Duration of burning period

The equation for the duration of the burning period t_d (hours) in the Eurocode simplifies to

$$t_d = 0.00013 e_t / F_v = \frac{0.00013E}{A_v\sqrt{H_v}} \tag{4.22}$$

where e_t is the fuel load (MJ/m^2 total surface area), and E is the total energy content of the fuel (MJ).

It is interesting to compare this duration with the theoretical duration for ventilation controlled burning given by Equations (4.2) and (4.3). Assuming a value for heat of combustion of $\Delta H_c = 15.5$ MJ/kg, the duration of the burning period given by Equation (4.22) is only 67% of the theoretical duration. The reason for this is not stated in the Eurocode, but it is probably intended to allow for some burning to take place during the decay period of the fire. If the heat release rate is constant during the burning period with a linear decay rate, the implied curve of heat release rate *versus* time is shown in Figure 4.15 where the duration of the decay period is equal to the duration of the burning period.

Decay rate

The Eurocode uses a reference decay rate $(dT/dt)_{ref}$ equal to 625°C per hour for fires with a burning period less than half an hour, decreasing to 250°C per hour for fires with a burning period greater than 2 hours, based on the ISO 834 testing standard (ISO, 1975). This decay rate is shown in Figure 4.16.

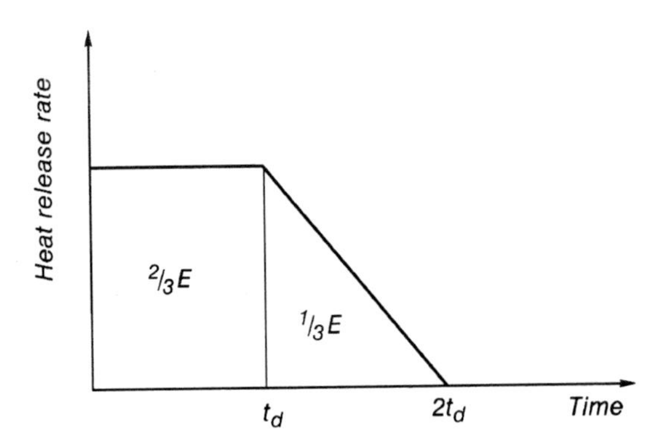

Figure 4.15 Heat release rate implied by Eurocode parametric fire

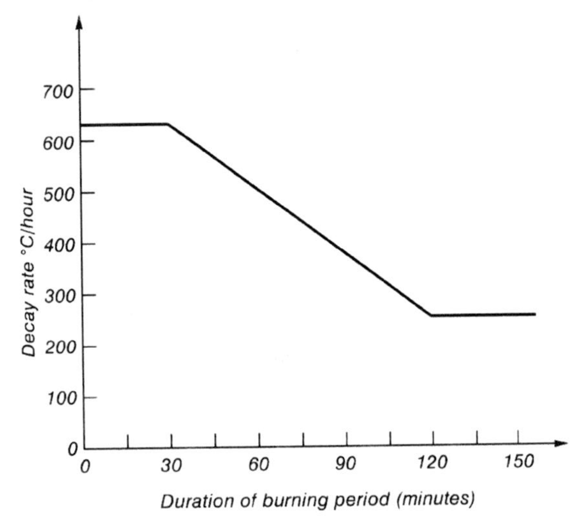

Figure 4.16 Rate of temperature decay in Eurocode parametric fires

In the Eurocode, the reference temperature is modified using fictitious time from Equation (4.18). This fictitious time, derived for use during the burning period, has not been justified for use in the decay period, and has been shown to give unsatisfactory results, with extremely fast decay rates for large openings in well-insulated compartments and extremely slow decay rates for small openings in poorly insulated compartments.

Feasey and Buchanan (2000) have shown that it is more accurate to modify the reference decay rate for ventilation factor and thermal insulation in a different way, with the resulting design decay rate given by

$$\frac{\mathrm{d}T}{\mathrm{d}t} = \left(\frac{\mathrm{d}T}{\mathrm{d}t}\right)_{\mathrm{ref}} \frac{\sqrt{F_{\mathrm{v}}/0.04}}{\sqrt{b/1900}} \qquad (4.23)$$

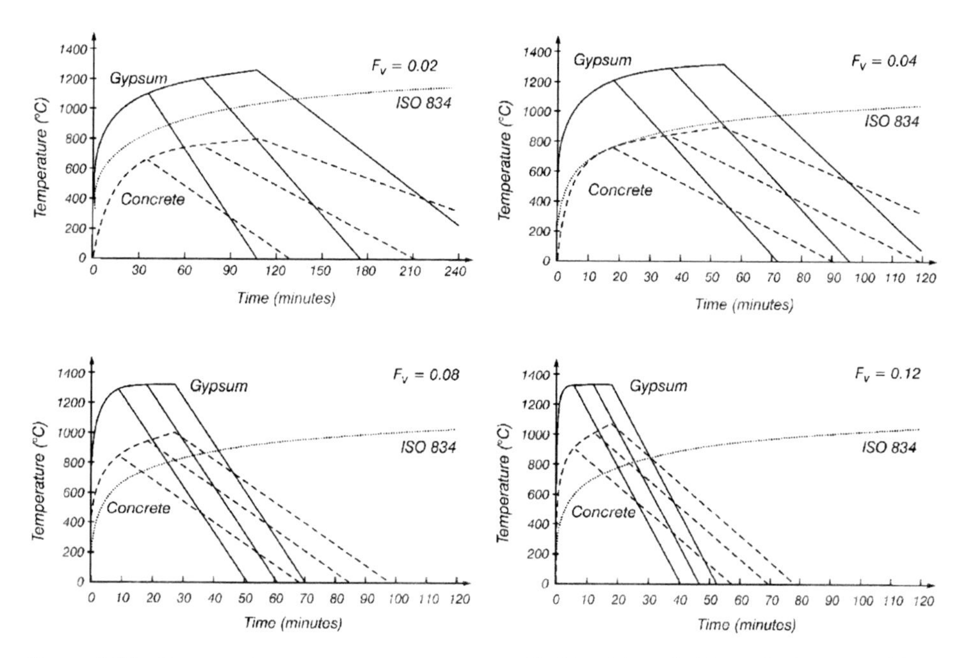

Figure 4.17 Parametric time–temperature curves (fuel load is 400, 800 and 1200 MJ/m² floor area)

This is equivalent to using a second fictitious time, similar to that in the growth period, but using square root rather than squared terms to give a much better fit to test results and computer simulations.

Time–temperature curves

Figure 4.17 shows the modified Eurocode time–temperature equation plotted for a range of ventilation factors, fuel loads and materials. The temperatures in the burning period have been calculated from Equations (4.17) and (4.20). The rate of temperature decay is from Equation (4.23), not from the Eurocode.

In each part of Figure 4.17, curves have been drawn for three fire loads and for two types of construction, showing the significant dependence of fire temperatures on the thermal properties of the bounding materials. The fire loads are 400, 800 and 1200 MJ/m² floor area, for a room 5×5 m in plan and 3 m high. The materials are normal weight concrete ($b = 1900\,\mathrm{W\,s^{0.5}/m^2\,K}$) and gypsum plaster board ($b = 410\,\mathrm{W\,s^{0.5}/m^2\,K}$). A typical commercial office building with a mixture of these materials on the walls and ceiling would give curves between these two, similar to a building made from lightweight concrete. The Eurocode suggestion of limiting the lower value of thermal inertia to $b = 1000$ $\mathrm{W\,s^{0.5}/m^2\,K}$ has not been followed, in order to give closer results to the European test fires.

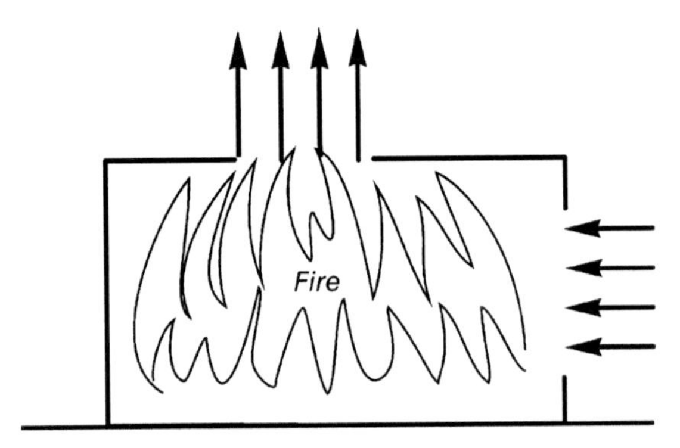

Figure 4.18 Vent flows for room with ceiling opening

4.6 OTHER FACTORS

4.6.1 Additional Ventilation Openings

Ventilation controlled fires are very sensitive to the size and location of openings. The presence of a ceiling opening allows combustion products to exit the ceiling opening while cool air enters the window. This significantly increases the ventilation to the fire as shown in Figure 4.18.

Magnusson and Thelandersson (1970) provide an approximate nomogram for allowing for ceiling vents, shown in Figure 4.19. If all of the lines shown are assumed to be straight lines through the origin, the nomogram can be simplified to give a fictitious ventilation factor

$$(A_v\sqrt{H_v})_{\text{fict}} = A_v\sqrt{H_v} + 2.3A_h\sqrt{h} \tag{4.24}$$

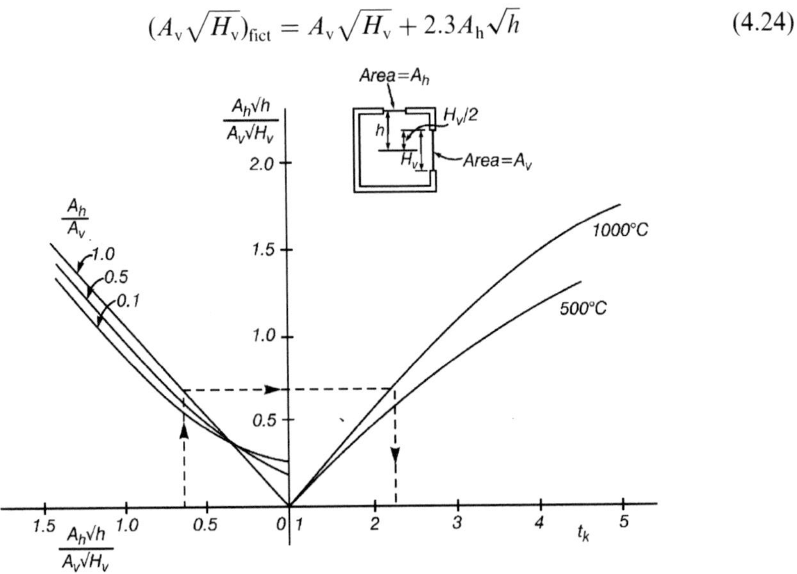

Figure 4.19 Nomogram for calculating ventilation factor for roof vents (Reproduced from (Magnusson and Thelandersson, 1970) by permission of Fire Safety Engineering Department, Lund University)

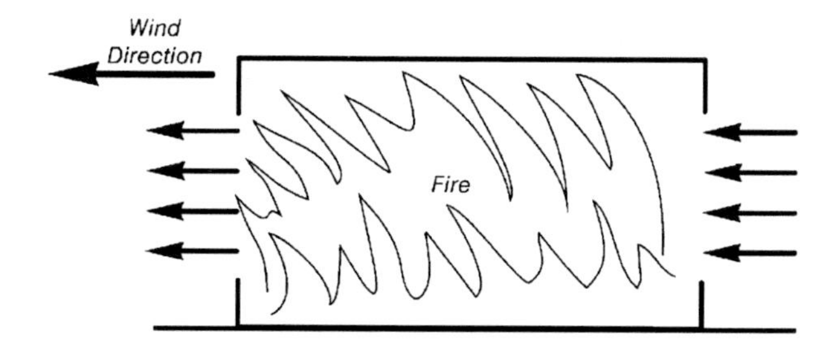

Figure 4.20 Vent flows for two windows, with wind blowing

where A_h is the area of the horizontal opening in the ceiling (m^2), and h is the vertical distance from mid-height of the window opening to the ceiling of the compartment. This approximate expression can only be used for values of $A_h\sqrt{h}/A_v\sqrt{H_v}$ in the range 0.3 to 1.5. Beyond the upper limit, the expression is not valid because the window opening no longer dominates the flow of gases. According to Magnusson and Thelandersson, their model has been shown to work up to this limit in tests reported by Thomas *et al.* (1963). The term $(A_v\sqrt{H_v})_{fict}$ from Equation (4.24) can be used in place of $A_v\sqrt{H_v}$ to calculate the burning rate in Equation (4.2), or to select a time–temperature curve from Figure 4.8.

A room with openings on two opposite walls may have cross ventilation, especially if there is a wind blowing as shown in Figure 4.20, producing increased rates of burning. No research has been done on this type of scenario, other than some estimates for external steel structures exposed to fires from windows (Law and O'Brien, 1981).

4.6.2 Progressive Burning

For large compartments such as open plan offices or industrial buildings, it is unlikely that a post-flashover fire will occupy the whole space at one time. All of the time–temperature curves presented on post-flashover fires relate to small rooms of sizes which have been tested, up to about 6 m by 6 m in floor area and 2.4 to 3.0 m in height. There is almost no test data for post-flashover fires in compartments with larger floor areas or taller ceilings. In general it is probably conservative to use the models described above, because the probability of flashover and full-room involvement is less as the size of the compartment increases, and the assumption of full-room involvement for the full burning period is most severe on the structure.

A recent series of tests in a long narrow room is described by Kirby *et al.* (1994). This room was approximately 6 m wide and 20 m deep with uniformly distributed wood cribs as fuel. Even when the fire was ignited at the end farthest from the window, burning moved quickly to the window end, then progressed slowly from that end towards the back as shown in Figure 4.21. Temperatures measured at points A, B and C in Figure 4.21 are shown in Figure 4.22 where it can be seen that the total duration of elevated temperatures is similar at all three points, but the peak

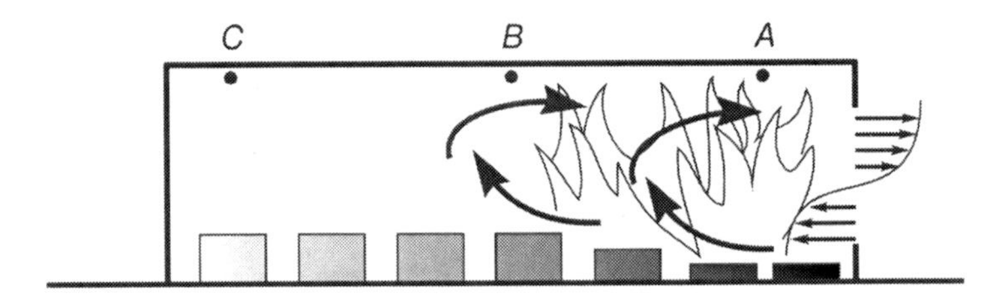

Figure 4.21 Progressive burning in a deep room with one window

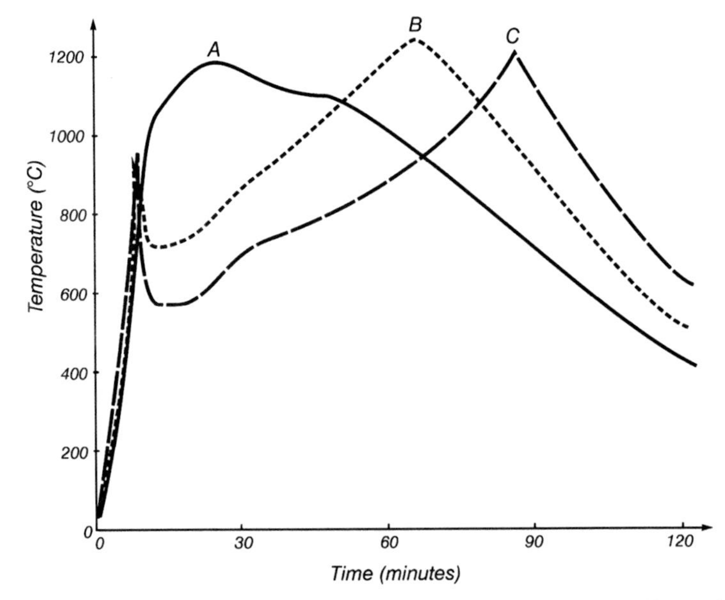

Figure 4.22 Temperatures during progressive burning in a deep room (Reproduced from Kirby *et al.* (1994) by permission of Corus UK Ltd)

temperature progresses back into the compartment as the fuel closest to the window is progressively burned.

Clifton (1996) has proposed a model for fire spreading within a large compartment. Thomas and Bennetts (1999) report similar behaviour in small-scale experiments, finding that most fires burn progressively with the object nearest the window burning first, and often delaying burning of other items in the room. These findings have significant implications for modelling of post-flashover fires, and are leading to renewed research efforts.

4.6.3 Localized Fires

The discussion of post-flashover fires in this chapter has been based on the assumption that a fully developed fire occurs and creates the same temperature conditions throughout the fire compartment. In some circumstances, possibly in a large space where there are no nearly combustibles, or in a fire partially controlled by sprinklers, there could be a localized fire which has much less impact on the building structure than a fully developed fire. Tests by Hasemi *et al.* (1995) have been used by Franssen *et al.* (1998) to calculate steel temperatures in steel beams above burning cars in car parking buildings. Structural design calculations can be made in such cases if the member temperatures are known, but it is always conservative to assume a fully developed fire. Bailey *et al.* (1996(a)) have investigated the structural response of a multi-bay steel frame to a spreading fire including the effects of cooling during the decay period.

WORKED EXAMPLE 4.1

Calculate the radiant heat flux at floor level in a room with a hot upper layer at 600°C. Assume that the smoke in the upper layer has an emissivity of 0.7. Assume that the area of the ceiling is large relative to the room height, so that the configuration factor is 1.0.

Emitter temperature	$T = 600°C = 873$ K
Configuration factor	$\varphi = 1.0$
Emissivity	$\varepsilon = 0.7$
Stefan–Boltzmann constant	$\sigma = 5.67 \times 10^{-8}$ W/m^{2}K^4
Radiant heat flux	$\dot{q}'' = \phi \varepsilon \sigma T^4$
	$= 1.0 \times 0.7 \times 5.67 \times 10^{-8} \times 873^4/1000$
	$= 23.1$ kW/m^2

WORKED EXAMPLE 4.2

Using Thomas' Flashover Criterion, calculate the heat release rate necessary to cause flashover in a room 6.0 m by 4.0 m in floor area, and 3.0 m high, with one window 2.0 m high by 3.0 m wide.

Length of room	$l_1 = 6.0$ m
Width of room	$l_2 = 4.0$ m
Height of room	$H_r = 3.0$ m
Area of internal surfaces	$A_t = 2\ (l_1\ l_2 + l_1\ H_r + l_2\ H_r)$
	$= 2\ (6 \times 4 + 6 \times 3 + 4 \times 3) = 108$ m^2
Height of window	$H_v = 2.0$ m
Width of window	$B = 3.0$ m
Area of window	$A_v = BH_v = 3.0 \times 2.0 = 6.0$ m^2
Heat release for flashover	$Q_{fo} = 0.007\ A_t + 0.378\ A_v \sqrt{H_v}$
	$= 0.007 \times 108 + 0.378 \times 6.0 \times \sqrt{2.0}$
	$= 3.96$ MW

WORKED EXAMPLE 4.3

Calculate the ventilation controlled heat release rate for a post-flashover fire in the room of Worked Example 4.2, if the burning wood has a heat of combustion of 16 MJ/kg.

Rate of burning $\dot{m} = 0.092\ A_v\sqrt{H_r}$

$= 0.092 \times 6.0 \times \sqrt{2.0} = 0.781$ kg/s

Heat of combustion $\Delta H_c = 16$ MJ/kg

Heat release rate $Q_{vent} = \dot{m}\Delta H_c = 0.78 \times 16 = 12.48$ MW

Calculate the duration of burning if the available fuel load energy density is 800 MJ/m^2 floor area.

Fuel load energy density $e_f = 800$ MJ/m^2

Floor area $A_f = 6.0 \times 4.0 = 24$ m^2

Total energy $E_f = e_f\ A_f = 800 \times 24 = 19\,200$ MJ

Duration of burning $t_b = E_f\ /Q_{vent} = 19\,200/12.5 = 1536$ s (25.6 minutes)

WORKED EXAMPLE 4.4

Calculate the ventilation controlled heat release rate and duration of burning for the room of the previous examples, using Law's equation. Assume that the window is in the long side of the room.

Opening factor $\Omega = (A_t - A_v)/A_v\sqrt{H_v}$

$= (108-6)/6\sqrt{2} = 12.0$ m$^{-1/2}$

Room width $W = 6.0$ m

Room depth $D = 4.0$ m

Rate of burning $\dot{m} = 0.18\ A_v\sqrt{H_v\ W/D}\ (1-e^{-0.036\Omega})$

$= 0.18 \times 6 \times \sqrt{2 \times 6/4}\ (1-e^{-0.036 \times 12})$

$= 0.657$ kg/s

Heat release rate $Q_{vent} = \dot{m}\Delta H_c = 0.657 \times 16 = 10.5$ MW

Duration of burning $t_b = E_f/Q_{vent} = 19\,200/10.5 = 1829$ s (30.5 minutes)

WORKED EXAMPLE 4.5

Calculate the fuel controlled heat release rate for wood fuel in a post-flashover fire if 30 m^2 of wood is exposed to a radiant heat flux of 60 kW/m^2. Assume that the wood has a heat of gasification of 6.0 MJ/kg.

Incident heat flux	$\dot{q}_i'' = 0.06\,\text{MW/m}^2$
Area of fuel	$A_{fuel} = 30\,\text{m}^2$
Heat of combustion	$\Delta H_c = 16\,\text{MJ/kg}$
Heat of gasification	$L_v = 6.0\,\text{MJ/kg}$
Heat release rate	$Q_{fuel} = \dot{q}_i''\, A_{fuel}\, \Delta H_c / L_v$
	$= 0.060 \times 30 \times 16/6 = 4.80\,\text{MW}$

WORKED EXAMPLE 4.6

Calculate the fuel controlled heat release rate for the wood in the previous example, using Babrauskas' equations for the regression rate, for thick slabs of wood and thin wood slabs 50 mm thick. Wood density is 400 kg/m^3

Thick slab of wood	
Regression rate	$v_p = 9.0 \times 10^{-6}\,\text{m/s}$
Density	$\rho = 400\,\text{kg/m}^3$
Area of fuel	$A_{fuel} = 30\,\text{m}^2$
Heat of combustion	$\Delta H_c = 16\,\text{MJ/kg}$
Heat release rate	$Q_{fuel} = v_p\, \rho\, A_{fuel}\, \Delta H_c$
	$= 9.0 \times 10^{-6} \times 400 \times 30 \times 16$
	$= 1.73\,\text{MW}$

Note that this is much less than in the previous example.

Wood 50 mm thick	
Thickness of slab	$D = 0.05\,\text{m}$
Regression rate	$v_p = 2.2 \times 10^{-6}\, D^{-0.6}$
	$= 2.2 \times 10^{-6} \times 0.05^{-0.6} = 13.3 \times 10^{-6}\,\text{m/s}$
Heat release rate	$Q_{fuel} = v_p\, \rho\, A_{fuel}\, \Delta H_c$
	$= 13.3 \times 10^{-6} \times 400 \times 30 \times 16$
	$= 2.55\,\text{MW}$

WORKED EXAMPLE 4.7

Recalculate the heat release rate from Worked Example 4.3 with a ceiling opening of 3.0 m^2.

Area of ceiling opening	$A_h = 3.0\,\text{m}^2$
Height above window mid-height	$h = 1.5\,\text{m}$ (assume window is mid way between floor and ceiling)

Window area	$A_v = 6.0\,m^2$
Window height	$H_v = 2.0\,m^2$
Modified opening parameter	$(A_v\sqrt{H_v})_{fict} = A_v\sqrt{H_v} + 2.3\ A_h\sqrt{h}$
	$= 6.0\ \sqrt{2.0} + 2.3 \times 3.0\ \sqrt{1.5}$
	$= 8.49 + 8.45 = 16.9\,m^{3/2}$
Rate of burning	$\dot{m} = 0.092\ (A_v\sqrt{H_v})_{fict}$
	$= 0.092 \times 16.9 = 1.55\,kg/s$
Heat of combustion	$\Delta H_c = 16.0\,MJ/kg$
Heat release rate	$Q_{vent} = \dot{m}\Delta H_c = 16.0 \times 1.55 = 24.8\,MW$

(Note that this ceiling opening almost doubles the rate of burning and the heat release rate, which will halve the duration of burning.)

WORKED EXAMPLE 4.8

Calculate the maximum temperature for the room of Worked Example 4.3, using Law's equations.

Duration of burning (from Worked Example 4.4)	$t_b = 24.8\,min$
Opening factor	$\Omega = 12.0\,m^{-1/2}$
Maximum temperature	$T_{max} = 6000\ (1 - e^{-0.1 \times \Omega})/\sqrt{\Omega}$
	$= 6000\ (1 - e^{-0.1 \times 12})/\sqrt{12} = 1210°C$
Check reduction factor for fuel load	
Total fuel load	$E_f = 19\,200\,MJ$
Calorific value of wood	$\Delta H_c = 16.0\,MJ/kg$
Fuel load (wood equivalent)	$L = E_f/\Delta H_c = 19\,200/16.0 = 1200\,kg$
Area of windows	$A_v = 6.0\,m^2$
Area of internal surfaces	$A_t = 108\,m^2$
Temperature parameter	$\psi = L/\sqrt{A_v(A_t - A_v)}$
	$= 1200/\sqrt{6\ (108 - 6)} = 48.5$
Reduced maximum temperature	$T = T_{max}\ (1 - e^{-0.05\psi})$
	$= 1210\ (1 - e^{-0.05 \times 48.5}) = 1103°C$

WORKED EXAMPLE 4.9

Estimate a time–temperature curve for the previous room using the Swedish curves.

Area of window	$A_v = 6.0\,m^2$
Height of window	$H_v = 2.0\,m$

Area of internal surfaces	$A_t = 108\,m^2$
Floor area	$A_f = 24.0\,m^2$
Ventilation factor	$F_v = A_v\sqrt{H_v}/A_t = 6.0 \times \sqrt{2.0}/108 = 0.079\ m^{-1/2}$
Fuel load (floor area)	$e_f = 800\,MJ/m^2$
Fuel load (total area)	$e_t = e_f\ A_f/A_t = 800 \times 24.0/108 = 178\,MJ/m^2$

The time–temperature curve can be roughly interpolated from the bottom left-hand graph in Figure 4.8, giving a maximum temperature of about 950°C after 20 minutes, dropping to 350°C after one hour.

WORKED EXAMPLE 4.10

Use the parametric fire equations to calculate the duration and the maximum temperature for a fire in a room 4.0 m×6.0 m in area, 3.0 m high, with one window 3.0 m wide and 2.0 m high. The fire load is 800 MJ/m² floor area. The room is constructed from concrete with the following properties:

Thermal conductivity	$k = 1.6\,W/mK$
Density	$\rho = 2300\,kg/m^3$
Specific heat	$c_p = 980\,J/kg\ K$
Thermal inertia of concrete	$b = \sqrt{k\rho c_p} = 1900\ W\,s^{0.5}/m^2\ K$
Length of room	$l_1 = 6.0\,m$
Width of room	$l_2 = 4.0\,m$
Floor area	$A_f = l_1\ l_2 = 6.0 \times 4.0 = 24.0\,m^2$
Height of room	$H_r = 3.0\,m$
Area of internal surfaces	$A_t = 2\ (l_1\ l_2 + l_1\ H_r + l_2\ H_r)$
	$= 2\ (6 \times 4 + 6 \times 3 + 4 \times 3) = 108\,m^2$
Window height	$H_v = 2.0\,m$
Window width	$B = 3.0\,m$
Window area	$A_v = H_v B = 2.0 \times 3.0 = 6.0\,m^2$
Ventilation factor	$F_v = A_v\sqrt{H_v}/A_t = 0.079\,m^{-1/2}$
Fuel load energy density	$e_f = 800\,MJ/m^2$
Total fuel load	$E = e_f\ A_f = 800 \times 24 = 19\,200\ MJ$
Duration of parametric fire	$t_d = 0.00013\ E/(A_v\sqrt{H_v})$
	$= 0.00013 \times 19200\ /\ (6.0\ \sqrt{2.0}) = 0.294$ hour (17.6 minutes)
Fictitious duration	$t^* = t_d\ (F_v/0.04)^2/(b/1900)^2$
	$= 0.294\ (0.079/0.04)^2/(1900/1900)^2$
	$= 1.15$ hours

Maximum temperature $T = 1325\,(1-0.324\,e^{-0.2t^*} - 0.204\,e^{-1.7t^*} - 0.472e^{-19t^*})$

$= 1325\,(1-0.257-0.029-0)$

$= 946°C$

Duration is less than 30 minutes, so reference decay rate is $(dT/dt)_{ref} = 625°C/hour$

Decay rate $dT/dt = (dT/dt)_{ref}\sqrt{F_v/0.04}/\sqrt{b/1900}$

$= 625\,\sqrt{0.079/0.04}/\sqrt{1900/1900}$

$= 878°C/hour$

The time to drop 946°C at this rate is 946/878 = 1.08 hours.

Total duration from flashover to extinction is 0.29 + 1.08 = 1.37 hours (1 hour 22 minutes).

These curves can be easily calculated and plotted using a computer spreadsheet.

WORKED EXAMPLE 4.11

Repeat Worked Example 4.10 if the concrete walls and ceiling are covered with a 12 mm thick layer of gypsum plaster board.

The thermal properties of the materials are as follows:

	Concrete	*Gypsum board*
Thermal conductivity k	1.6 W/mK	0.20 W/mK
Density ρ	2300 kg/m^3	700 kg/m^3
Specific heat c_p	980 J/kg K	1700 J/kg K
$b = \sqrt{k\rho c_p}$	1900 W s$^{0.5}$/m^2 K	488 W s$^{0.5}$/m^2 K

Length of room $l_1 = 6.0\,m$

Width of room $l_2 = 4.0\,m$

Floor area $A_f = l_1\,l_2 = 6.0 \times 4.0 = 24.0\,m^2$

Height of room $H_r = 3.0\,m$

Area of internal surfaces $A_t = 2\,(l_1\,l_2 + l_1\,H_r + l_2\,H_r)$

$= 2\,(6\times4 + 6\times3 + 4\times3) = 108\,m^2$

Window height $H_v = 2.0\,m$

Window width $B = 3.0\,m$

Window area $A_v = H_v B = 2.0\times3.0 = 6.0\,m^2$

Ventilation factor $F_v = A_v\sqrt{H_v}/A_t = 0.079\,m^{-1/2}$

Fuel load energy density $e_f = 800\,MJ/m^2$

Total fuel load $E = e_f\,A_f = 800\times24 = 19\,200\,MJ$

Duration of parametric fire $t_d = 0.00013\,E/(A_v\sqrt{H_v})$

$= 0.00013\times19200/(6.0\,\sqrt{2.0})$

$= 0.294$ hour (17.6 minutes) (1060 seconds)

The limiting thickness, $s_{lim,1}$, of the fire exposed material is calculated from

$$\begin{aligned} s_{lim,1} &= \sqrt{tk/\rho c_p} \\ &= \sqrt{(1060 \times 0.2)/(700 \times 1700)} \\ &= 0.0133\,\text{m}\ (13.3\,\text{mm}) \end{aligned}$$

since s_1 $(= 12\,\text{mm}) < s_{lim,1}$ (13.3 mm),

then
$$\begin{aligned} b &= (s_1/s_{lim,1})b_1 + (1 - s_1/s_{lim,1})b_2 \\ &= (12/13.3) \times 488 + (1 - 12/13.3) \times 1900 \\ &= 626\ \text{s}^{0.5}/\text{m}^2\ \text{K} \end{aligned}$$

Fictitious duration
$$\begin{aligned} t^* &= t_d\ (F_v/0.04)^2/(b/1900)^2 \\ &= 0.294\ (0.079/0.04)^2/(626/1900)^2 \\ &= 10.56\ \text{hours} \end{aligned}$$

Maximum temperature
$$\begin{aligned} T &= 1325\ (1 - 0.324\ e^{-0.2t^*} - 0.204\ e^{-1.7t^*} - 0.472^{e-19t^*}) \\ &= 1325\ (1 - 0.0392 - 0 - 0) \\ &= 1273°\text{C} \end{aligned}$$

Duration is less than 30 minutes, so reference decay rate is $(dT/dt)_{ref} = 625°\text{C/hour}$

Decay rate
$$\begin{aligned} dT/dt &= (dT/dt)_{ref}\ \sqrt{F_v/0.04}/\sqrt{b/1900} \\ &= 625\ \sqrt{0.079/0.04}/\sqrt{626/1900} \\ &= 1530°\text{C/hour} \end{aligned}$$

The time to drop 1273°C at this rate is 1273/1530 = 0.832 hours.

Total duration from flashover to extinction is 0.29 + 0.83 = 1.12 hours (1 hour 8 minutes).

5
Fire Severity

5.1 OVERVIEW

This chapter gives an overview of basic methods for designing structures for fire safety. It describes methods for quantifying the severity of post-flashover fires, for comparison with the provided fire resistance. This chapter describes the standard fire used for fire-resistance testing and approvals, and the concept of equivalent fire severity which is used for comparing real fires with the standard time–temperature curve.

5.2 FIRE SEVERITY AND FIRE RESISTANCE

5.2.1 Verification

The fundamental step in designing structures for fire safety is to verify that the fire resistance of the structure (or each part of the structure) is greater than the severity of the fire to which the structure is exposed. This verification requires that the following design equation be satisfied:

$$\text{fire resistance} \geqslant \text{fire severity} \tag{5.1}$$

where *fire resistance* is a measure of the ability of the structure to resist collapse, fire spread or other failure during exposure to a fire of specified severity, and *fire severity* is a measure of the destructive impact of a fire, or a measure of the forces or temperatures which could cause collapse or other failure as a result of the fire. There are several different definitions of fire severity and fire resistance, leading to different ways of comparing them using different units. These comparisons can be confusing if not made correctly, so it is important for designers to understand the alternatives clearly.

As shown in Table 5.1, there are three methods for comparing fire severity with fire resistance. The verification may be in the *time* domain, the *temperature* domain or the *strength* domain, as discussed below.

Table 5.1 Three methods for comparing fire severity with fire resistance

Domain	Units	Fire Resistance	$\geqslant$ Fire Severity
Time	minutes or hours	Time to failure	$\geqslant$ Fire duration as calculated or specified by code
Temperature	°C	Temperature to cause failure	$\geqslant$ Maximum temperature reached during the fire
Strength	kN or kNm	Load capacity at elevated temperature	$\geqslant$ Applied load during the fire

Time domain

By far the most common procedure is for fire severity and fire resistance to be compared in the *time domain* such that:

$$t_{fail} \geqslant t_s \tag{5.2}$$

where t_{fail} is the time to failure of the element, and t_s is the fire duration, as specified by a code or calculated, both times having units of minutes or hours.

The time to failure of a building element is usually a *fire-resistance rating*, which may be obtained from a published listing of ratings or by calculation, as described in Chapter 6. The fire duration, or fire severity, is usually a time of standard fire exposure specified by a building code, or the equivalent time of standard fire exposure calculated for a real fire in the building, as described later in this chapter.

Temperature domain

It is sometimes necessary to verify design in the *temperature domain* by ensuring that the maximum temperature (°C) in a part of the structure is no greater than the temperature (°C) which would cause failure. Failure in this context could be thermal failure of a separating element or structural collapse of a load-bearing member. Verification in the temperature domain requires that:

$$T_{fail} \geqslant T_{max} \tag{5.3}$$

where T_{fail} is the temperature which would cause failure of the element, and T_{max} is the maximum temperature reached in the element during the fire, or the temperature at a certain time specified by the code.

The temperature reached in the element can be calculated by a thermal analysis of the structural assembly exposed to the design fire. For a separating element, the failure temperature is the temperature on the unexposed face which would allow fire to spread into the next compartment. For a structural element, the temperature which would cause collapse can be calculated from knowledge of the loads on the element, the load capacity at normal temperatures, and the effect of elevated temperatures on the structural materials, as described in Chapter 7.

The temperature domain is most likely to be used for an element which serves an insulating or containing function. The temperature domain is less suitable for structural elements because it does not adequately consider internal thermal gradients or structural behaviour.

Strength domain

Verification in the *strength domain* is a comparison of the applied load at the time of the fire with the load capacity of structural members throughout the fire, such that

$$R_f \geqslant U_f \tag{5.4}$$

where R_f is the minimum load capacity reached during the fire, or the load capacity at a certain time specified by the code, and U_f is the applied load at the time of the fire.

These values may be expressed in units of force and resistance for the whole building, or as internal member actions such as axial force or bending moment in individual members of the structure. The load capacity during the fire can be calculated from a thermal analysis and a structural analysis at elevated temperatures, as described later in this chapter. The load capacity must be obtained by calculation because almost no structural test results are available for full burnout fires. The loads at the time of the fire can be calculated using load combinations from national loadings codes as described in Chapter 7.

No safety factors are shown in Equations (6.2) to (6.4). This is because the required level of safety is obtained by using conservative values for the individual terms.

Example

The comparison of fire severity with fire resistance described above can be rather confusing, so the three different domains of verification are illustrated with a simple example. Figure 5.1(a) shows the temperature of a steel beam during fire exposure. Calculations show that the beam will fail when the steel temperature reaches T_{fail} at time t_{fail}. The building code requires that the beam should have a fire resistance of t_{code} or in other words the required fire severity is t_{code}.

Verification in the time domain requires checking that the beam does not fail prematurely, so that the time to failure t_{fail} is greater than the fire severity specified by the code t_{code} (check 1 in Figure 5.1(a)). Verification in the temperature domain requires checking that the steel temperature which would cause failure T_{fail} is greater than T_{code} which is the temperature reached in the beam at time t_{code} (check 2 in Figure 5.1(a)). These two checks will give identical results because they are both based on the same process.

Figure 5.1(b) shows the load capacity of the same steel beam during the fire. The imposed load at the time of the fire is U_f. The load capacity before the fire is R_{cold} and the graph shows how this decreases during the fire. At the time t_{code} the load capacity of the beam has reduced to R_{code}. Verification in the strength domain simply requires checking that the reduced load capacity is greater than the applied load (check 3 in Figure 5.1(b)). All three of these verification checks give identical results.

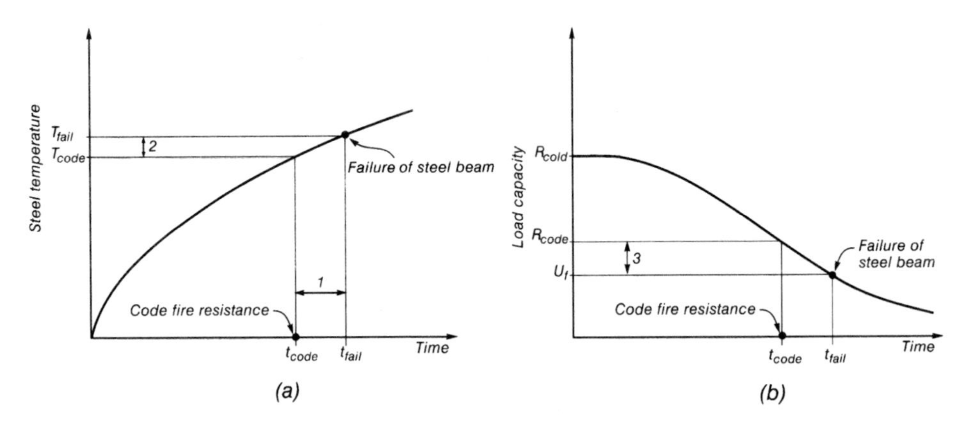

Figure 5.1 Behaviour of a steel beam in fire: (a) temperature increase, (b) loss of strength

5.2.2 Fire Exposure Models

Figure 5.2 illustrates a range of different design situations (from CIB, 1986). The left-hand column shows three fire exposure models which represent three different ways in which a design fire can be specified.

Fire exposure H_1 represents exposure to a standard test fire for a specified period of time, t_{code} as prescribed by a building code. This is the most common specification of fire exposure. Traditional prescriptive codes specify the required fire resistance directly, leaving little opportunity for fire engineers to calculate a specific fire severity for any particular building. Prescriptive codes usually require fire resistance to be somewhere between half an hour and four hours, in half hour or one hour steps, with little or no reference to the severity of the expected fire.

Fire exposure H_2 represents a modified duration of exposure to the standard test fire. The equivalent time, t_e is the time of exposure to the standard test fire considered to be equivalent to a complete burnout of a real fire in the same room. Methods of calculating equivalent fire severity are described later in this chapter. Many performance-based codes allow the use of time equivalent formulae as an improvement on simple prescriptive fire-resistance requirements.

Fire exposure H_3 represents a realistic fire which would occur if there was a complete burnout of the room, with no intervention or fire suppression. This is the type of fire described by the time–temperature curves in Chapter 4.

The other columns of Figure 5.2 show that assessment of fire resistance may consider a single element, a sub-assembly or a whole structure. The words in the lower boxes show that test results are only likely to be used for single elements exposed to H_1or H_2 fires, with calculations becoming necessary in most other cases.

Verification that a member or structure has sufficient fire resistance will be by comparison of times, temperatures, or strength as described above. With reference to Figure 5.2, verification to fire exposures H_1 and H_2 is likely to be in the time domain, where an assigned fire resistance (in hours) is compared with the required fire resistance (also in hours). Verification using exposure to a complete burnout (H_3) is

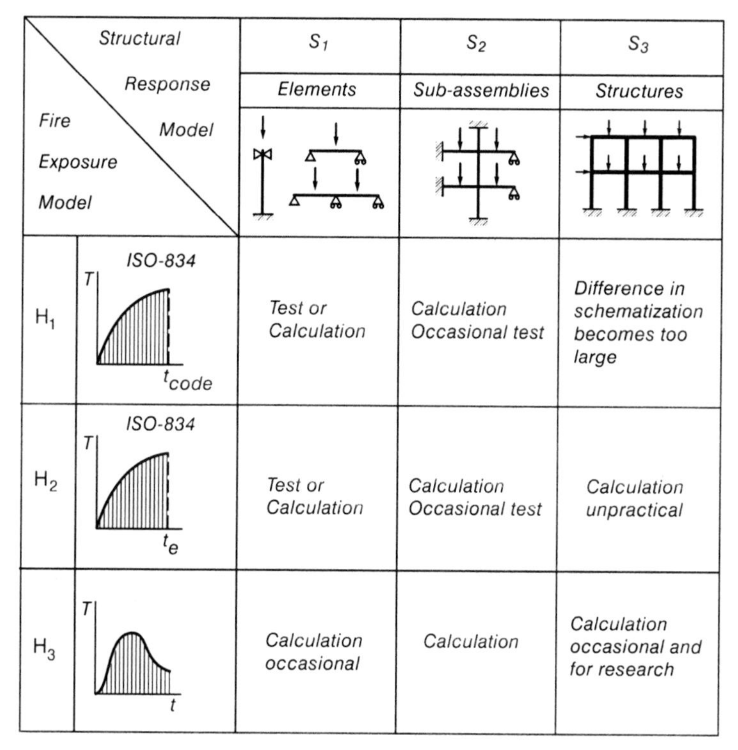

Figure 5.2 Fire models and structural response models (Reprinted from CIB (1986) with permission from Elsevier Science)

more likely to be a comparison of temperatures for insulating elements or a comparison of strength for structural elements.

5.2.3 Design Combinations

The above options illustrate that several alternative methods can be used for verifying fire resistance requirements. Because of the large number of possible combinations, it is essential for designers to specify clearly which combination of exposure and resistance is being used. Both the design and the assessment of the design can become very confusing if the selected combination is not clearly stated and used accordingly.

Table 5.2 shows a list of the most common combinations, to help designers select a combination for a particular design. In very general terms, both the accuracy of the prediction and the amount of calculation effort increase downwards in the table.

5.3 FIRE SEVERITY

Fire severity is a measure of the destructive potential of a fire. Fire severity is most often defined in terms of a period of exposure to the standard test fire, but this is not appropriate for real fires which have very different characteristics.

Table 5.2 Design combinations for verifying fire resistance

Combination	Fire exposure model	Assessment of fire resistance	Verification domain
1	Prescriptive code (H_1)	Listed rating or calculation	Time
2	Time-equivalent formula (H_2)	Listed rating or calculation	Time
3	Predicted real fire (H_3)	Calculation	Temperature or strength

The fire severity used for design depends on the legislative environment and on the design philosophy. In a prescriptive code environment, the design fire severity is usually prescribed with little or no room for discussion. In a performance-based code environment, the design fire severity is usually a complete burnout fire or the equivalent time of a complete burnout fire. In some cases the design fire may be a shorter time which only allows for escape, rescue or fire-fighting. The equivalent time of a complete burnout is the time of exposure to the standard test fire that would result in an equivalent impact on the element, as described later in this chapter.

Damage to a structure is largely dependent on the amount of heat absorbed by the structural elements. Heat transfer from post-flashover fires is mainly radiative which is proportional to the fourth power of the absolute temperature. Hence the severity of a fire is largely dependent on the temperatures reached and the duration of the high temperatures. Some damage such as phase changes or melting are temperature-dependent rather than heat-dependent, so the maximum temperature as well as the duration of the fire is also important.

5.4 STANDARD FIRE

Most countries around the world rely on full-size fire-resistance tests to assess the fire performance of building materials and structural elements. The time–temperature curve used in fire-resistance tests is called the 'standard fire'. Full-size tests are preferred to small-scale tests because they allow the method of construction to be assessed, including the effects of thermal expansion and deformation under load. Babrauskas and Williamson (1978(c) and 1978(d)) and Cooper and Steckler (1996) describe the origin of the standard fire-resistance test.

The most widely used test specifications are ASTM E119 (ASTM 1988(a)) and ISO 834 (ISO, 1975). Other national standards include British Standard BS 476 Parts 20–23 (BSI 1987), Canadian Standard CAN/ULC-S101-M89 (ULC, 1989) and Australian Standard AS 1530 Part 4 (SAA, 1990(c)). Most national standards are based on either the ASTM E119 test or the ISO 834 test, which are compared below. This chapter concentrates on the fire temperatures in test furnaces. Other aspects of fire-resistance furnaces are described in more detail in Chapter 6.

5.4.1 Time–temperature Curves

The standard time–temperature curves from ASTM E119 and ISO 834 are compared in Figure 5.3. They are seen to be rather similar. All other international fire resistance test standards specify similar time–temperature curves (Lie, 1995).

In the ISO 834 specification (ISO, 1975) the temperature T (°C) is defined by

$$T = 345\log_{10}(8t + 1) + T_0 \tag{5.5}$$

where t is the time (minutes) and T_0 is the ambient temperature (°C).

The ASTM E119 curve is defined by a number of discrete points, which are shown in Table 5.3, along with the corresponding ISO 834 temperatures. Several equations approximating the ASTM E119 curve are given by Lie (1995), the simplest of which gives the temperature T (°C) as

$$T = 750[1 - e^{-3.79553\sqrt{t_h}}] + 170.41\sqrt{t_h} + T_0 \tag{5.6}$$

where t_h is the time (hours).

Figure 5.3 also shows two alternative design fires from the Eurocode (EC1, 1994). The upper curve is the hydrocarbon fire curve, intended for use where a structural member is engulfed in flames from a large pool fire. The temperature T (°C) in the hydrocarbon fire curve is given by

$$T = 1080(1 - 0.325e^{-0.167t} - 0.675e^{-2.5t}) + T_0 \tag{5.7}$$

The lower curve is intended for the design of structural members located outside a burning compartment. Unless they are engulfed in flames, exterior structural members will be exposed to lower temperatures than members inside a compartment. The temperature T (°C) for external members is given by

$$T = 660(1 - 0.687e^{-0.32t} - 0.313e^{-3.8t}) + T_0 \tag{5.8}$$

where for Equations (5.7) and (5.8), t is the time (minutes) and T_0 is the ambient temperature (°C).

5.4.2 Furnace Parameters

Fire severity in a test environment depends on a number of characteristics of the testing furnace. Even if two furnaces are operated according to the same time–temperature curve, they may not impact the test specimen with the same severity of fire exposure, depending on various parameters.

Temperatures are not always uniform throughout the furnace. Even if the average temperature follows the specified curve precisely, this may be an average of lower and higher temperatures which could have a severe local impact on the test specimen.

It is clear from Figure 5.3 that the ASTM E119 and ISO 834 curves are similar. The tests can be considered to give roughly equivalent thermal exposure, but there are some significant differences. Heating of furnaces is controlled to ensure that the

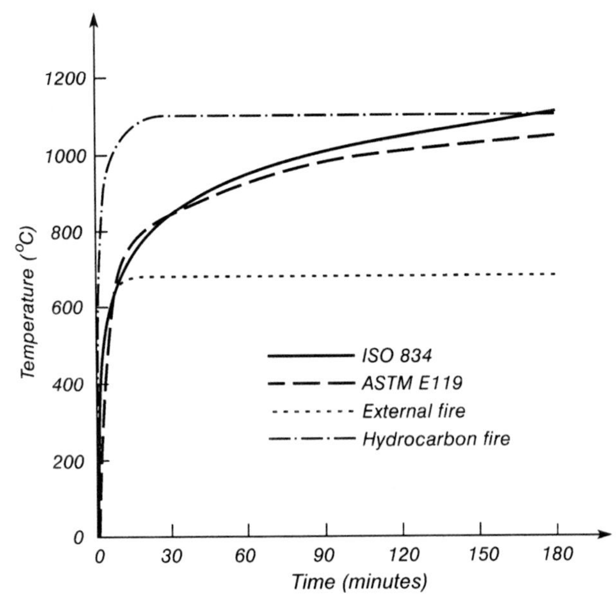

Figure 5.3 Standard time–temperature curves

temperatures measured by thermocouples follow the prescribed curve given by Equation (5.5) or Table 5.3. The ASTM E119 specification requires furnace temperatures to be measured with thermocouples located in heavy steel pipes with capped ends, which heat up more slowly than the exposed thin wire thermocouples specified in the ISO 834 test, so even for the same temperature curve, the furnace gas temperatures will be higher in the furnace with capped thermocouples than in a furnace with bare wire thermocouples. Babrauskas and Williamson (1978(b)) have shown that the temperature difference is most significant during the first five minutes of the test.

For the same method of measuring temperature, and the same time–temperature curve, there can be significant differences between the heating conditions in various furnaces, depending on the size of the furnace, the type of fuel and the furnace lining material. Fire-resistance furnaces can be fuelled with either oil or gas. Some

Table 5.3 ASTM E119 and ISO 834 time–temperature curves

Time (minutes)	ASTM E119 Temperature (°C)	ISO 834 Temperature (°C)
0	20	20
5	538	576
10	704	678
30	843	842
60	927	945
120	1010	1049
240	1093	1153
480	1260	1257

gas-fired furnaces have pre-mixed burners, others have diffusion burners. These differences in fuel and fuel mixing can affect the luminosity of the flames, which affects the emissivity, hence the heat transfer to the furnace walls and to the test specimen. The most common wall lining materials are fire bricks or ceramic fibre blankets, which have very different thermal properties, hence different rates of heat transfer to the test specimen. Temperatures will increase less rapidly in furnaces lined with bricks.

The differences between furnaces has been a particular problem in Europe, where harmonization of testing standards between many countries is in progress. As a solution it is being proposed that furnace conditions should be controlled by replacing the usual small thermocouples with a 'plate thermometer' which is designed to measure the exposure of the test sample rather than the temperature of the furnace gases. Most European countries are supporting the adoption of a European standard for the plate thermometer, which is expected to greatly reduce the differences in severity of exposure between furnaces in different countries.

5.5 EQUIVALENT FIRE SEVERITY

5.5.1 Real Fire Exposure

The concept of *equivalent fire severity* is used to relate the severity of an expected real fire to the standard test fire. This is important when designers want to use published fire-resistance ratings from standard tests with estimates of real fire exposure. The behaviour of post-flashover fires has been described in Chapter 4. This section describes methods of comparing real fires to the standard test fire.

5.5.2 Equal Area Concept

Early attempts at time equivalence compared the area under time–temperature curves. Figure 5.4 illustrates the concept, first proposed by Ingberg (1928), by which two fires are considered to have equivalent severity if the areas under each curve are equal, above a certain reference temperature. This has little theoretical significance because the units of area are not meaningful. Even though Ingberg was aware of its technical inadequacy he used the equal area concept as a crude but useful method of comparing fires. After carrying out furnace tests, he developed a relationship between fire load in a room and the required fire resistance of the surrounding elements. This approach, subsequently used by US code writers to specify fire-resistance ratings, has been useful, but ignores the effects of ventilation and fuel geometry on fire severity.

The equal area concept is used for correcting the results of standard fire-resistance tests if the standard curve is not exactly followed within the tolerances specified in the standard (ASTM, 1988(a)).

The impact of a fire on a surrounding structure is a function of the heat transfer into the structure. A problem with the equal area concept is that it can give a very poor comparison of heat transfer for fires with different shaped time–temperature curves. Heat transfer from a fire to the surface of a structure is mostly by radiation,

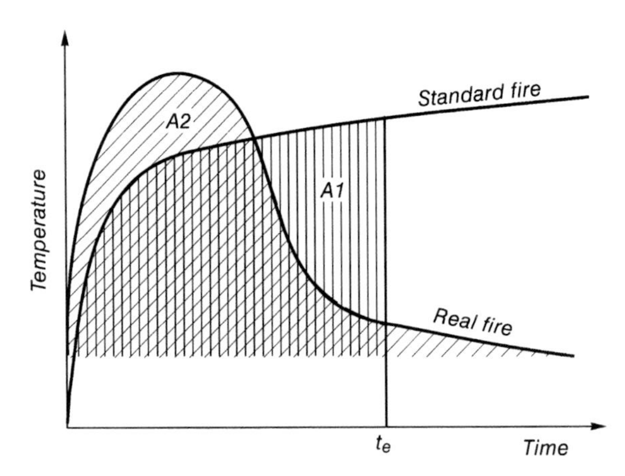

Figure 5.4 Equivalent fire severity on equal area basis

the balance by convection. Because radiative heat transfer is proportional to the fourth power of the absolute temperature, heat transfer to the surface in a short hot fire may be much greater than in a long cool fire, even if both have equal areas under the time–temperature curves.

Babrauskas and Williamson (1978(a) and 1978(b)) also point out that there could be a critical difference between a short hot fire and a longer cool fire if the maximum temperature in the former is sufficient to cause melting or some other critical phase change in a material which would be much less affected in the cooler fire.

5.5.3 Maximum Temperature Concept

A more realistic concept, developed by Law (1971), Pettersson *et al.* (1976) and others, is to define the equivalent fire severity as the time of exposure to the standard fire that would result in the same maximum temperature in a protected steel member as would occur in a complete burnout of the fire compartment. This concept is shown in Figure 5.5 which compares the temperatures in a protected steel beam exposed to the standard fire with those when the same beam is exposed to a particular real fire.

In principle, this concept is applicable to insulating elements if the temperature on the unexposed face is used instead of the steel temperature, and is also applicable to materials which have a limiting temperature, such as the 300°C temperature at which charring of wood generally begins.

The maximum temperature concept is widely used, but it can give misleading results if the maximum temperatures used in the derivation of a time-equivalent formula are much greater or lower than those which would cause failure in a particular building.

5.5.4 Minimum Load Capacity Concept

In a similar concept based on load capacity, the equivalent fire severity is the time of exposure to the standard fire that would result in the same load bearing capacity as

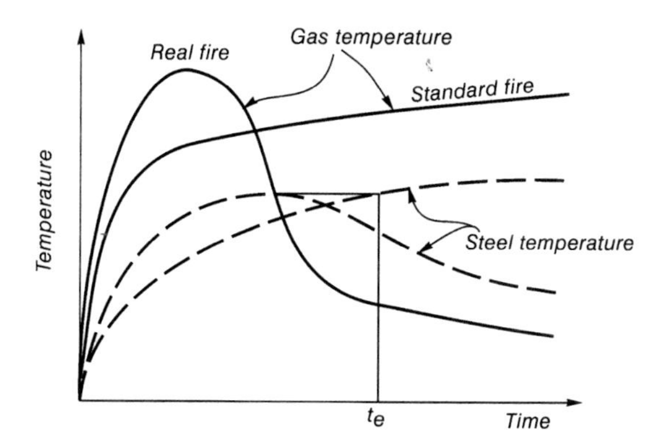

Figure 5.5 Equivalent fire severity on temperature basis

the minimum which would occur in a complete burnout of the firecell. This concept is shown in Figure 5.6 where the load bearing capacity of a structural member exposed to the standard fire decreases continuously, but the strength of the same member exposed to a real fire increases after the fire enters the decay period and the steel temperatures decrease. This approach is the most realistic time equivalent concept for the design of load bearing members. The minimum load concept is difficult to implement for a material which does not have a clearly defined minimum load capacity, for example with wood members where charring can continue after the fire temperatures start to decrease.

5.5.5 Time-equivalent formulae

A number of time-equivalent formulae have been developed by fitting empirical curves to the results of many calculations of the type shown conceptually in Figure 5.5. The resulting formulae are based on the maximum temperature of protected steel members exposed to realistic fires.

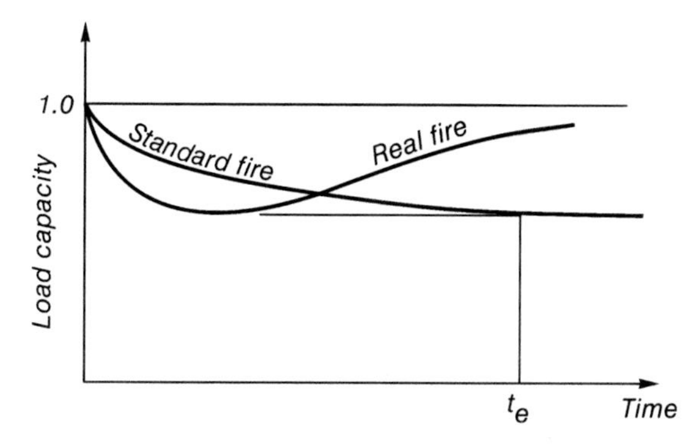

Figure 5.6 Equivalent fire severity on load capacity basis

CIB formula

The most widely used time equivalent formula is that published by the CIB W14 group (CIB, 1986), derived by Pettersson (1973) based on the ventilation parameters of the compartment and the fuel load. The equivalent time of exposure to an ISO 834 test t_e (min) is given by:

$$t_e = k_c w\, e_f \tag{5.9}$$

where e_f is the fuel load (MJ/m^2 of floor area), k_c is a parameter to account for different compartment linings, and w is the ventilation factor (m$^{-0.25}$) given by:

$$w = \frac{A_f}{\sqrt{A_v A_t \sqrt{H_v}}} \tag{5.10}$$

where A_f is the floor area of the compartment (m^2), A_v is the total area of openings in the walls (m^2), A_t is the total area of the internal bounding surfaces of the compartment (m^2), and H_v is the height of the windows (m).

Law formula

A similar formula was developed by Margaret Law on the basis of tests in small-scale compartments (Thomas and Heselden, 1972) and larger-scale compartments (Law, 1973). The formula is given by:

$$t_e = \frac{A_f\, e_f}{\Delta H_c \sqrt{A_v (A_t - A_v)}} \tag{5.11}$$

where ΔH_c is the calorific value of the fuel (MJ/kg).

The CIB formula and the Law formula are only valid for compartments with vertical openings in the walls. They cannot be used for rooms with openings in the roof. The Law formula gives similar results to the CIB formula, generally with slightly larger time equivalent values.

Eurocode formula

These formulae were later modified and incorporated into the Eurocode (EC1, 1994), referred to often as the 'Eurocode Formula' giving t_e (minutes) as

$$t_e = k_b w\, e_f \tag{5.12}$$

where k_b replaces k_c and the ventilation factor w is altered to allow for horizontal roof openings. The ventilation factor is given by

$$w = \left(\frac{6.0}{H_r}\right)^{0.3} \left[0.62 + \frac{90(0.4 - \alpha_v)^4}{1 + b_v a_h}\right] > 0.5 \tag{5.13}$$

where H_r is the compartment ceiling height (m),

$$\alpha_v = A_v/A_f \qquad 0.025 \leqslant \alpha_v \leqslant 0.25 \tag{5.14}$$

$$\alpha_h = A_h/A_f \qquad \alpha_h \leqslant 0.20 \tag{5.15}$$

$$b_v = 12.5\ (1 + 10\ \alpha_v - \alpha_v^2) \tag{5.16}$$

A_f is the floor area of the compartment (m^2), A_v is the area of vertical openings in the walls (m^2), and A_h is the area of horizontal openings in the roof (m^2).

The derivation of the Eurocode formula and a similar formula in the German Standard DIN, 18230-1 (DIN, 1996) is based on work by Schneider *et al.* (1990). It is understood to have come from an empirical analysis of calculated steel temperatures in a large number of fires simulated by a German computer program called Multi-Room-Fire-Code (U. Schneider, personal communication).

An important difference from the CIB formula is that the Eurocode equivalent time is independent of opening height, but depends on the ceiling height of the compartment, so the two formulae can give different results for the same room geometry. The results are similar for small compartments with tall windows, but the Eurocode formula gives much lower fire severities for large compartments with tall ceilings and low window heights.

Values of the terms k_c and k_b are given in Table 5.4, where they are shown to depend on the compartment materials (roughly inversely proportional to the thermal inertia). The 'general' case is that recommended for compartments with unknown materials. Note that k_c and k_b have slightly different numerical values and dimensions, because of the different ventilation factors in the respective formulae. The bottom line marked 'large compartments' is a modification to the Eurocode formula recommended by Kirby *et al.* (1999) for large spaces, after several experimental fires in a large compartment measuring 23 × 5.5 m by 2.7 m high.

Using typical thermal properties of materials from Table 3.1, a building constructed from steel is in the 'high' category, normal and lightweight concrete are 'medium', and gypsum plaster and any materials with better insulating properties are in the 'low' category. The Eurocode formula has been used to create the tables in the Approved Documents to the New Zealand Building Code (BIA, 1992).

Table 5.4 Values of k_c or k_b in the time equivalent formulae

			$b = \sqrt{(k\rho c_p)}$			
Formula	Term	Units	high >2500	medium 720–2500	low <720	General
CIB W14	k_c	min m$^{2.25}$/MJ	0.05	0.07	0.09	0.10
Eurocode	k_b	min m^2/MJ	0.04	0.055	0.07	0.07
Large compartments	k_b	min m^2/MJ	0.05	0.07	0.09	0.09

k = thermal conductivity (W/mK), ρ = density (kg/m^3), c_p = specific heat (J/kg K)

Validity

Time-equivalent formulae are empirical. They have generally been derived by calculation, using a particular set of design fires for small rooms and the maximum temperature concept for certain protected steel members with various thicknesses of insulation. As such they may not be applicable to other shapes of time–temperature curve, to larger rooms, to other types of protection, or to other structural materials. None of the formulae described above have well documented derivations which describe their limitations.

It is generally accepted that the time equivalent formulae can be applied to protected steelwork (for which they were derived) and reinforced concrete members. They are not intended for unprotected steelwork or for timber construction. Eurocode 1, Part 2-2 states that the formula can be used for all types of materials, but the Swedish national application document does not permit it to be used for timber structures. Thomas *et al.* (1997) compared the formulae with the results of calculations for typical concrete, steel and timber structures, using design fires calculated by the COMPF2 computer program, finding that there are many situations where the formulae do not give a good prediction of actual behaviour, usually on the unsafe side. Law (1997) compares several time-equivalent formulae, including those shown above, concluding that the CIB formula and the Law formula give much more accurate results than the Eurocode formula.

In conclusion, these time-equivalent formulae are a crude approximate method of introducing real fire behaviour into fire engineering calculations. It is much more accurate to make designs from first principles with the use of estimated post-flashover fire temperatures such as those described in Chapter 4.

WORKED EXAMPLE 5.1

Calculate the equivalent fire severity using the Eurocode formula for a room 4.0 m × 6.0 m in area, 3.0 m high, with one window 3.0 m wide and 2.0 m high. The fire load is 800 MJ/m² floor area. The room is constructed from concrete.

Length of room	$l_1 = 6.0\,\mathrm{m}$
Width of room	$l_2 = 4.0\,\mathrm{m}$
Floor area	$A_f = l_1\, l_2 = 6.0 \times 4.0 = 24.0\,\mathrm{m}^2$
Height of room	$H_r = 3.0\,\mathrm{m}$
Fuel load energy density	$e_f = 800\,\mathrm{MJ/m^2}$

For concrete

Thermal conductivity	$k = 1.6\,\mathrm{W/mK}$
Density	$\rho = 2200\,\mathrm{kg/m^3}$
Specific heat	$c_p = 880\,\mathrm{J/kg\,K}$
Thermal inertia	$\sqrt{k\rho c_p} = 1760\,\mathrm{W\,s^{0.5}/m^2\,K}$ (medium)
Conversion factor	$k_b = 0.055$

Window height	$H_v = 2.0\,\text{m}$
Window width	$B = 3.0\,\text{m}$
Window area	$A_v = H_v B = 2.0 \times 3.0 = 6.0\,\text{m}^2$
Horizontal vent area	$A_h = 0$ (no ceiling opening)
	$\alpha_v = A_v/A_f = 6.0/24.0 = 0.25$
	$\alpha_h = A_h/A_f = 0$
	$b_v = 12.5(1 + 10\,\alpha_v - \alpha_v^2) = 43.0$
Ventilation factor	$w = (6.0/3.0)^{3.0}\,[0.62 + 90\,(0.4 - 0.25)^4/1(1 + 43.0 \times 0)] = 0.820\,\text{m}^{-0.3}$
Equivalent fire severity	$t_e = e_f k_b w = 800 \times 0.055 \times 0.820 = 36.1$ minutes

WORKED EXAMPLE 5.2

Repeat Worked Example 5.1 with an additional ceiling opening of $3.0\,\text{m}^2$.

Ceiling opening area	$A_h = 3.0$
	$\alpha_h = A_h/A_f = 3.0/24.0 = 0.125$
Ventilation factor	$w = (6.0/3.0)^{0.3}\,[0.62 + 90\,(0.4 - 0.25)^4/(1 + 43.0 \times 0.125)] = 0.772\,\text{m}^{-0.3}$
Equivalent fire severity	$t_e = e_f k_b w = 800 \times 0.055 \times 0.772 = 34.0$ minutes

WORKED EXAMPLE 5.3

Repeat Worked Example 5.1 using the CIB formula and the Law formula.

CIB formula

Length of room	$l_1 = 6.0\,\text{m}$
Width of room	$l_2 = 4.0\,\text{m}$
Floor area	$A_f = l_1 l_2 = 6.0 \times 4.0 = 24.0\,\text{m}^2$
Height of room	$H_r = 3.0\,\text{m}$
Fuel load energy density	$e_f = 800\,\text{MJ/m}^2$
Total area of the internal surface	$A_t = 2(l_1 l_2 + l_1 H + l_2 H) = 2(6 \times 4 + 6 \times 3 + 4 \times 3) = 108\text{m}^2$

For concrete

Thermal conductivity	$k = 1.0\,\text{W/mK}$
Density	$\rho = 2200\,\text{kg/m}^3$
Specific heat	$c_p = 880\,\text{J/kg K}$
Thermal inertia	$\sqrt{k\rho c_p} = 1391\,\text{W s}^{0.5}/\text{m}^2\,\text{K}$ (medium)
Conversion factor	$k_c = 0.07\,\text{min.m}^{2.25}/\text{MJ}$
Window height	$H_v = 2.0\,\text{m}$
Window width	$B = 3.0\,\text{m}$
Window area	$A_v = H_v B = 2.0 \times 3.0 = 6.0\,\text{m}^2$

Ventilation factor

$$w = A_f/(A_v\, A_t\, H_v^{0.5})^{0.5}$$
$$= 24/(6 \times 108 \times 2^{0.5})^{0.5} = 0.793\,\text{m}^{-0.25}$$

Equivalent fire severity

$$t_e = e_f\, k_c\, w = 800 \times 0.07 \times 0.793$$
$$= 44.4 \text{ minutes}$$

Law formula

Net calorific value of wood $\Delta H_c = 16\,\text{MJ/kg}$

Equivalent fire severity

$$t_e = e_f\, A_f/[\Delta H_c\,(A_v\,(A_t - A_v))^{0.5}]$$
$$= 800 \times 24/[16 \times (6(108-6))^{0.5}]$$
$$= 48.6 \text{ minutes}$$

6

Fire Resistance

6.1 OVERVIEW

This chapter describes fire resistance, the standard fire resistance test and the ways in which it is used for achieving fire-resistance ratings of building elements. This chapter also describes methods for calculating the fire resistance of structural members and discusses the importance of fire resistance of components and assemblies in real buildings.

6.2 FIRE RESISTANCE

Fire resistance is a measure of the ability of a building element to resist a fire. Fire resistance is most often quantified as the time for which the element can meet certain criteria during exposure to a standard fire-resistance test. Structural fire resistance can also be quantified using the temperature or load capacity of a structural element exposed to a fire.

It is important to recognize that *individual materials do not possess fire resistance.* Fire resistance is a property assigned to building elements which are constructed from a single material or a mixture of materials. Some building elements may be simple elements such as a single steel column or a concrete floor slab. Other building elements may be complex assemblies of several layers of different materials such as a composite floor and suspended ceiling system.

A *fire-resistance rating* is the fire resistance assigned to a building element on the basis of a test or some other approval system. Some countries use the terms *fire rating*, *fire-endurance rating* or *fire-resistance level.* These terms are usually interchangeable. Fire-resistance ratings are most often assigned in whole numbers of hours or parts of hours, in order to allow easy comparison with the fire-resistance requirements specified in building codes. For example, a wall that has been shown by test to have a fire resistance of 75 minutes will usually be assigned a fire resistance rating of one hour.

6.3 ASSESSING FIRE RESISTANCE

Building elements need to be assigned fire-resistance ratings for comparison with the fire severity specified by codes. The most common method of assessing fire resistance is to carry out a full-scale fire-resistance test. It is becoming increasingly possible to assess fire resistance by calculation in lieu of full-scale tests, as permitted explicitly by codes such as the Uniform Building Code (UBC, 1997). The results of assessments of fire resistance obtained from tests, calculations or expert opinions are listed in various documents maintained by testing authorities, code authorities or manufacturers. These listings of fire-resistance ratings are in three main categories; *generic ratings*, which apply to typical materials, *proprietary ratings*, which are linked to particular manufacturers, and approved *calculation methods*. Generic and proprietary ratings are obtained directly or indirectly from full-scale fire-resistance tests. This chapter describes all these methods of assessing and listing fire resistance.

Fire resistance of any building element depends on many factors, including the severity of the fire test, the material, the geometry and support conditions of the element, restraint from the surrounding structure and the applied loads at the time of the fire. Many building codes and manufacturers' documents simply list fire resistance of one, two or four hours with little or no reference to these factors which are discussed further in this book.

6.4 FIRE-RESISTANCE TESTS

Almost all countries have building codes that specify fire-resistance ratings for building elements. Fire-resistance ratings are most often specified in hours or minutes, with typical values ranging from half an hour to two, three or four hours. The time–temperature curves followed in the tests have been described in Chapter 5. Fire-resistance tests are not intended to simulate real fires. Their purpose is to allow a standard method of comparison between the fire performance of structural assemblies.

Many countries require that fire resistance be based on the results of *full-scale* fire-resistance tests. The required sizes for full-scale tests are given below. Full-scale tests are expensive, but for many years it has been considered essential to test elements of building construction at a large scale because cheaper small-scale tests are not able to assess the effects of potential problems caused by connections, shrinkage, deflections, and gaps between panels of lining materials.

Full-scale testing is the most common method of obtaining fire-resistance ratings, but fire-resistance tests are very expensive, so are only undertaken when considered necessary. The high expense of full-scale fire-resistance testing is encouraging manufacturers to share test results within trade organizations, and is hastening the development of new calculation methods to predict fire resistance by calculation rather than by test. All calculations should be based on the results of full-scale tests to avoid the potential problems described above.

Fire-resistance tests are carried out on representative specimens of building elements. For example, if a representative sample of a flooring system has been exposed to the standard fire for at least two hours while meeting the specified failure criteria, a similar assembly can be assigned a two hour fire resistance rating for use in a real building. The implication is that the built assembly will behave at least as well in a real fire as the tested assembly did in the full-scale fire test. Obvious difficulties are that there are many differences between the tested and the built assemblies. The tested assemblies nearly always have different sizes and shapes, and different loads or boundary conditions than in real buildings, and the test fire may be very different from a real fire. These problems are addressed later in this book.

6.4.1 Standards

For fire-resistance testing, many countries use the International Standard ISO 834 (ISO, 1975) or have national standards based on ISO 834 (for example AS 1530 Part 4 (SAA, 1990(c)). Most European countries have standards similar to ISO 834. The standard used in the United States and some other countries is ASTM E119 (ASTM, 1988(a)), first published in 1918. The Canadian standard (ULC, 1989) is based on ASTM E119. The relevant British Standards are BS 476 Parts 20-23 (BSI, 1987).

6.4.2 Test Equipment

A typical fire-test furnace consists of a large steel box lined with fire bricks or a ceramic fibre blanket. The furnace will have a number of burners, most often fuelled by gas but sometimes by fuel oil. There must be an exhaust chimney, several thermocouples for measuring gas temperatures and usually a small observation window.

National and international standards for fire-resistance testing do not specify the construction of the furnace in detail, which sometimes causes problems when making comparisons between tests from different furnaces. The standards are more concerned with the fire temperatures to be followed during the test and the failure criteria. As stated in Chapter 5, most national standards are based on either the ASTM E119 test or the ISO 834 test, which have some minor but important differences. Fortunately, despite minor differences, fire-resistance test methods are very similar around the world, so that international comparisons are always possible. It is exceedingly difficult for any country to make a major change to standard test procedures because of the cost of re-testing and re-classifying the large number of assemblies which have been tested in the past.

In a typical test, a wall or floor assembly is constructed in a frame remote from the furnace, then brought to the furnace in its frame, and used to close off the furnace opening before the test begins. The burners are ignited at the start of the test and controlled to produce the time–temperature curve specified by the testing standard. Temperatures, deformations and applied loads are monitored during the test. The essential temperature measurements are those in the furnace itself and on the unexposed face of the specimen. In some tests, temperatures are measured at other locations within the test specimen or inside the furnace for research and development purposes.

The most common apparatus for full-scale fire-resistance testing is the vertical wall furnace (Figure 6.1). The minimum size specified by most testing standards is $3.0 \times 3.0\,m^2$ (ISO 834 or ASTM E119). Some furnaces are 4.0 m tall. Floors, roofs or beams are tested in a horizontal furnace (Figures 6.2, 6.3). ASTM E119 specifies a

Figure 6.1 Typical furnace for full-scale fire-resistance testing of walls

minimum size of 16 m² with a span at least 3.7 m. ISO 834 recommends a size of 3 m×4 m. Some furnaces can be tipped from horizontal to vertical orientation to test both walls and floors. Special furnaces are available in some laboratories for testing individual columns or beams, or other non-standard items.

Test specimens are intended to represent actual construction as closely as possible. The moisture content of the test specimen is important because high moisture content can increase fire resistance considerably, especially delaying the temperature rise on the unexposed surface of floors or walls. Most testing standards specify conditions of relative humidity and temperature for conditioning of specimens and also methods of correcting test results for non-standard moisture content.

6.4.3 Failure Criteria

The three failure criteria for fire resistance testing are *stability*, *integrity* and *insulation*.

To meet the *stability* criterion, a structural element must perform its load-bearing function and carry the applied loads for the duration of the test, without structural collapse. Many testing standards have a limitation on deflection or rate of deflection for load-bearing tests, so that a test can be stopped before actual failure of the test

Table 6.1 Failure criteria for construction elements

	Stability	Integrity	Insulation
Partition		X	X
Door		X	X
Load-bearing wall	X	X	X
Floor/ceiling	X	X	X
Beam	X		
Column	X		
Fire-resistant glazing		X	

specimen which would damage the furnace. Commonly specified failure criteria are a deflection of 1/20 of the span, or a limiting rate of deflection when the deflection is 1/30 of the span.

The integrity and insulation criteria are intended to test the ability of a barrier to contain a fire, to prevent fire spreading from the room of origin. To meet the *integrity* criterion, the test specimen must not develop any cracks or fissures which allow smoke or hot gases to pass through the assembly. The ASTM E119 specification requires that there be no passage of flame or hot gases sufficient to ignite cotton waste. To meet the *insulation* criterion the temperature of the cold side of the test specimen must not exceed a specified limit, usually an average increase of 140°C and a maximum increase of 180°C at a single point. These temperatures represent a conservative indication of the conditions under which fire might be initiated on the cool side of the barrier.

All fire-rated construction elements must meet one or more of the three criteria as shown in Table 6.1, depending on their function. Note that fire-resistant glazing need only meet the integrity criterion because it is not load bearing and it cannot meet the insulation criterion as glass has very little resistance to radiant transfer of heat.

An increasing international trend is for fire codes to specify the required fire resistance separately for stability, integrity and insulation. For example, a typical load-bearing wall may have a specified fire-resistance rating of 60/60/60, which means that a one hour rating is required for stability, integrity and insulation. If the same wall was non-load-bearing, the specified fire-resistance rating would be –/60/60. A fire door with a glazed panel may have a specified rating of –/30/–, which means that this assembly requires an integrity rating of 30 minutes, with no requirement for stability or insulation.

An additional integrity criterion for walls and partitions in some testing standards is the hose-stream test. This test requires that no water should pass through the wall when it is subjected to water from a standard fire-fighting hose immediately after the fire test. The ASTM E119 standard allows a duplicate test specimen to be subjected to the hose-stream test after fire exposure of half of the time of the fire-resistance rating. The ISO 834 standard does not include a hose-stream test.

6.4.4 Standard of Construction

The standard of construction of fire-test specimens is sometimes of concern. Fire tests are supposed to be carried out on representative samples typical of normal construction, but some manufacturers may want to use their very best materials and workmanship for the fire test. Such problems can be overcome with accurate reporting by independent testing agencies, including good descriptions of the details of materials and fastenings. Many unsuccessful fire tests are never reported, so published test results may only represent the very best of the specimens actually tested.

6.4.5 Furnace Pressure

The pressure inside the test furnace is important. The furnace pressure affects the *integrity* criterion, because positive pressure will force flames or hot gases through any cracks. There are no requirements for pressure in the ASTM E119 test specification, hence many fire resistance tests in the United States are conducted at a low negative pressure to give the most favourable test result. The ISO 834 test method specifies a positive pressure of 10 Pa under a horizontal test specimen such as a floor system. For vertical test specimens such as walls the pressure gradient must be linear, with 10 Pa at the top and at least two thirds of the specimen subjected to positive pressure. The British Standard BS 476 Part 20 (BSI, 1987) specifies a pressure gradient of 8.5 Pa/m of height with the neutral axis 1 m above the floor level and a maximum pressure of 20 Pa. These pressures are sufficient to force hot gases through small openings near the top of a wall, but are too low to have a significant effect on load-bearing capacity.

Standard fire resistance tests do not assess resistance to blast forces. Special design is necessary for walls, partitions, or other barriers which may be subjected to impact loading or forces resulting from explosions or blast. This problem was highlighted in the Piper Alpha oil platform disaster, where walls with otherwise adequate fire resistance were blown out by explosions, leading to rapid and catastrophic spread of fire.

6.4.6 Applied Loads

All elements which are required to meet the *stability* criterion (see Table 6.1) should be tested under an applied load. As an exception, the ASTM E119 test method permits steel columns and beams to be tested without an applied load provided that the average temperature does not exceed 538°C. At this temperature, a fire-exposed steel member would have approximately half of its normal temperature load capacity (see Chapter 8). Testing of unloaded specimens can produce unsafe results because the test does not assess the effect of deformations during the fire which could affect the restraint from surrounding structure, the behaviour of connections or the 'stickability' of the applied fire protection materials.

The level of load on a structural element during a real fire can have a large effect on its structural performance, as discussed in the next chapter. Similarly, the level of applied load during a fire test can have a significant effect on the level of fire resistance achieved. It is not easy to decide what level of load to apply to a test specimen during a fire-resistance test, because low applied loads will give a high fire-resistance rating but may limit the use of the product in other applications, whereas higher applied loads will lead to a lower fire-resistance rating and may result in an additional test being required if the assembly just fails to achieve a particular rating. Such difficulties can be overcome with the use of calculations to predict fire-resistance ratings based on a small number of full-scale tests.

Both ASTM E119 and ISO 834 recommend that the applied load in a fire test should be the maximum permitted load under nationally accepted design rules, but they also permit lower loads to be used provided that they are clearly identified. The definition of 'maximum permitted load' or 'maximum design load' will vary greatly from country to country depending on whether the relevant structural design code is in 'working stress' format or 'limit states' format. See Chapter 7 for more discussion of loads and loading standards.

When deciding what level of loads to apply during a fire resistance test, it is best to apply those loads which produce stresses in the tested element similar to those expected in the actual building at the time of an unwanted fire. This may be possible if a specific prototype is being tested, but may be very difficult if a proprietary product is being tested for possible use in a multitude of different situations. The precise level of the loads and stresses is not important, but it is essential that the loads applied during the test be described accurately in the test report. Designers can then ensure that the corresponding stresses are not exceeded in the fire design. Loading devices and levels of structural restraint vary significantly from furnace to furnace, so it is important that these are also described in the test report.

The loads in some real members may be very low at the time of a fire, especially if the member was designed for deflection rather than strength, so it may be decided to use higher stresses in the fire-test specimen, to allow subsequent use in a wider variety of applications. If the designed assembly has a larger span than that tested, the comparison with the test results should be on the basis of stresses rather than loads, but it is often difficult to keep shear and flexural stresses in the same ratio.

6.4.7 Restraint and Continuity

Flexural continuity and axial restraint have a significant effect on fire resistance, as described in the next chapter. For this reason, the support and restraint conditions are important in fire tests of structural elements, especially floors and beams.

Figure 6.2 Floor furnace with a heavy surrounding beam for providing axial restraint to the test specimens (Reproduced from Lie, 1972)

Most national testing standards require test specimens to be supported in a condition similar to that expected in actual buildings (Figure 6.2). Since 1970, ASTM E119 has had separate requirements for 'restrained ratings' and 'unrestrained ratings', but does not clearly define these. For floors, ASTM E119 requires that restrained specimens shall be 'reasonably restrained in the furnace', whereas restrained beams are to be tested 'simulating the restraint in the construction represented'. Not all furnaces have been designed to provide restraint during fire tests.

ISO 834 states that the test specimens should be installed in the furnace in such a way that the boundary conditions provide the degree and the type of restraint to which they will be subjected in practice. Where details of the use are not available, ISO 834 specifies that the support conditions should be clearly pinned or be fully fixed against rotation, and be clearly free to expand or be fully restrained against axial movement, with the actual support conditions documented in the test report. See Chapter 7 for more discussion of restraint and continuity.

6.4.8 Small-scale Furnaces

Many laboratories have small-scale furnaces which are used for research and development. Intermediate-scale furnaces (1.0 to 2.5 m^2) are used for standard testing of small items such as fire doors, using the same standard time–temperature curve as in a full-scale furnace. Even if the same time–temperature curve is followed, the fire severity may not be identical to a large furnace because the different geometry may result in different heat transfer coefficients from the walls and hot gases of the furnace to the test specimen. Tests in small- and intermediate-scale test

Figure 6.3 Detail of loading arrangement for the fire testing of floors

furnaces may not pick up potential problems such as shrinkage, deflections or connection problems which could occur in real fires or in full-scale tests. Despite these limitations, they can give useful information in many situations, particularly regarding thermal transmission.

6.5 APPROVED FIRE-RESISTANCE RATINGS

6.5.1 Listings

Most countries require that fire-resistance tests be certified by a recognized testing laboratory or approvals agency. In North America, independent testing organizations such as Underwriters Laboratories (UL, 1996) and Southwest Research Institute (SWRI, 1996) maintain registers of approved assemblies to which they have assigned fire-resistance ratings. Most of these ratings are based on tests which they have carried out in accordance with recognized testing standards. Generic ratings based on these approvals are listed in national building codes (e.g. (BCA 1996, NBCC, 1995, UBC, 1997). In small countries, approvals are often needed for tests conducted in other countries, so that in New Zealand, for example, a register of approved listings is maintained by the national standards organization (SNZ, 1991) containing many listings derived in other countries. Some trade organizations (e.g. ASFPCM, 1988, Gypsum Association, 1994) maintain industry listings of approvals for products manufactured or used by their members.

Listings generally fall into three categories: *generic ratings*, *proprietary ratings*, or *calculation methods*.

Generic ratings

Generic fire resistance ratings, or 'tabulated ratings' are listings which assign fire resistance to typical materials with no reference to individual manufacturers or detailed specifications. For example, many national codes list tables of generic ratings for fire protection of structural steel members by encasement in a certain thickness of concrete, with no details of concrete quality or reinforcing. Generic ratings are derived from many full-scale fire resistance tests carried out over many years. Generic ratings are widely used because they can be applied to commonly available materials in any country. Generic ratings are usually very conservative, and they are inadequate because they apply only to standard fire exposure, and make no allowance for the size and shape of the fire-exposed member or the level of load.

Proprietary ratings

Proprietary fire-resistance ratings apply to proprietary products made by specific manufacturers. Proprietary ratings are based on the results of full-scale fire tests commissioned by the manufacturer of the materials. Proprietary ratings are usually accompanied by an approved specification detailing the materials and construction methods, and it is the assembly rather than the materials which has the approved rating. Unless they are covered by a suitable agreement, proprietary ratings cannot be applied to similar products from other manufacturers because there may be differences in materials or installation methods, and the rating may legally be the property of the manufacturer.

Proprietary ratings may be less conservative than generic ratings because they relate to more closely defined products. Proprietary ratings are usually based on standard fire exposure and make no allowance for the level of applied load, but they sometimes include reference to the size and shape of the fire-exposed member in a more accurate way than generic ratings.

Calculation methods

As the science of fire engineering develops, it is becoming more feasible to assess fire resistance by calculation as well as by test. Many listing agencies and national design codes now include approved calculation methods for assessing fire resistance. Many of these methods are described in this book. Calculations should be based on full-scale fire resistance test results of similar assemblies.

6.5.2 Expert Opinion

Most of the listings described in the above documents are based directly on the results of full-scale fire resistance tests. Such fire tests are very expensive, so testing

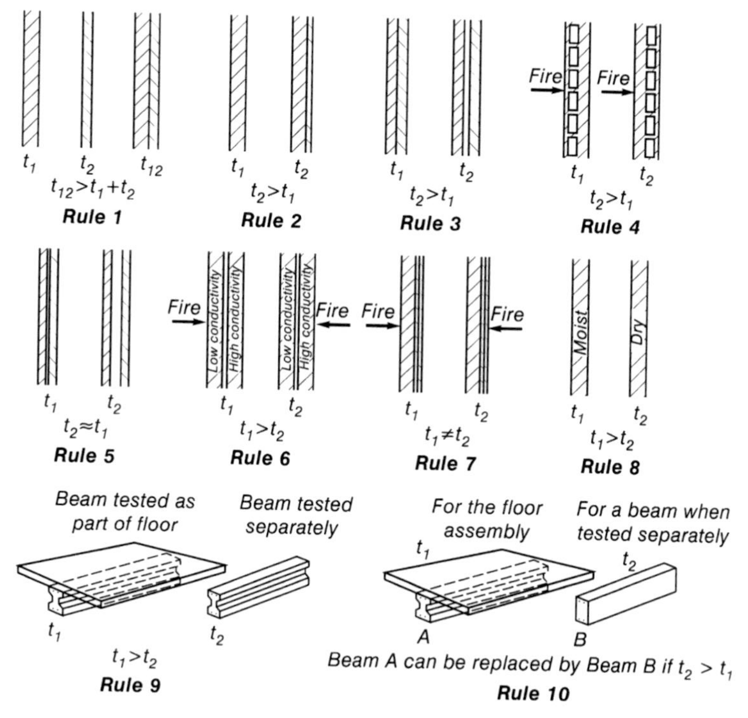

Figure 6.4 Harmathy's ten rules of fire resistance (Reproduced from Harmathy (1965) by permission of National Fire Protection Association)

and approving authorities are increasingly asked to give written expert opinions on assemblies which are similar to, but different from, those which have passed a test. An increasing number of listed fire-resistance ratings are based on such expert opinions. The opinion will state whether the assembly would be considered likely to pass a test, based on observations of similar successful tests and the considered experience of the testing and approving personnel.

As an indication of the factors to be considered in making an opinion, Figure 6.4 illustrates a useful set of empirical 'rules' for comparing fire resistance of similar assemblies (Harmathy, 1965). These 'rules' of fire endurance have stood the test of time and are applicable in almost all situations. These rules have been expanded and explained in more detail by Lie (1992).

6.6 FIRE RESISTANCE BY CALCULATION

Figure 6.5 shows a flow chart for the process of calculating the strength of a structural assembly exposed to a complete burnout of a fire compartment. The resulting load capacity can be compared with the expected applied load at the time of a fire, to verify whether the design is satisfactory. This is design in the strength domain (see Chapter 5).

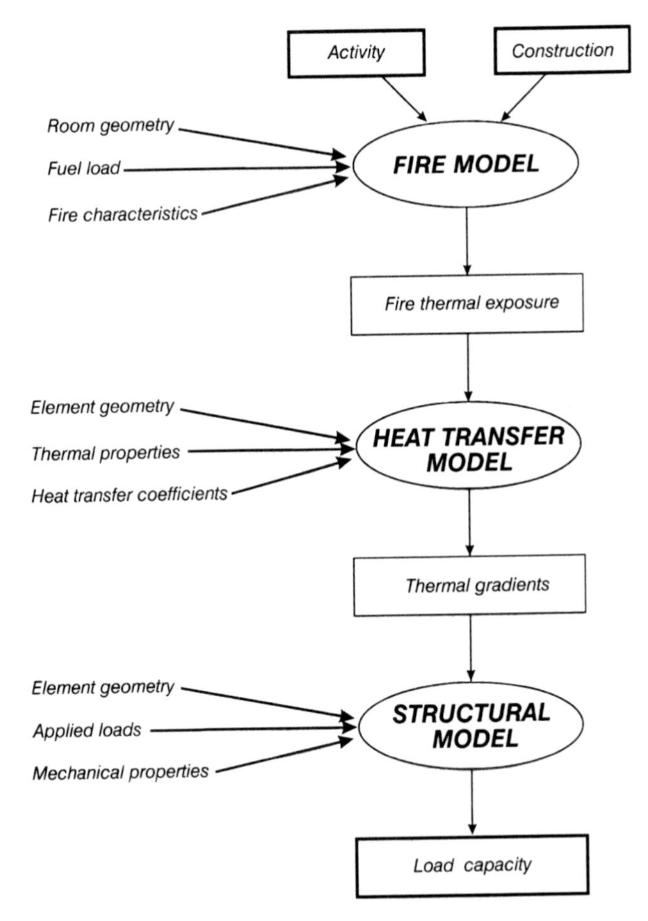

Figure 6.5 Flow chart for calculating the load capacity of a structure exposed to fire

The process of calculating structural fire behaviour has three essential component models: a fire model, a heat transfer model and a structural model.

6.6.1 Fire Model

Fire models have been discussed in Chapter 5. Input can be any selected time–temperature curve including the standard fire, a measured real fire or a parametric fire curve.

6.6.2 Heat transfer model

The heat transfer model is an essential component of calculating fire resistance because the load capacity or the containment ability of a fire-exposed element or structure depends on the internal temperatures. The temperature of any material exposed to a fire increases as heat is conducted from the hot fire-exposed surface to the cooler interior. The temperature gradients depend on the radiative and convective heat transfer coefficients at the surface, and the conduction of heat within the member.

For non-load bearing elements designed to contain fires, the output from a heat transfer model can be used directly to assess whether the time to critical temperature rise on the unexposed face is acceptable. For simple structural elements with a single limiting temperature, the output from a heat transfer model can be used directly to assess whether the critical temperature is exceeded. These situations do not require the application of a structural model. They are examples of verification in the temperature domain.

For more complicated structural elements or assemblies, the output from the heat transfer model is essential input to a structural model for calculating load-bearing capacity. Temperature gradients within a member may or may not be significant. When a material such as steel with a high thermal conductivity is heated slowly, as in a protected assembly, it may be sufficiently accurate to disregard temperature gradients and assume that all the material is at the same temperature. For materials with low thermal conductivity like concrete, it becomes very important to know the thermal gradients during the fire because these affect the temperature of the reinforcing steel. Heat transfer calculations are less important for large timber members because the rate of charring is more dependent on the thermal diffusivity of the wood than on the fire environment.

Calculation of heat transfer requires knowledge of the geometry of the element, thermal properties of the materials and heat transfer coefficients at the boundaries. Practical difficulties are that some thermal properties are very temperature dependent, and heat transfer coefficients are not well established. Heat transfer to surfaces of the element is a combination of convection and radiation. Heat transfer through solid materials is by conduction. Heat transfer through voids is a combination of convection and radiation. Heat transfer calculations have been described briefly in Chapter 3. Specific examples are given in later chapters.

6.6.3 Structural Model

Models for calculating the performance of structural elements exposed to fire are described in the next chapter. Hand calculation methods can be used for simple elements but sophisticated computer models are necessary for the analysis of frames or larger structures. Computer-based structural analysis models must be able to include the effects of thermal expansion, loading and unloading, large deformations and non-linear material properties which are temperature dependent, all for a framework of interconnected members of different materials.

Hand calculation methods for the main structural materials are given later in this book.

6.7 FIRE RESISTANCE OF ASSEMBLIES

Most of the above discussion relates to fire resistance of individual elements of building construction. Real buildings are more than just a collection of elements, so the fire resistance of the whole building must be considered by considering the fire resistance of its component parts and their location in the building. Many elements in a real building may be of different sizes, shapes and with different fixing details than those tested. Any assessment of fire resistance must consider the three failure criteria of stability, insulation and integrity. Fire resistance of a few representative assemblies are discussed in this section.

6.7.1 Walls

Most fire-resistance furnaces are specifically designed for the testing of walls, as briefly described earlier in this chapter. Non-load-bearing partitions are easier to test than load-bearing walls, because only the integrity and insulation criteria are important, and large movement at the edges of the walls is not expected. When testing load-bearing walls, it is necessary to have suitable loading devices such as hydraulic jacks at the top or bottom of the wall. These need to be protected from the furnace temperatures, and be installed in such a way that the loaded edge of the wall is free to move as the load is applied or as subsequent deflections occur. For framed systems such as stud walls, it is important to ensure that no applied load is carried by the end studs which are bolted to the edge of the furnace. When assessing fire-test results for real buildings, it is important to find out whether the height, load or end fixity conditions are different from those in the fire test.

6.7.2 Floors

Floors are more difficult to test than walls because a larger furnace is necessary and the loading equipment is much more extensive. All floors are load bearing, so all must be loaded during testing. Because of furnace size limitations, most floor systems are tested at spans less than those commonly used in buildings. A tested floor system should have similar stresses and similar deformations to those expected in longer spans in a real building, to enable the test results to be extrapolated. Floors designed to span in only one direction must be free to behave in that way in the test furnace. Deformations are particularly important because the flexural curvature of a floor will affect the integrity of any fire-resisting membrane or applied fire protection.

Many floor–ceiling assemblies rely on a ceiling membrane as an essential part of the fire-resisting system, so these assemblies must be tested as a complete system. Flexural continuity and axial restraint can have a large influence on the results of fire-resistance testing of floors, as described previously.

Traditionally, floors have only been fire tested from below. This is because most fires tend to spread up, not down, and the most vulnerable part of the structure is usually on the underside. There has been some recent concern about the possibility of fires burning downwards through light timber floors clad on the top surface with particle board or plywood, but observations at real fires show that there is often a lot of debris on the floors, from collapsed ceilings, fittings and contents, which provide some protection to the top surface of the floor system.

Figure 6.6 A special furnace for fire resistance testing of columns, with an unprotected steel column ready for testing (Reproduced by permission of Corus UK Ltd)

6.7.3 Beams

Beams are always tested for fire resistance as part of a floor or roof assembly, with fire exposure from below. The floor or roof may be structurally part of the beam, with composite action, or it may simply be a non-structural component to seal off the top of the furnace. When assessing fire resistance of beams it is essential to know whether there is composite action. A major difficulty in fire testing of beams is the limited span available in almost all furnaces. As with floors, a tested beam should have similar stresses and similar deformations to those expected in longer spans in a real building, but this can be very difficult to achieve over a small span. Flexural continuity and axial restraint can be difficult to provide because of the large forces involved, but these can have a very large influence on the results. Effects such as these may be better assessed by calculation.

6.7.4 Columns

Columns are usually tested in special furnaces which expose the column to fire from all sides (Figure 6.6). There are only a small number of column furnaces in the world, so fire resistance of columns is often achieved by calculation rather than by test, or by using conservative generic ratings. In real buildings, columns are often built into walls which protect one or more sides of the column from fire exposure. This may reduce the load-bearing capacity during the fire because of temperature gradients through the cross section leading to thermal distortion and non-uniform load capacity. A column built into a wall can be tested in a wall furnace, but very few such tests have been done and the fire resistance of such columns is poorly understood.

6.7.5 Penetrations

Penetrations through walls and floors can severely reduce the ability of these barriers to contain a fire. If fire enters a cavity in a wall or floor assembly through a penetration, it can also severely reduce the load capacity. Walls and floors in typical buildings have numerous penetrations for electrical, plumbing and air handling services. The fire resistance of all of these must be considered as part of the fire safety design. There are standard methods for testing fire stops at penetrations through walls and floors (e.g. ASTM, 1988(b)). Methods for protecting 'poke-through' penetrations are given by Gustaferro and Martin (1988). Particular problems can occur if unprotected penetrations are not visible, such as hidden penetrations through walls above suspended ceilings, and penetrations made at a later date as part of alterations to the building. Many proprietary products are available for sealing gaps and openings in buildings, including fire-resisting boards, paints, mastic sealants, intumescent strips and pillows. Products such as these can make the difference between success and failure of passive fire protection strategies.

Fire test performance of plastic pipe penetrations is described by England *et al.* (2000).

The fire resistance of walls and floors can also be reduced if the barrier is penetrated by a heat-conducting member such as an unprotected steel beam. Fire on one side of the barrier can cause the member to heat up, conducting heat to the unexposed side, where ignition can occur if combustible materials are in contact. Some codes prohibit this type of penetration. The danger of fire spread can be reduced by insulating the beam for a certain distance on either side of the wall. Deformations in steel beams passing through walls can also damage the wall unless the structure is specifically designed to prevent such damage.

6.7.6 Junctions and Gaps

Junctions between walls and floors are seldom fire tested. It is important for designers to assess the connecting details to ensure that the junctions do not give weaknesses in barriers that otherwise have excellent fire resistance. There are many proprietary products for providing fire resistance at junctions. Most common materials, including concrete, steel and wood, can be used for preventing fire spread through junctions, provided that the materials have sufficient thickness and are well-detailed. Aluminium and plastic materials are not suitable because they melt at low temperatures. Gaps between precast concrete panels can be fire rated using a ceramic fibre blanket (Gustaferro and Martin, 1988, UBC, 1997). Section 11.9.7 describes fire resistance of junctions in light timber frame construction.

6.7.7 Seismic Gaps

In seismic regions, buildings are provided with seismic gaps to allow differential movements to occur in the event of an earthquake. The seismic gaps can be within a building (to separate non-structural items and prevent damage when the structure moves), or between buildings (to allow separate parts of the building to move independently). Expected movement on one floor within a building can be 50 mm or more, and expected movement between parts of multi-storey buildings can be up to 0.5 m. It is very difficult to provide details and flexible filling materials to accommodate these movements, and also provide fire resistance before, during and after an earthquake. A review of this problem is given by James and Buchanan (2000). Many proprietary products are available for filling seismic gaps, but their fire performance after large movements is often not proven.

6.7.8 Fire Doors

Doors are a very important part of the passive fire protection in many buildings. There are many proprietary fire-resisting doors on the market, but they are usually expensive and have to meet different requirements in different countries. If a door is to match the fire resistance of the wall in which it is installed, the whole door assembly must be able to meet the integrity and insulation requirements for the specified fire-resistance period. Solid core doors can easily be made with sufficient fire resistance, but weaknesses occur at the handle, hinges and all around the door edges. Many countries require that fire-rated doors should be tested with exactly the same hardware as will be used in practice (Figure 6.7). The edge of the door or the frame is often fitted with a strip of intumescent material that swells into a foam when heated, to prevent flames penetrating the gap around the door.

Glazed doors are only required to meet integrity requirements because most glazed panels cannot meet insulation requirements. Various codes have different limitations on the maximum size of glass panel in fire doors. To meet the integrity requirements the glass must be special fire-resistant glass or be wired glass. There are an increasing number of proprietary fire-resistant glazing products on the market.

Fire safety requirements for doors are very different in different countries. Some fire doors are required to prevent spread of smoke, in which case they must pass an air leakage test as well as a fire test. Many aspects of the performance of fire doors under test are described by England *et al.* (2000), who also propose an improved test method.

Real fire experience has shown that steel roller-shutter doors maintain excellent integrity in severe fires. No insulation rating is possible because the thin steel of the door heats up very rapidly, but a roller-shutter door can restrict the spread of fire provided that there are no combustible materials near the unexposed face of door.

Figure 6.7 Fire-resistance test of two doors. The door on the left has had an integrity failure, as shown by penetration of flames and hot gases (Reproduced by permission of Building Research Association of New Zealand)

6.7.9 Ducts

Air-handling ducts are potential paths for fire spread in buildings. Some authorities require ducts to have be provided with fire resistance. Typical steel ducts can only provide an integrity rating, which can be improved to an insulation rating with insulating material such as a ceramic fibre blanket internally or externally. More solid fire-resistant ducts can be made from multiple layers of material such as gypsum board. There is no standard test method for ducts, but some systems have been tested successfully in non-standard tests.

An air-handling duct passing through a barrier can cause a serious reduction in the fire resistance of the barrier. This can be prevented by placing a "fire damper" inside the duct where it passes through the barrier. Some dampers are also designed to control smoke movement. The dampers are designed to close automatically. Small dampers operate when a spring-loaded blade or curtain inside the duct is released by melting of a heat-activated fusible link. Dampers in large ducts may have motorised closers which are activated by the fire detection system in the building. Another type of system has blades covered with intumescent material which swells up to close the duct at high temperatures. Testing requirements are described by England *et al.* (2000).

When there is a severe fire on one side of a wall penetrated by a duct, the collapsing duct on the hot side of the wall may cause damage to the wall itself, reducing the fire resistance. To prevent such damage, the fire damper should be firmly attached to the wall, and the duct should be constructed with joints which allow the duct to pull away from the damper, leaving the damper intact as part of the wall. This approach cannot be applied to fixed services such as cable trays.

6.7.10 Glass

Glass is a vitreous solid material with crystal structure similar to a liquid. On heating, it goes through a series of phases of decreasing viscosity. Most typical glass softens or melts at temperatures from 600 to 800°C, but it will crack or break if exposed to thermal shock at much lower temperatures, due to differential temperatures within the glass or because of expansion of the surrounding frame. Normal window glass is assumed to break and fall out of the windows at the time of flashover, although tests have shown that this does not always occur. Double glazing tends to remain in place much longer than single layers of glass.

Glass is sometimes used in fire-resisting barriers, where it can only provide an integrity rating, because it has no structural capability at elevated temperatures and cannot provide an insulation rating unless it is coated with some sort of intumescent coating. If glazing is to be used in a fire-resisting barrier, it must be assembled with special glass, either wired glass (reinforced with fine wires in both directions) or specially formulated toughened glass. Fire-resisting glazing is usually installed in steel frames which clamp the glass and prevent it from deforming excessively when it gets hot. Aluminium cannot be used because of its low melting temperature. Glazed assemblies can be tested in full-scale fire resistance tests, where the assessment is only for the integrity criterion.

A number of proprietary insulated glazing systems have recently been developed, consisting of alternating layers of glass or sodium silicate with transparent intumescent materials. These products are transparent at room temperatures, but become opaque at high temperatures, achieving fire resistance of up to two hours.

Glass walls and windows can provide resistance to fire spread if they are sprayed continuously with water from a properly designed sprinkler system (England *et al.*, 2000, Kim *et al.*, 1998).

6.7.11 Historical construction

Fire engineers are sometimes asked to report on the fire safety of historical buildings. This often requires information on the fire resistance of old materials and obsolete building systems. This information can often be obtained from many current listings, and calculations can be made using the information in this book. A useful reference is Appendix L to NFPA 909 (NFPA, 1997) which gives extensive lists of fire ratings of elements such as masonry walls, hollow clay tile floors, old-style doors and cast iron columns which are no longer used in new construction.

7

Design of Structures Exposed to Fire

7.1 OVERVIEW

This chapter describes the overall process of designing structures for fire exposure. It also describes the available tools for making structural calculations, and explains the importance of loads and support conditions in estimating load capacity under fire conditions. These calculations are used for verifying fire performance in the strength domain.

Most building structures are made up of a number of elements such as walls, floors and roofs, often supported by members such as beams and columns. To avoid the collapse of a building structure, the combination of elements and their supporting members must perform their load-bearing function for the duration of the fire.

In many simple structures, collapse of one member will result in total collapse of the structure. Structural failure occurs if the applied load exceeds the load capacity at any time during the fire. In more complex structures it may be possible for the structure to survive a fire even if one or more members loses its load-carrying capacity. As individual members try to expand under heating, they create interactions with the surrounding structure which do not occur under normal temperature conditions.

7.2 STRUCTURAL DESIGN AT NORMAL TEMPERATURES

Before describing the procedure for structural design under fire conditions, it is important to review design at normal temperatures, in order to define terms and maintain consistency. The basic steps in making a structural design are:

(1) establish the functional requirements for the building,

(2) make a conceptual design of the structural system,

(3) guess the sizes of the main structural members,

(4) estimate the loads on the structure,

(5) make a structural analysis to determine internal forces and stresses,

(6) check whether the guessed initial sizes have sufficient strength and stiffness, and

(7) repeat steps as necessary.

The steps (4) through (6) will be described in more detail. These apply equally to new or existing buildings.

7.2.1 Loads

Types of load

Loads on structures are usually differentiated as 'dead loads' and 'live loads'. Dead loads are loads which are always present, being the self-weight of the building materials and any permanent fixtures. Live loads are loads which may or may not occur at any time, from a wide variety of sources including the following:

- Human occupancy loads are from the weight of people. These may vary from zero to very high levels, especially where crowds can gather. Day-to-day loads are usually much less than the loads specified by structural design codes.

- Non-human occupancy loads come from equipment, goods, and other moveable objects. The weight of objects may be very low and variable in spaces like office buildings, or heavy and semi-permanent in warehouses and libraries.

- Snow loads are seasonal, with large geographic differences. Some areas may expect heavy snow to remain for several months every year, whereas others may expect no snow, or very infrequent snow loads.

- Wind loads are experienced by most buildings. The probability of extreme wind loads varies greatly, depending on location and topography. Wind loads are usually lateral loads on walls or uplift loads on roofs.

- Major earthquakes are extreme events which have a very long return period. Some areas expect no earthquakes, others may have many small earthquakes and others have a low but significant probability of a rare major earthquake. Earthquake loads are inertial loads acting at the centre of mass, mostly in the horizontal plane.

Load combinations

The above loads never occur all at the same time. Structural design at normal temperatures requires the investigation of several alternative load combinations as specified by national codes. At the time of a fire the most likely load is the dead load and a part of the occupancy loading.

7.2.2 Structural Analysis

Structural analysis is the process of understanding the way in which applied loads on the floors or roofs or walls of a building 'flow' through the beams, columns and other structural members to the foundations. The building structure resists the applied loads by deforming slightly under the load. The flow of loads through the structure is accompanied by deformations and the development of internal forces in each structural member. Internal forces may be bending moments, axial forces, or shear forces. In European terminology (used in the structural Eurocodes, developed over the past ten years) loads are referred to as 'actions' and internal forces as 'effects of actions'.

Structural analysis is used to calculate the deformations of the structure under the applied loads and the internal actions in every member. The structural analysis of simple structures is performed by hand calculation from first principles of statics and mechanics, often with reference to standard formulae. Computer programs are widely used for the structural analysis of more complex structures.

If member sizes are known, internal forces can be converted to stresses, usually expressed as a combination of normal stresses and shear stresses.

Non-linear analysis

Most structural analysis assumes that the structure is linear and elastic. A linear-elastic structure is one where deformations are directly proportional to the applied loads and the structure reverts to its original shape when all loads are removed. The linear-elastic assumption is good for most structures at low levels of load.

There are two main sources of non-linearities in structural analysis. Geometrical non-linearities occur when deformations become so large that they induce additional internal actions, resulting in even larger deformations. Column buckling is the most common case of geometrical non-linearity. Material non-linearities occur when materials are stressed beyond the elastic range causing yielding or 'plastic' behaviour, in which case the structure has permanent deformations after the loads are removed. Understanding of non-linear behaviour becomes important if the ultimate strength of the structure is to be understood.

The most simple computer programs for structural analysis consider only linear-elastic behaviour. More advanced programs can include both geometrical and material non-linear analysis. Non-linear behaviour can be very important under fire conditions because deformations are larger and material strength is less than in normal temperature conditions. Computer programs for structural analysis in fire conditions are discussed later.

7.2.3 Design Format

The specification of design loads and material strength depends on the format of the design code, which varies from country to country.

The traditional design format, still used in many countries, is *working stress* design or *allowable stress* design where member stresses under the actual loads expected in service are compared with the allowable or permissible stresses which are considered safe for the material under long-term loads. There is usually a large safety factor built into the safe working stresses.

Modern design codes use an *ultimate strength* design format in which internal actions resulting from the maximum likely values of load ('characteristic loads') are compared with the expected member strength using the short-term strength of the weakest likely materials ('characteristic strength'). This design format is known as *limit states design* in Europe and *load and resistance factor design* (*LRFD*) in North America. There are minor differences between these formats.

Limit states design clearly differentiates between the strength limit state (or ultimate limit state) and the serviceability limit state. The 'strength' or 'ultimate' limit state is concerned with preventing collapse or failure whereas the 'serviceability' limit state is concerned with controlling deflection or vibration which may affect the service of the building. Structural design for fire is mainly concerned with the ultimate limit state because it is strength and not deflection which is critical to prevent collapse of buildings in fire.

Many national codes are in transition from working stress design to ultimate strength design. It is possible to make a rough straight comparison (or soft conversion) between the two formats. For simple structural members, both formats should result in similar member sizes.

Working stress design format

The loads in *working stress design* or *allowable stress design* are the typical loads expected in service. The dead load is the self weight of the structure calculated by the designer, and the live loads are specified by national codes.

Considering dead loads and occupancy loads, most codes specify only one load combination for the design load L_w given by

$$L_w = G + Q \tag{7.1}$$

where G is the dead load and Q is the live load

In the structural design process, the load L_w is used to calculate internal forces (bending moment, axial force and shear force) in each structural member, then the resulting stresses are calculated and these are compared with the allowable design strength for the material, which is considered to be the safe stress for long-term loads.

For example, in the design of a tension member, the axial tensile force N_w (Newtons) in the member is calculated from the above load combination. The resulting tensile stress f_t^* (MPa) is calculated from

$$f_t^* = N_w / A \tag{7.2}$$

where A is the cross sectional area of the member (mm^2).

The design equation which must be satisfied is given by

$$f_t^* \leqslant f_a \tag{7.3}$$

where f_a is the allowable design stress in the code (MPa). The actual level of safety is not clearly known in this format because the loads are not the worst loads that could occur, and the allowable stresses are known to be safe, but are not directly related to the failure stresses.

Ultimate strength design format

In *ultimate strength design*, the characteristic dead load is the self weight of the structure calculated by the designer, the same as in working stress design. Characteristic occupancy loads are specified by national codes for various uses, usually being estimates of loads which have a 5% probability of being exceeded in a 50 year period.

Considering only dead loads and occupancy loads for the strength limit state, most codes specify two load combinations, where each load is increased by a 'load factor' giving factored loads L_u of

$$L_u = 1.4G_k \quad \text{or} \quad L_u = 1.2G_k + 1.6Q_k \tag{7.4}$$

where G_k is the characteristic dead load and Q_k is the characteristic live load.

The load factors of 1.2, 1.4 and 1.6 in Equation (7.3) are called *partial safety factors* (γ_G, γ_Q) in the structural Eurocodes. The first combination in Equation (7.3) is for situations where dead loads dominate, and the second is for most situations where there is a mixture of dead load and occupancy loads. The load factors in Equation (7.4) are from the recommended US code (ASCE, 1995) and the New Zealand Loading Code (SNZ, 1992). The Eurocode combinations are similar (EC1, 1994). Similar combinations are specified for use with wind, snow or earthquake loads. The values from the New Zealand Loading Code (SNZ, 1992) are

$$\begin{aligned} L_u &= 1.2G_k + 0.4Q_k + W_k \quad \text{or} \quad L_u = 0.9G_k + W_k \\ L_u &= 1.2G_k + 1.2S_k \\ L_u &= G_k + 0.4Q_k + E_k \end{aligned} \tag{7.5}$$

where W_k is the characteristic wind load, S_k is the characteristic snow load, and E_k is the characteristic earthquake load.

In the structural design process, the worst of these load combinations is used to calculate the internal forces (bending moment, axial force and shear force) in each structural member, which are compared with the load capacity of the member. The load capacity is obtained from the short-term characteristic strength specified in the material code. The characteristic stress is an estimate of the fifth percentile stress at failure (usually calculated with 75% confidence). The nominal load capacity is reduced by a *strength reduction factor* Φ which is intended to allow for uncertainty in the estimates of material strength and section size. The value of Φ is normally in the range 0.7 to 0.9. In the European system, the strength reduction factor Φ is replaced by $1/\gamma_M$ where γ_M is the partial safety factor, analogous to the inverse of the strength reduction factor Φ, for each material.

Hence, verification of the design for strength requires that

$$U^* \leqslant \Phi R \tag{7.6}$$

where U^* is the internal force resulting from the applied load, R is the nominal load capacity, and Φ is the strength reduction factor ($1/\gamma_M$).

The internal force U^* may be axial force N^*, bending moment M^* or shear force V^* occurring singly or in combination. The load capacity R will be the axial strength, flexural strength or shear strength, in the same combination.

For example, in the design of a tension member, the axial force N^* obtained from the worst factored load combination in Equation (7.3) must not exceed the design capacity ΦN_n so the design equation is

$$N^* \leqslant \Phi N_n \tag{7.7}$$

where N_n is the nominal axial load capacity (Newtons) given by

$$N_n = f_t A \tag{7.8}$$

where f_t is the characteristic tensile strength (MPa), and A is the cross-sectional area (mm^2).

When comparing working stress design with ultimate strength design, note that the characteristic tensile strength f_t is larger than the long-term allowable tensile strength f_{tw} with a corresponding difference between the loads N^* and N_w.

7.2.4 Material Properties

Derivation of material design values for strength depend on the format of the design system in use. In the traditional system of *working stress design*, the design strength (or permissible stress) represents the stress which can be sustained safely under long duration loads. In the more modern systems of *limit states design* or *LRFD*, the characteristic strength (or design strength) represents the stress at which the material will fail under short duration loads. In most countries the characteristic strength is the fifth percentile short-term failure stress (estimated with 75% confidence) for a typical population of material of the size and quality under consideration. For the modulus of elasticity two values are needed: the fifth percentile value for buckling strength calculations, and the mean value for deflection calculations.

The normal temperature properties of steel, concrete and timber are compared briefly as an introduction to elevated temperature design in subsequent chapters, using Figures 7.1 and 7.2. Figure 7.1 shows a simply supported beam with two point loads. The bending moment at mid-span produces the internal strain distribution as shown with tensile strains at the bottom and compressive strains at the top. Figure 7.2 shows typical stress–strain relationships for the three materials. These are not drawn at the same scale because the yield strength of steel is much greater than the crushing strength of concrete or wood which are similar.

It can be seen that steel has the same properties in both compression and tension, with elastic behaviour to a well-defined yield point, followed by very ductile behaviour. Concrete has very little dependable tensile strength, but is strong in compression, with limited ductility. Ductility can be substantially increased by

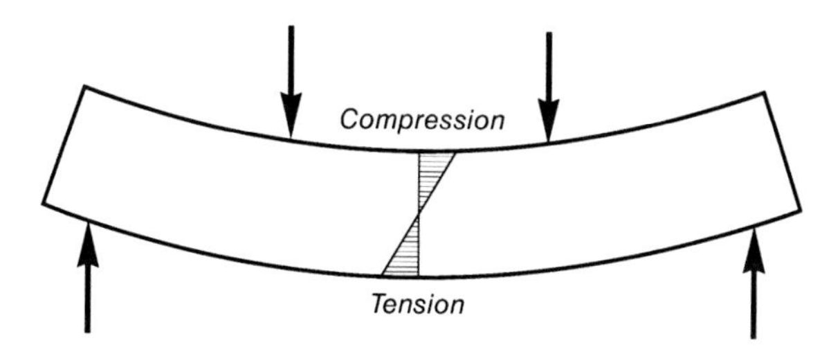

Figure 7.1 Internal strains in a simply supported beam

confining the compression zone with stirrups. Wood is ductile in compression but exhibits brittle failure in tension. The solid line shows parallel to grain behaviour where the tensile strength is very high, and the dotted line shows perpendicular to grain behaviour where the tensile strength is very weak (in splitting).

These stress–strain relationships are used to show flexural stresses in beams of typical beam cross sections, in the lower part of Figure 7.2. Internal stresses are shown for beams lightly stressed in the elastic range and stressed near failure in the inelastic range. When approaching its ultimate flexural strength, the steel beam develops plastic yielding over most of the cross section, depending on the amount of curvature. At ultimate strength, the reinforced concrete beam has a parabolic stress distribution in compression, with the resulting compressive force equal to the yield force of the reinforcing bars yielding in tension. The parabolic compression block is approximated by the dotted rectangle for design purposes (Park and Paulay, 1975).

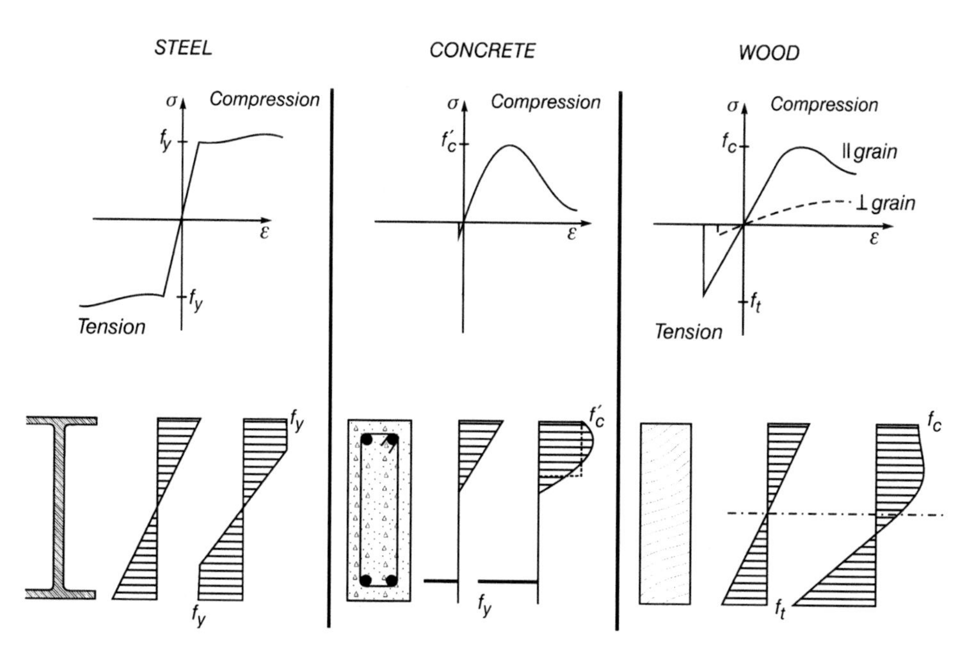

Figure 7.2 Stress–strain relationships and internal flexural stresses for steel, concrete and timber

The strength of timber depends on the material properties. Commercial quality timber usually has low tensile strength due to defects, so it fails when the stress distribution is in the linear-elastic range. For high quality timber with no defects in the tension zone, ductile yielding occurs in compression as shown, lowering the neutral axis, and causing very high tensile stresses.

7.2.5 Probability of Failure

The objective of structural design is to provide buildings with an acceptably low probability of failure under extreme loading conditions. Probabilities of failure are not usually stated in design codes, but they have been used by the writers of ultimate strength format codes, to establish the necessary strength reduction factors to give a target level of safety for all anticipated conditions, using characteristic values of load and strength.

The sketch in Figure 7.3(a) shows schematically that load U and resistance R are both probabilistic quantities, with a distribution of values about a mean. There is a small probability of failure which can be calculated from the area of overlap between the two curves if their distributions are known. The characteristic values of member resistance usually represents the lower fifth percentile tail of the strength distribution, and the design load represents a high percentile of likely loads for a given return period.

When considering Figure 7.3(a) for fire design, the load and resistance curves can be quantified using any one of time, temperature or strength (as shown in Table 5.1). If Figure 7.3(a) represents load and resistance at room temperature, both the curves will shift to the left under fire conditions as the expected loads are less and the strength decreases due to elevated temperatures.

The ultimate limit state representing failure occurs if $R < U$, so the likelihood of failure is related to the difference $R - U$. Figure 7.3(b) shows the frequency distribution of $R - U$. The probability $R - U < 0$ is given by the shaded area under the distribution. Limit-state design codes are usually calibrated to give a certain reliability index β, which is the number of standard deviations of the mean value of $R - U$ above zero, as shown in Figure 7.3(b). For given distributions, the strength reduction factor Φ is derived by code writers to give a target reliability index β, in the

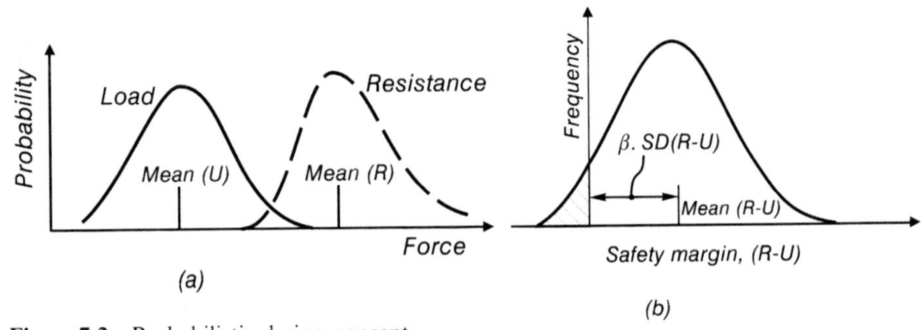

Figure 7.3 Probabilistic design concept

range between 2 and 3. A target value of 2.4 is used in the United States, as described by ASCE (1995).

The above discussion shows that although there is a probabilistic framework behind the ultimate strength code formats, day-to-day design is a deterministic process. Design for fire has far more uncertainty than design for normal temperature conditions. This book considers structural fire safety in a deterministic framework, rather than a probabilistic framework. The science of structural reliability is rapidly developing, but applications to fire safety are still in their infancy despite pioneering work more than 20 years ago by Magnusson (1972). Recent European studies are described by Schleich (1999).

7.3 STRUCTURAL DESIGN IN FIRE CONDITIONS

Structural design for fire is conceptually similar to structural design for normal temperature conditions. Before making any design it is essential to establish clear objectives, and determine the severity of the design fire. The design can be carried out using either working stress or ultimate strength format, but only the ultimate-strength design format will be illustrated here. The main differences of fire design compared with normal temperature design are that, at the time of a fire:

- the applied loads are less,
- internal forces may be induced by thermal expansion,
- strengths of materials may be reduced by elevated temperatures,
- cross-sectional areas may be reduced by charring or spalling,
- smaller safety factors can be used, because of the low likelihood of the event,
- deflections are not important (unless they affect strength),
- different failure mechanisms need to be considered.

The above factors may be different for different materials. For example, Figure 7.4(a) shows how failure of a simply-supported steel beam occurs when the yield strength of the material drops so low that it is exceeded by the actual stress in the member at the time of the fire. The stress in the member does not change during the fire because the loads are constant and the section properties do not change. In contrast, Figure 7.4(b) shows the same situation for a timber beam, where the stresses increase steadily (under constant load) due to loss of cross section by charring. The material strength only decreases very slightly due to elevated temperatures within the beam. As before, failure occurs when the stress in the member exceeds the material strength.

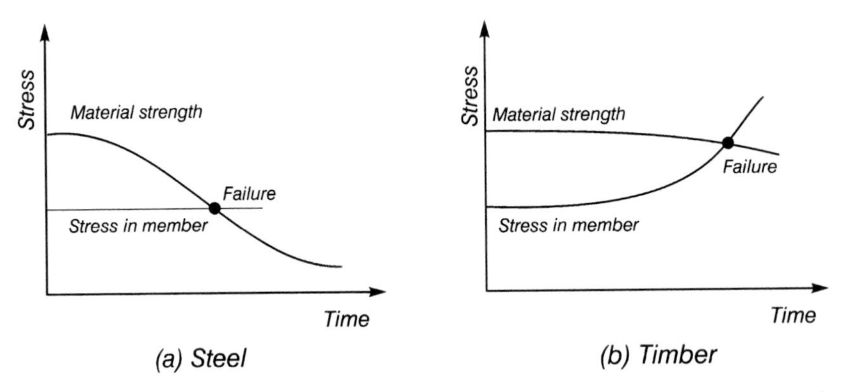

Figure 7.4 Member failure in fire, due to internal stresses exceeding material strength

7.3.1 Design Equation

Verification of design for strength during fire requires that the applied loads are less than the load capacity of the structure, for the duration of the fire design time. This requires satisfying the design equation given by

$$U^*_{\text{fire}} \leqslant \Phi_f R_{\text{fire}} \tag{7.9}$$

where U^*_{fire} is the design action from the applied load at the time of the fire, R_{fire} is the nominal load capacity at the time of the fire, and Φ_f is the strength reduction factor for fire design. The design force U^*_{fire} may be axial force N^*_{fire}, bending moment M^*_{fire}, or shear force V^*_{fire} occurring singly or in combination, with the load capacity calculated accordingly.

The strength reduction factor Φ_f accounts for uncertainty in the estimates of material strength and section size. Fire design is based on the most likely expected strength, so most national and international codes specify a strength reduction factor of $\Phi_f = 1.0$. In the Eurocodes, the partial safety factor γ_M is also equal to 1.0 for fire design. In both the North American and European formats the design equation for fire conditions now becomes

$$U^*_{\text{fire}} \leqslant R_{\text{fire}} \tag{7.10}$$

This is the equation that will be used in the following chapters for the design of steel, concrete and timber structures.

7.3.2 Loads for Fire Design

Load combinations

In the event of a fire, the most likely applied loads are much lower than the maximum design loads specified for normal temperature conditions, so different load combinations are used. Most codes refer to an 'arbitrary point-in-time load' to be used for the fire-design condition. As an example, in the New Zealand code (SNZ, 1992) the design load L_f for fire is

$$L_f = G_k + 0.6Q_k \quad \text{or} \quad L_f = G_k + 0.4Q_k \tag{7.11}$$

the first for storage occupancies with semi-permanent loads, and the second for all other occupancies. The Eurocode and the USA recommendations are slightly different as shown in Table 7.1. The differences between these codes are minor. In all cases, the member actions under fire conditions are much less than in normal temperature conditions. This is especially true for members which have been designed for load combinations including wind, snow or earthquake, or for members sized for deflection control or for architectural reasons.

Some codes such as the New Zealand code (SNZ, 1992) do not consider wind or snow or earthquake at the same time as the fire. The omission of snow may be appropriate if snow loads are small and infrequent, but there should be a nominal level of lateral load to ensure that sway modes are included in the analysis. In contrast, the Eurocode has additional load combinations including both snow and wind. In the situation where the main action is considered to be wind, the load combination for fire is

$$L_f = G_k + 0.5W_k + 0.3Q_k \tag{7.12}$$

In the situations where the main action is considered to be snow, the load combination for fire is

$$L_f = G_k + 0.2S_k + 0.3Q_k \tag{7.13}$$

where W_k is the characteristic value of wind load, and S_k is the characteristic value of snow load.

The fire itself may induce forces in a structure. These are most likely as a result of restraint from the surrounding structure preventing thermal expansion, or from a flexural member becoming a tension member after large deformations have occurred. Such loads are most likely to occur in steel structures because steel members tend to heat more rapidly than other materials. Restraint forces are often significant in concrete structures, but fire-induced forces are less important in timber structures.

Table 7.1 Dead and live load factors for fire design

	Dead load	Permanent live load	Other live load
New Zealand (SNZ, 1992)	G_k	$0.6Q_k$	$0.4Q_k$
Eurocode (EC1, 1994)	G_k	$0.9Q_k$	$0.5Q_k$
USA (ASCE, 1995)	$1.2\ G_k$	$0.5Q_k$	$0.5Q_k$
Ellingwood and Corotis (1991)	G_k	$0.5Q_k$	$0.5Q_k$

Load ratio

Under normal day-to-day conditions, all buildings have an extremely low probability of failure. The term 'load ratio' r_{load} is the ratio of the expected loads on the structure during a fire to the loads that would cause collapse at normal temperatures, given by

$$r_{load} = U^*_{fire}/R_{cold} \tag{7.14}$$

Most buildings have a load ratio of 0.5 or less at most times, so that the strength of any member could drop by half or more before collapse would be expected. The load ratio is far less than 0.5 for buildings or parts of buildings designed to resist extreme events such as rare snowstorms, hurricanes or earthquakes. The lower the load ratio, the greater the fire resistance, because of the large loss in load carrying capacity which can occur before failure occurs. This is a most important concept for structural fire design.

Working stress design

Most modern loading codes, which specify load combinations for fire design, are in limit states (LRFD) format with loads similar to those described above. Loading codes which are in working stress format do not usually include load combinations for fire design, so designers have to use the normal temperature design load combination of $L_w = G + Q$ for fire design. The Australian timber code (SAA, 1990) allows the design load for fire to be 75% of the normal temperature design load, giving $L_{w,f} = 0.75(G + Q)$ which will give more realistic design results.

7.3.3 Structural Analysis for Fire Design

Structural analysis for fire design is essentially the same process as structural analysis for normal temperature design, but it is complicated by the effects of elevated temperatures on the internal forces and the properties of materials.

For many simple structural elements exposed to fire, load carrying capacity can be calculated with simple hand calculation methods, using the same techniques as for cold conditions. Examples for steel, concrete and timber structures are given in later chapters. The major changes from cold conditions are the use of lower applied loads and temperature-reduced material properties. For some materials such as wood, an alternative approach is to use reduced section properties with no change in the material properties.

Hand calculations are most appropriate for single elements with simple supports, especially where internal temperatures are uniform or where the temperature of one part of the member is critical. Structural analysis must consider the possibility of instability failures as well as strength failures.

Many tools are available for calculating the structural fire performance of a load-bearing construction. These can be categorized into hand calculations, design charts, proprietary computer programs and generic computer programs. Hosser *et al.* (1994) and Sullivan *et al.* (1994) review many alternative methods for calculating structural performance.

7.3.4 Computer Calculations

Computer models

A number of proprietary computer programs are available for calculating the load capacity of structural members or frameworks. A review of 13 programs by Sullivan *et al.* (1994) concluded that none of them is sufficiently user friendly or well-documented for routine use as a design tool. Most of the programs are specifically designed for steel or concrete structures. None has been designed for use with timber structures. More recent programs which have been used extensively for analysis of steel and concrete structures are SAFIR (Franssen *et al.*, 2000) and VULCAN (Rose *et al.*, 1998).

There are many generic finite-element stress-analysis programs which can be used for structures exposed to fires, including such programs as NASTRAN and ANSYS. For example, Thomas (1997) used the ABAQUS program to predict the structural performance of light timber-frame walls and floors, with good results. Rotter *et al.* (1999) used ABAQUS for analysis of fire behaviour of the eight-storey Cardington steel frame. The use of these programs is suitable for research but not for daily design, because of the difficulties of managing large quantities of input and output data, and the large number of elements required to model even simple structures.

Calculation procedures

The general procedure for making computer calculations of the internal forces in a member exposed to fire is not simple, because of the difficulty of combining the various components of strain from Equation (7.16), involving non-linear material properties and geometric non-linearities associated with large deflections. There are two basic types of computer-based calculation procedure. These are the finite element method and the moment–curvature method.

Finite-element method The finite-element programs are the most powerful, and are the most often used for large structures. These programs are based on the matrix solution of a very large number of simultaneous equations. For example, a program such as SAFIR (Franssen *et al.*, 2000) uses a fine grid of elements over each cross section, with long elements ('fibres') between the nodes along the length of each member. The analysis consists of setting up and solving the following matrix equation

$$\{F\} = [K]\,\{Q\} \qquad (7.15)$$

where $\{F\}$ is a vector containing the generalized forces of the structure, [K] is the stiffness matrix of the structure, and $\{Q\}$ is a vector containing the generalized displacements of the structure.

These matrices are obtained by assembling the individual matrices for each element, where $\{f\}$ is the element force vector, $[k]$ is the element stiffness matrix, and $\{q\}$ is the displacement vector for the element, respectively. To allow for geometric non-linearities, the stiffness matrix of each element is composed of two separate matrices. A textbook on the finite-element method should be consulted for further information on this method of analysis.

Moment–curvature method The moment–curvature method is based on the development of moment–curvature–thrust relationships for the heat-affected cross section. These relationships can be used to build a curvature diagram for the whole member, to check equilibrium and compatibility of the deformed shape at each time-step in the analysis. This type of analysis is most suitable for individual members, such as for steel columns (Poh and Bennetts, 1995) or concrete walls (O'Meagher and Bennetts, 1991).

It is useful to go through the typical step-by-step procedure for writing such a computer program. For example, Figure 7.5 shows a reinforced concrete wall exposed to fire from one side. The fundamental assumption is that 'plane sections remain plane' so that the strain profile through any cross section of the structural element remains linear throughout the time of the fire exposure. The calculation procedure involves finding the strain profile for which the internal forces are in equilibrium with the applied load effects at one cross section. This must be repeated at each node along the length of the member, in order to achieve equilibrium and deformation compatibility for the whole member. For a member with non-linear geometrical behaviour, such as a long column, the second-order effects must be included in the analysis, usually with a trial-and-error approach to find the shape of the deflected member at which equilibrium can be reached. The calculations may involve matrix manipulations or a trial-and-error approach. It is necessary to find a complete solution at each time step. The whole process must be repeated at increasing time steps throughout the growth and decay of the fire.

For example, Figure 7.5(a) shows the thermal gradient through a reinforced concrete wall with central reinforcing, as calculated by a thermal analysis computer program at a number of segments through the thickness of the wall. Figure 7.5(b) shows the assumed total strain profile for the first trial. Figure 7.5(c) shows the thermal strains calculated from the temperature profile at each segment, knowing the relationship of thermal expansion as a function of temperature for this material. Figure 7.5(d) shows the transient strains, calculated in the same way. Transient strains apply only to concrete structures heated for the first time, as described in Chapter 9.

The stress-related strain is now obtained by a process of elimination, subtracting the thermal and transient strains at each segment through the wall from the assumed total strain in accordance with (Equation 7.16). Creep strain is usually not included explicitly because it is too difficult to specify accurately. Figure 7.5(e) shows the resulting stress-related strain where it can be seen that the hot face of the wall is in tension, with a compression region behind that, followed by a second tensile region

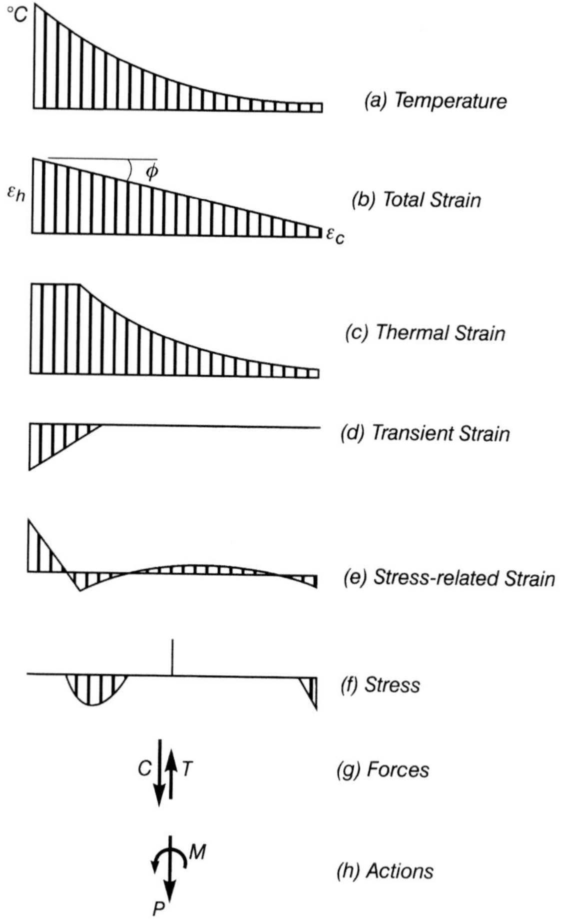

Figure 7.5 Internal stresses and forces in a concrete wall exposed to fire on one side (Reproduced from (O'Meagher and Bennetts, 1991) by permission of Elsevier Science)

in the centre of the wall and a small compression block near the cold face. It can be seen that despite the simple thermal gradient and linear strain diagram, the strain profile through the wall is not simple. The stress-related strains can be used with the stress–strain relationships for the materials (Figure 7.2) to calculate the stress associated with the strain in each segment, plotted in Figure 7.5(f). Note that by entering with the strain to output the stress allows the use of any shape of non-linear stress–strain relationship, including a falling curve after maximum stress has been reached, as shown for concrete and timber in Figure 7.2. In Figure 7.5(f) it has been assumed that the concrete has no tensile strength, so concrete stresses only appear in the compressive regions. The vertical line at the centre of the wall represents the tensile stress in the single reinforcing bar in the centre of the wall.

The stresses at each segment can be summed to give the total tensile and compressive force on the cross section, as shown in Figure 7.5(g). These can be resolved into an axial force and a bending moment (Figure 7.5(h)) for comparison with the internal forces at that cross section. If equilibrium is not achieved, another total strain profile must be used in the next trial. Repetitive calculations such as these can be used to construct moment–curvature relationships for each member, leading to calculations of individual members and whole structures exposed to fires.

7.4 MATERIAL PROPERTIES IN FIRE

Material properties at normal temperatures have been briefly described with reference to Figure 7.2. The strength and modulus of elasticity of all materials change with elevated temperature. Methods of deriving material properties at elevated temperature are discussed below. Details for specific materials are given in the following chapters.

7.4.1 Testing Regimes

When structural elements are exposed to fire, they experience temperature gradients and stress gradients, both of which vary with time. Mechanical properties of materials for fire design purposes must be determined and published in a way that is consistent with the anticipated fire exposure.

Constant temperature tests of materials can be carried out in four possible regimes.

(1) The most common test procedure to determine stress–strain relationships is to use a testing machine to impose a constant rate of increase of strain (by controlling the rate of travel of the machine loading head) measuring the load, from which the stress can be derived.

(2) A similar regime is to control the rate of increase of load (or stress) and measure the deformation (hence the strain).

(3) A creep test is one in which the load is kept constant and the deformations over time are measured.

(4) A relaxation test is one in which a constant initial deformation is imposed and the reduction in load over time is measured.

When the effects of changing temperatures are added, there are two more possible testing regimes.

(5) In a transient creep test, the specimen is subjected to an initial load, then the temperature is increased at a constant rate while the load is maintained at a constant level and deformations are measured.

(6) The final test regime is similar except that the applied load is varied throughout the test in order to maintain a constant level of strain as the temperature is increased at a constant rate.

These six regimes are illustrated in Figure 7.6 derived from Anderberg (1988) and Schneider (1988). The most common of these are regimes (1) and (5). The results of regime (1) tests depend on the rate of loading, because of the influence of creep. The results of regime (5) tests depend on the rate of temperature increase. All these regimes present some difficulties because the effect of creep influences all of the test results, and a difficulty with transient tests on large specimens is that the rate of temperature increase may not be uniform over the cross section. Figure 7.6 does not consider the effect of changing moisture content which can be another important variable, especially for timber structures, making testing for material properties even more difficult.

For most materials, stress–strain relationships at elevated temperatures can be obtained directly from steady-state tests at certain elevated temperatures (Regime (1)), or they can be derived from the results of transient tests. Anderberg (1988) compares stress–strain relationships obtained in both ways and points out that there are differences due to the effect of creep. For most materials, yield strength and modulus of elasticity both decrease with increasing temperature.

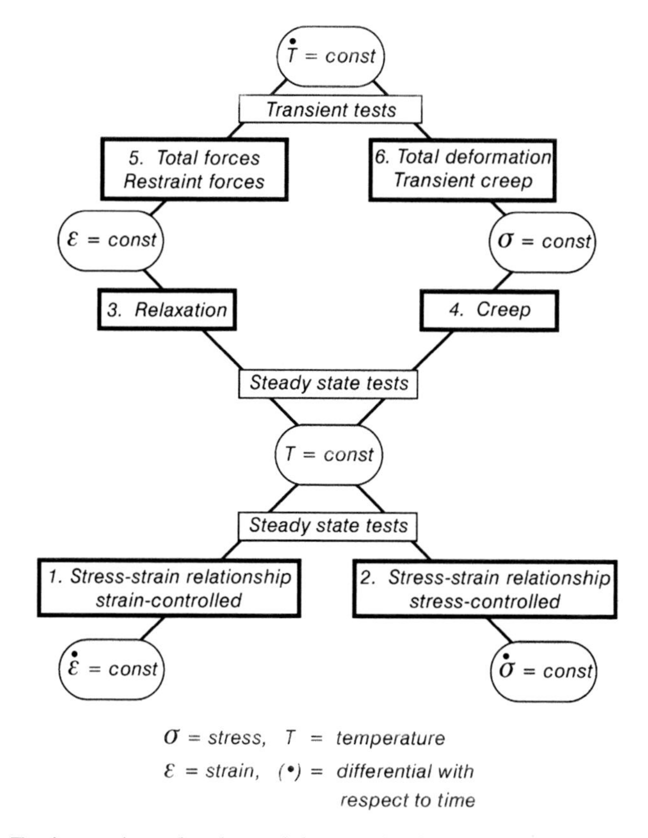

Figure 7.6 Testing regimes for determining mechanical properties of materials at elevated temperatures (Reprinted from Anderberg (1988) with permission from Elsevier Science)

7.4.2 Components of Strain

Analysis of a structure exposed to fire requires consideration of the deformation of the structure under the applied loads. The deformation of materials at elevated temperature is usually described by assuming that the change in strain $\Delta\varepsilon$ consists of four components, being

$$\Delta\varepsilon = \varepsilon - \varepsilon_i = \varepsilon_\sigma(\sigma,T) + \varepsilon_{th}(T) + \varepsilon_{cr}(\sigma,T,t) + \varepsilon_{tr}(\sigma,T) \qquad (7.16)$$

where ε is the total strain at time t, ε_i is the initial strain at time $t=0$, $\varepsilon_\sigma(\sigma,T)$ is the mechanical, or stress-related strain, being a function of both the applied stress σ and the temperature, $\varepsilon_{th}(T)$ is the thermal strain being a function only of temperature, T, $\varepsilon_{cr}(\sigma,T,t)$ is the creep strain, being additionally a function of time, and $\varepsilon_{tr}(\sigma,T)$ is the transient strain which only applies to concrete.

Stress-related strain

The stress-related strain (or mechanical strain) refers to the strain which results in stresses in the structural members. These stresses are based on the stress–strain relationships shown in Figure 7.2, used for the structural design of all materials. For the fire design of individual structural members such as simply supported beams which are free to expand on heating, the stress-related strain is the only component of strain that needs to be considered. If the reduction of strength with temperature is known, member strength can easily be calculated at elevated temperatures using simple formulae such as those given in this chapter.

The stress-related strains in fire exposed structures may be well above yield levels, resulting in extensive plastification, especially in steel buildings with redundancy or restraint to thermal expansion. Computer modelling of fire exposed structures requires knowledge of stress–strain relationships not only in loading, but also in unloading, as members deform and as structural members cool in real fires (Franssen, 1990, El-Rimawi *et al.*, 1996).

Thermal strain

Thermal strain is the well-known thermal expansion that occurs when most materials are heated, with expansion being related to the increase in temperature. Thermal strain is not important for fire design of simply supported members, but must be considered for frames and complex structural systems, especially where members are restrained by other parts of the structure and the thermal strains can induce large internal forces.

Creep strain

Creep is the term which describes long-term deformation of materials under constant load. Under most conditions, creep is only a problem for members with very high permanent loads. If the load is removed there will be slow recovery of some of the creep deformations, as shown in Figure 7.7(a). Creep becomes more important at elevated temperatures because creep can accelerate as load capacity reduces, leading to secondary and tertiary creep as shown in Figure 7.7(b). 'Relaxation' is the

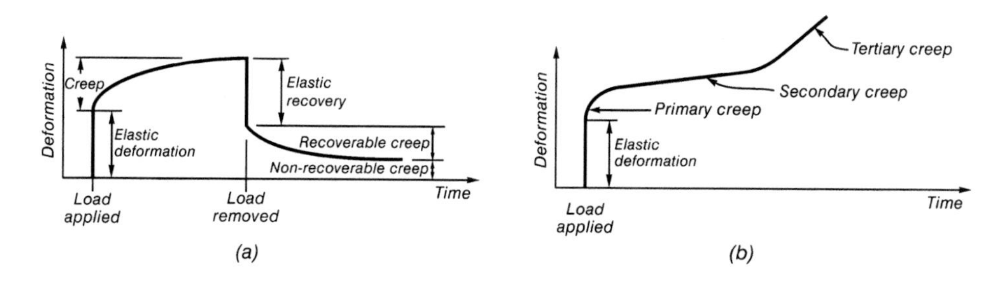

Figure 7.7 Creep in structural materials: (*a*) creep under normal conditions, (*b*) creep at elevated temperatures

complementary term which describes the reduction of stress in materials subjected to constant deformation over a long period of time.

Creep is relatively insignificant in structural steel at normal temperatures. However, it becomes very significant at temperatures over 400 or 500°C and is highly dependent on stress level. At higher temperatures the creep deformations in steel can accelerate rapidly, leading to plastic behaviour and 'runaway' failure. Creep in wood is complicated by changes in moisture content such that creep deformations tend to be larger in environments where the moisture content of the wood fluctuates over time, hence creep can become a major concern in fire-exposed wood which is at temperatures around 100°C.

Creep strain is not usually included explicitly in fire engineering calculations because of the added complexity and the lack of sufficient input data. This applies to both hand and computer methods. Any structural analysis computer program for elevated temperature is already very complex without having to explicitly include the effects of time-dependent behaviour. The effects of creep are usually allowed for implicitly by using stress–strain relationships which include an allowance for the amount of creep that might usually be expected in a fire-exposed member.

Transient strain

Transient strain is caused by expansion of cement paste when it is heated for the first time under load. Transient strain is often included in analytical models for predicting the behaviour of reinforced concrete structures exposed to fire, as described in Chapter 9.

Effect of strain components

Equation (7.16) can be simplified, ignoring the last two terms to give

$$\varepsilon_{\text{total}} = \varepsilon_{\text{mechanical}} + \varepsilon_{\text{thermal}} \tag{7.17}$$

where $\varepsilon_{mechanical}$ is the stress-related strain and $\varepsilon_{thermal}$ is the thermal strain resulting from thermal expansion. This is a key relationship for understanding the structural behaviour of fire-exposed structures, because structural engineers are interested in stresses and deformations in structures. The deformations in the structure depend on the total strain ε_{total} and the stresses in the structure depend on the stress-related strain $\varepsilon_{mechanical}$.

Rotter *et al.* (1999) explain this further by considering two contrasting types of structure. For a lightly loaded structure in which there is no resistance to thermal expansion, the total strain is dominated by the thermal strain and the mechanical strain is very low, and hence most deformations (bowing or elongation) result from the thermal strain which is only a function of temperature. In a very different type of structure where there is severe restraint to thermal expansion, there can be no elongation, so $\varepsilon_{total} = 0$, hence the thermal and mechanical strains are approximately equal and opposite. Both may be very large, resulting in high levels of plastification and high stresses (with much yielding) because of the high mechanical strains.

These aspects of structural behaviour under fire conditions are not what would be intuitively expected by most structural engineers. The design of simple members exposed to fire is not difficult, as described in this book, but highly redundant structures must be analysed with sophisticated computer programs in order to quantify these effects, as described below.

7.5 DESIGN OF INDIVIDUAL MEMBERS EXPOSED TO FIRE

This section outlines the principles of structural design for individual members exposed to fire conditions. This is the 'simplified' design method as described in the Eurocodes. The following sections describe the different approach needed for members which form part of larger structural assemblies.

7.5.1 Tension Members

To continue the example of a simple tension member, the design process for fire is essentially the same as for normal temperature conditions. In the event of a fire, the factored axial force will reduce to N^*_{fire} (N) and the strength will reduce due to elevated temperatures or reduction of the cross section. To prevent failure, the expected loads must be compared with the expected strength N_f at the time of the fire. The strength reduction factor Φ_f is taken as 1.0 for fire design, so that the design equation becomes

$$N^*_{fire} \leqslant N_f \qquad (7.18)$$

where $N_f = f_{t,f}\ A_{fi}$, A_{fi} is the area of the cross section (possibly reduced by fire exposure) (mm^2), and $f_{t,f}$ is the characteristic tensile strength of the material at elevated temperature (MPa).

7.5.2 Compression Members

Design for compression follows similar principles, except that the factored axial force N^*_{fire} is now compressive. The design equation is the same as for tension

$$N^*_{fire} \leqslant N_f \tag{7.19}$$

Calculation of N_f requires understanding of the basic behaviour of compression members. A 'short column' will fail when the applied stress reaches the crushing strength of the column material. A 'long column' is a compression member whose load capacity is limited by lateral buckling, leading to an instability failure at an average stress less than the crushing strength of the material. Figure 7.8 shows the relationship between axial load capacity and length for a pin ended long column where the crushing strength N_c (N) of a short column is

$$N_c = f_c A \tag{7.20}$$

where f_c is the crushing strength of the material (MPa), and A is the cross sectional area of the column (mm^2).

The theoretical axial load capacity of a perfectly straight long column is known as the critical buckling strength N_{crit} (N) given by the Euler buckling formula

$$N_{crit} = \frac{\pi^2 EI}{L^2} \quad \text{or} \quad N_{crit} = \frac{\pi^2 EA}{(L/r)^2} \tag{7.21}$$

where E is the modulus of elasticity (MPa), A is the cross sectional area of the column (mm^2), I is the moment of inertia of the cross section in the direction of buckling (mm^4), L is the length of the column (mm), r is the radius of gyration of the cross section, $r = \sqrt{I/A}$ (mm), and L/r is the slenderness ratio. If the modulus of elasticity E is reduced during a fire, a column can buckle as shown in Figure 7.8(b).

The dotted line in Figure 7.8(a) shows the behaviour of real columns, as obtained from testing, with a gradual transition from short- to long-column behaviour. The deviation from the theoretical curve, which is allowed for in design codes, results from initial out-of-straightness, residual stresses and other factors. Column design may be by separate consideration of crushing and buckling behaviour, but most codes provide a 'buckling factor' for reducing the crushing strength to allow for the possible effects of buckling. The buckling factor is usually a function of the slenderness ratio (L/r), with a value of 1.0 for low slenderness, decreasing as the slenderness increases.

When compression members are exposed to fire, strength decreases because of reductions in the crushing strength and the modulus of elasticity. For some materials such as timber, the cross section is reduced by charring, leading to smaller section properties and an increase in the slenderness ratio. Design for fire is covered in the following chapters.

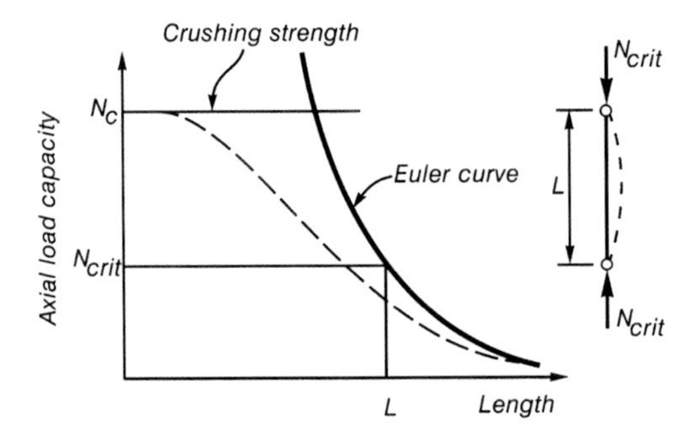

Figure 7.8 Column buckling. (a) Effect of member length on compressive load capacity. (b) Steel column which has buckled during fire exposure (Reproduced from HMSO (1961) by permission of Her Majesty's Stationery Office)

7.5.3 Beams

Flexural design

The process shown above for tension members can be applied to bending members. This section refers only to simply supported beams. Beams with continuous supports, axial restraint, or larger frames are described in later sections.

Figure 7.9 shows loads and bending moments for a simply supported roof beam designed to support both dead load and snow load. Note that in this book, bending moment diagrams are always plotted on the tension side of flexural members, following the European convention which is opposite to that used in North America. A positive bending moment is one which causes tension on the underside of the beam (a 'sagging' moment). A negative bending moment is one which causes tension on the top of the beam (a 'hogging' moment).

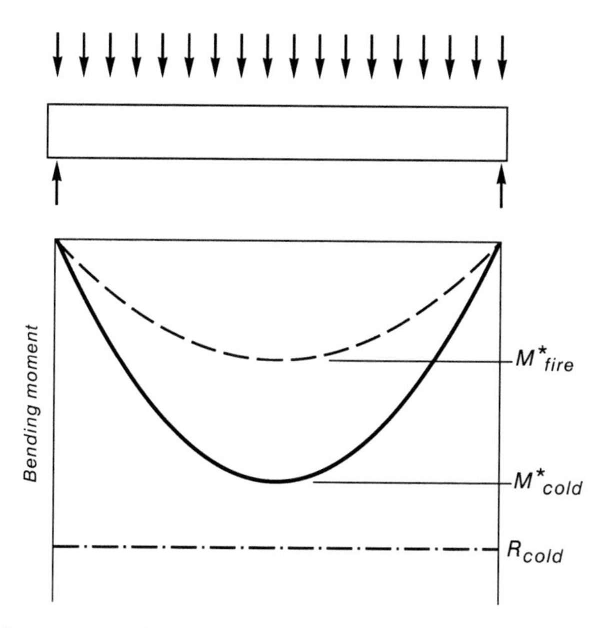

Figure 7.9 Bending moment diagrams for a simply supported beam

Under factored design loads of self weight and snow load, the bending moment diagram is shown by the solid curve, where the mid-span bending moment M^*_{cold} (kNm) is

$$M^*_{\text{cold}} = w_c L^2/8 \qquad (7.22)$$

where w_c is the uniformly distributed factored load (kN/m), and L is the span (m).

Under normal temperature conditions a member size must be selected with a sufficiently large section modulus Z to satisfy the design equation

$$M^*_{\text{cold}} \leqslant M_n \qquad (7.23)$$

where $M_n = \Phi f_b Z$, Φ is the strength reduction factor, f_b is the characteristic flexural strength at normal temperatures, and Z is the section modulus of the cross section.

For materials such as wood, where design is based on elastic behaviour, Z is the elastic section modulus. For rectangular sections $Z = bd^2/6$. For ductile materials like steel, the plastic section modulus S may be used instead of Z, giving slightly higher design strengths. (Note that the symbols Z and S are used with reversed meanings in some countries.)

The resulting short-term flexural capacity under cold conditions is $R_{\text{cold}} = f_b Z$, shown by the lower straight line on Figure 7.9. R_{cold} is greater than M^*_{cold} because of the following factors in the design process.

(1) The strength reduction factor Φ is always less than 1.0 in normal temperature conditions.

(2) The size of the selected member may be larger than exactly needed (because of steps in available sizes or because the size was chosen to control deflections, or for architectural reasons).

(3) For some materials such as timber, the strength depends on the duration of the load, so there will be a difference between the short-term capacity and long-term load demand.

If a fire occurs when there is no snow on the roof, the bending moment at mid-span will be less, as shown by the dotted curve. The mid-span bending moment M^*_{fire} (kNm) is given by

$$M^*_{fire} = w_f L^2/8 \tag{7.24}$$

where w_f is the uniformly distributed factored load for fire conditions (kN/m).

It can be seen that for failure to occur as a result of the fire, the flexural capacity would have to drop from R_{cold} to M^*_{fire}. In this case the design equation becomes

$$M^*_{fire} \leqslant M_f \tag{7.25}$$

where $M_f = f_{b,f} Z_f$, $f_{b,f}$ is the characteristic flexural strength of the material at elevated temperature, and Z_f is the appropriate section modulus of the cross section (possibly reduced by fire exposure).

The ratio M^*_{fire}/R_{cold} is the 'load ratio'. This example demonstrates the important principle that if the load ratio is low, the necessary drop in strength for failure to occur is large, hence the greater the fire resistance.

Lateral buckling of beams

The above equations for bending assume that the beam will fail by flexural yielding. Slender beams with no lateral restraint may fail by lateral torsional buckling at a load less than the flexural load capacity. This can only happen if the compression edge of the beam is free to buckle sideways. Lateral buckling is more of a problem for slender beams of materials like steel than for members of compact materials like concrete. At normal temperatures the critical buckling load can be calculated by using formulae from structural design codes. Chapter 8 gives guidance for checking lateral buckling of steel beams under fire exposure.

Any members providing lateral bracing to beams or columns must have at least the same fire resistance as the main members. This can be difficult to calculate if the bracing members are not actually load bearing, but only provide bracing. A common rule-of-thumb is that bracing members should be designed to resist $2\frac{1}{2}$% of the axial force in the braced member, in addition to any applied loads and self weight. This can also be used in fire design. The hierarchy of lateral support must be followed through carefully. For example, main roof beams may rely on secondary beams for lateral stability, the secondary beams may rely on purlins and the purlins on the roofing material. If the main beams are to resist a design fire, then all the related materials must remain in place for the duration of the fire exposure.

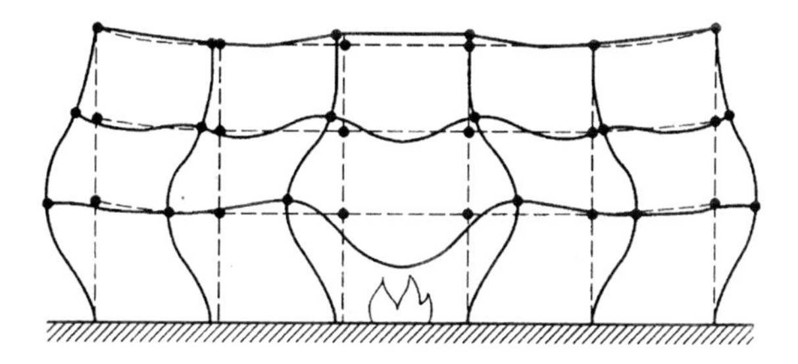

Figure 7.10 Frame deformations in the lower floors of a multi-storey frame resulting from a fire on the ground floor (Reproduced from Bresler and Iding (1982) by permission of Wiss, Janney Elstner Associates, Emeryville, California, USA

Shear

Design for shear can be handled in the same way if sufficient information is available on the shear resistance of materials and members at elevated temperatures.

7.6 DESIGN OF STRUCTURAL ASSEMBLIES EXPOSED TO FIRE

All the discussion above has referred to individual members. This section describes how the structural behaviour of a member exposed to fire can be enhanced by consideration of the whole structural assembly.

7.6.1 Frames

The behaviour of moment-resisting frames is more complex than the behaviour of individual members, because of continuity and axial restraint, and because fire-induced elongations and rotations affect other areas of the building which are not subjected to heating. For example, Figure 7.10 shows calculated deformations resulting from a fire in one bay of a multi-storey building (Bresler and Iding, 1982). In general, the continuity of moment-resisting connections enhances fire resistance of members in frames, so that design of individual members using the methods described above is conservative. Special purpose computer programs should be used when assessing the expected fire performance of large or special structures including multi-bay frames. Design of unbraced frames is more difficult than braced frames because lateral deformations and the resulting P-Δ effects must also be considered.

7.6.2 Redundancy

Many structures have very little structural *redundancy*, so that failure of a single element can cause failure of the whole structure. On the other hand, *redundant* structures have the capacity for considerable load sharing, so that when one element fails its load can be redistributed to other stiffer and stronger elements. This process is conceptually similar to moment redistribution, but the process is redistribution of internal forces from member to member rather than within a member. Many structures have many alternative load paths for load sharing between frames or roof trusses. In such structures a localized fire may cause structural failure of one or more individual elements without resulting in the collapse of the building.

The effects of redundancy are also related to the load ratio, because if the total loads on a structure at the time of a fire are much less than the full-design load, then fewer structural members may be necessary to support the structure, provided that there are sufficient load paths for the applied loads to get to the undamaged members. Redundancy is most often quoted as a benefit of steel construction, but it can apply to buildings of all materials, especially if the materials are ductile and there are alternative load paths.

If a fire-exposed structure is very redundant with many alternative load paths, as in a modern multi-storey steel-framed building, Rotter *et al.* (1999) have shown how large deformations can develop without any significant loss in load-carrying capacity, provided that the structure has sufficient ductility to accommodate the large deflections. In such buildings, failure must not be defined as loss of load capacity or excessive deflection of any single member because loads can be carried by other members.

7.6.3 Disproportionate Collapse

Disproportionate collapse is conceptually the opposite to redundancy. Whereas a redundant structure can suffer the failure of some parts without structural collapse, disproportionate collapse refers to a situation where failure of one element causes a major collapse, with a magnitude disproportionate to the initial event. This is a major concern in the UK largely as a result of the Ronan Point disaster in 1968 (Figure 7.11) where an explosion in one room of a multi-storey building caused a whole section of the building to collapse with considerable loss of life (HMSO, 1968). Design against disproportionate collapse requires the provision of some structural toughness with redundant load paths.

Disproportionate collapse can also occur if elements providing lateral restraint to main beams or columns are destroyed in a fire, allowing subsequent collapse of the main member. As an example, Comeau (1999) describes the collapse of a timber truss roof during a fire, resulting in the deaths of three fire-fighters. In this case the collapse was not a result of fire damage, but was caused when the fire-fighters cutting a hole in the roof removed the lateral restraint to the compression chord of the main roof trusses, causing collapse by buckling. For this reason, members providing lateral restraint to fire-rated members must also have appropriate fire resistance.

7.6.4 Continuity

Continuous beams

Flexural continuity can improve the fire resistance of a bending member. A member with flexural continuity is a statically indeterminate member which would need to

Figure 7.11 Multi-storey apartment building after a gas explosion caused disproportionate collapse to upper floors [Ronan Point, U.K., 1968] (Reproduced by permission of Building Research Establishment, UK)

fail at more than one cross section to lose its load carrying capability. A simply supported beam has no continuity, so failure at mid-span will cause collapse. Beams which are continuous over several supports or built into rigid frames have continuity which is beneficial in fire design, because collapse does not occur when the ultimate strength is reached at only one cross section.

In fire-exposed structures, the benefits of continuity allow load to be resisted in alternative ways through moment redistribution as heat-affected areas lose strength. The benefits are greatest in ductile concrete or steel members which can undergo large rotations at 'plastic hinges'. Additional benefits can occur in materials like reinforced concrete or composite structures where the flexural strength may be different in positive and negative bending.

A plastic hinge is a segment of a beam where large rotations occur with no significant increase in bending moment. Figure 7.12 shows the moment–curvature relationship for a beam of a ductile material like steel, for which the stress–strain relationship is shown in Figure 7.2. At low levels of bending moment there is a linear relationship between moment and curvature, with the slope of the line given by the product EI. The elastic curvature κ of a beam is given by

$$\kappa = M/EI \tag{7.26}$$

where M is the bending moment, E is the modulus of elasticity, and I is the moment of inertia (second moment of area) of the cross section. The relationship becomes

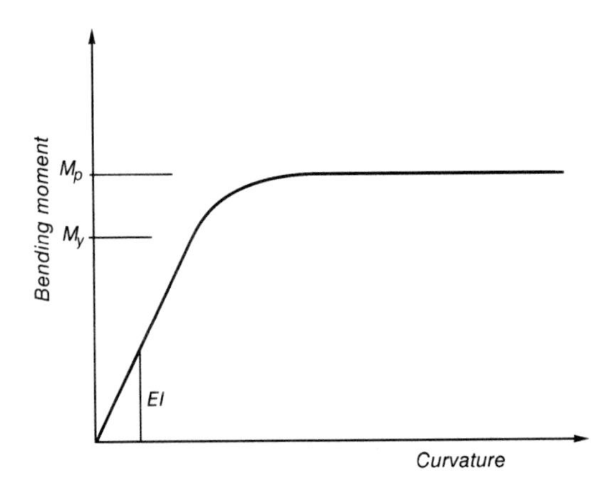

Figure 7.12 Moment–curvature relationship for a beam of ductile material

non-linear after first yielding occurs in the cross section at a bending moment M_y. The maximum bending moment which can be resisted by the cross section is the plastic moment M_p.

To assess the possible benefits of moment redistribution, the collapse mechanism causing failure must be considered. Figure 7.13 shows three different support conditions for a beam with a span L and uniformly distributed load w. The three support conditions are simply supported, continuous with built-in supports at both ends, and continuous at one end representing the end span of a continuous beam or a two-span beam. The bending moment diagrams and elastic deflected shapes for these and other combinations of load and support conditions can be obtained from structural engineering textbooks, or from structural analysis computer programs.

Figure 7.13 shows the elastic bending moment diagram and deflected shape for each of the three conditions under service loads. Figure 7.13 also shows the mechanisms that would occur if the loads were to increase towards the failure loads (or the strength decreases due to fire exposure). Note that the deflections are sketched with similar magnitudes for the elastic deflection and the failure mechanism, although the actual deflections at failure will be much greater than the elastic deflections.

Observation of the failure mechanism in Figure 7.13(a) shows that the simply supported beam will fail when its strength is exceeded at mid-span and a plastic hinge occurs at that point. There is no benefit from flexural continuity for a simply supported beam.

In Figure 7.13(b) it can be seen that the continuous beam will not reach its maximum load capacity until plastic hinges form at three points. If the flexural capacity is the same along the full length of the beam, and the same in both positive and negative directions, the final bending moment at all three plastic hinges will be identical, and the bending moment diagram at failure will be different from the elastic bending moment diagram. This shift in the bending moments is known as moment redistribution.

The end span beam in Figure 7.13(c) is an intermediate case between the simply supported and continuous beams, requiring two hinges to cause a failure mechanism.

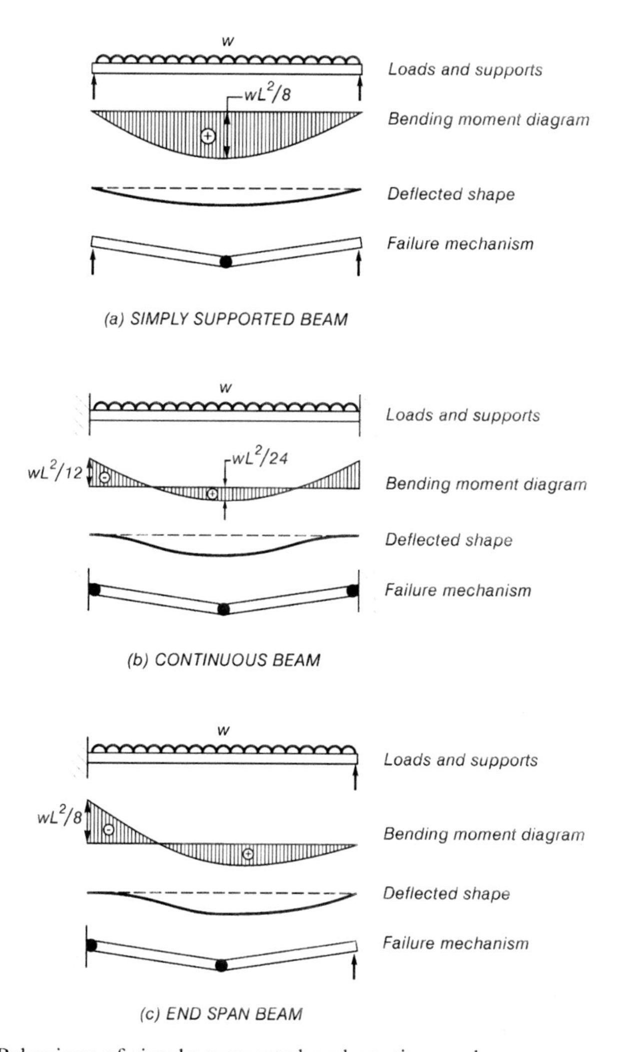

Figure 7.13 Behaviour of simply supported and continuous beams

Consider the behaviour as the load increases steadily. Because the elastic bending moment at the support is greater than that near mid-span, the plastic moment M_p will occur first at the support, then the bending moment near mid-span will increase as the load increases, accompanied by plastic rotation in the plastic hinge at the support. Eventually the plastic moment will be reached near mid-span causing the failure mechanism shown. The final shape of the bending moment diagram will again have changed due to moment redistribution.

Moment redistribution

Moment redistribution is discussed in more detail with reference to Figures 7.14 and 7.15 which show one span of a beam which is continuous over several supports. First

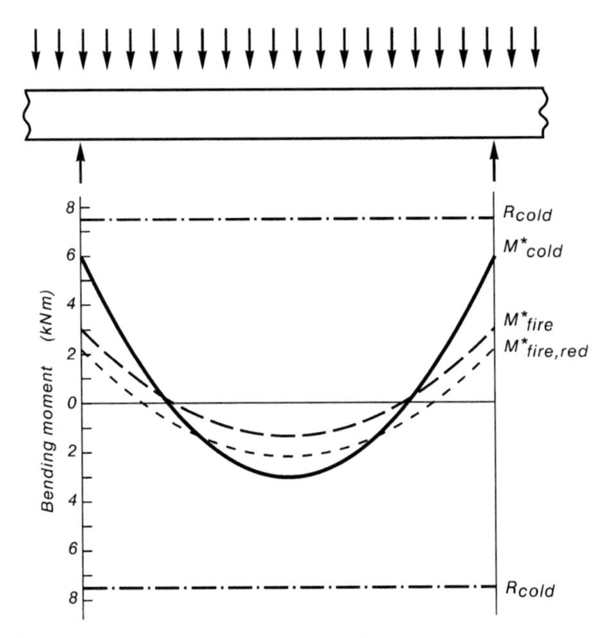

Figure 7.14 Moment redistribution to equal positive and negative moments

consider Figure 7.14 which is the same situation as shown in Figure 7.13(b). Under cold conditions with full factored dead and snow loads, the elastic bending moment diagram for a continuous beam is the solid line marked M^*_{cold}. This curve has exactly the same shape as the solid curve in Figure 7.9, but the continuity causes it to be re-positioned such that the end moment is double the mid-span moment. Any introductory book on structural mechanics will give the derivation of this elastic bending moment diagram.

With the reduced loads expected in fire conditions, the bending moments reduce to those shown by the curve M^*_{fire}. If this beam has a symmetrical cross section, such as a steel I-beam, the positive and negative flexural capacities at normal temperatures are equal, and the capacities are as shown by the horizontal lines marked R_{cold} with a larger safety margin against failure in positive bending (mid-span) than in negative bending (at the supports). As the flexural capacity drops from R_{cold} under fire exposure, a plastic hinge will occur at the support when the flexural capacity reaches M^*_{fire}. As the flexural capacity drops even further due to elevated temperatures, plastic rotation will occur at the support and the mid-span bending moment will increase due to moment redistribution. Failure will occur when a plastic hinge occurs at mid-span, and the bending moments are as shown by $M_{fire,\ red}$ with equal positive and negative moments both equal to the plastic moment capacity of the heated beam.

This situation will change if the beam has different flexural capacities in positive and negative bending, more common in reinforced concrete structures, or where the fire causes non uniform heating in the cross section. In Figure 7.15 the solid line M^*_{cold} and the dotted line M^*_{fire} are exactly the same as in Figure 7.14. The lines marked R_{cold} have been shown with different values in positive and negative bending, assuming that this is a reinforced concrete beam where the number of reinforcing bars has been

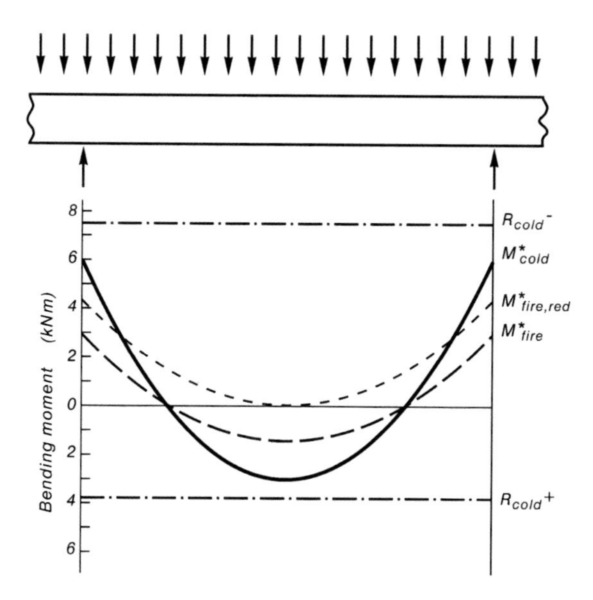

Figure 7.15 Moment redistribution to unequal positive and negative moments

selected to match the M^*_{cold} bending moment diagram. If the fire exposure causes the positive flexural capacity of the beam to drop to zero at mid-span, the beam will not fail provided that the negative flexural capacity R_{cold}^- does not drop below the value of M^*_{fire} at the supports. The beam now carries its entire load by cantilevering out from the supports. The bending moment diagram M^*_{fire} is the same as the diagram M^*_{fire} except that it has been lifted as moment distribution occurred. The value of $M^*_{\text{fire,red}}$ at the supports is now equal to the mid-span simply-supported value of M^*_{fire} in Figure 7.9. A serious consequence of moment redistribution in reinforced concrete is the need to re-evaluate the locations where reinforcing bars are terminated within the span of the beam. Numerical values are put on these moment redistributions in the worked examples at the end of Chapter 7.

Redistribution of bending moments can give significant advantages in the fire design of continuous beams in all materials. For a reinforced concrete beam, the final shape of the redistributed bending moment diagram depends on the relative amount of positive and negative reinforcing provided, and the expected influence of the fire on the flexural strength in both hogging and sagging moments. Depending on the strength of the beam when plastic hinges develop at the ends or in the centre the bending moment diagram M^*_{fire} may move up or down to any statically admissible position as plastic hinges occur. This can be done visually with sketched bending moment diagrams, or it can be calculated.

It is useful to know the shape of the bending moment diagram for a propped cantilever or the end bay of a multi-span beam, such as shown in Figure 7.16. If the positive plastic moment capacity is known, the negative moment M^- (kNm) at the support is given by

$$M^- = wL^2/2 - wL^2\sqrt{2M_p^+/wL^2} \tag{7.27}$$

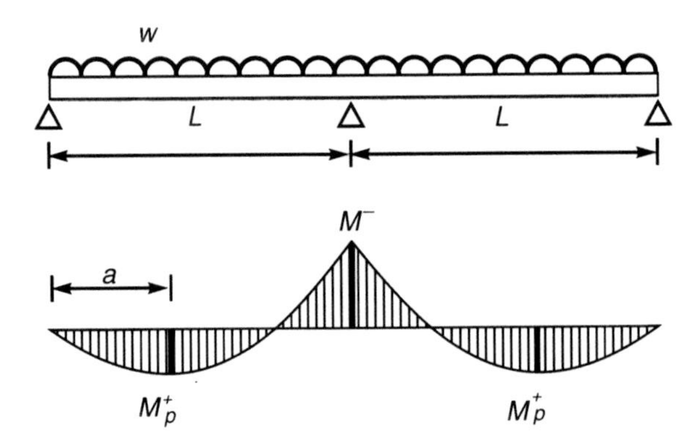

Figure 7.16 Bending moment diagram for a two-span continuous beam

where w is the uniformly distributed load on the beam (kN/m), L is the span of the beam (m), and M_p^+ is the positive plastic moment capacity (kNm).

The distance a (m) of the maximum positive moment from the pinned support is given by

$$a = \sqrt{2M_p^+/L} \tag{7.28}$$

Derivations of these equations can be found in structural mechanics textbooks.

7.6.5 Plastic Design

Some methods of calculating the benefits of moment redistribution in statically indeterminate beams have been described above. An alternative approach which is more versatile is to use plastic theory and the simple equations of virtual work.

Figure 7.17 shows a fixed-end beam of span L with uniformly distributed load w in the undeformed state and after large plastic deformations. In the plastic deformed state the beam can be considered to be two rigid bars each with a rotation θ, which produces a mid-span deflection of $\delta = \theta\, L/2$. The virtual work equation is based on the principle that the external work done by vertical movement of the applied load is equal to the internal work done by plastic rotation at all the plastic hinges. In this case the total load on the beam is wL and the average vertical displacement is $\delta/2$. The two plastic hinges at the supports have a rotation θ with plastic moment M_p^- and the central plastic hinge has a rotation 2θ with plastic moment M_p^+. The virtual work equation now gives

$$\begin{aligned} \text{external work} &= \text{internal work} \\ wL\delta/2 &= 2M_p^-\theta + 2\theta M_p^+ \end{aligned} \tag{7.29}$$

Substituting $\delta = \theta L/2$ gives

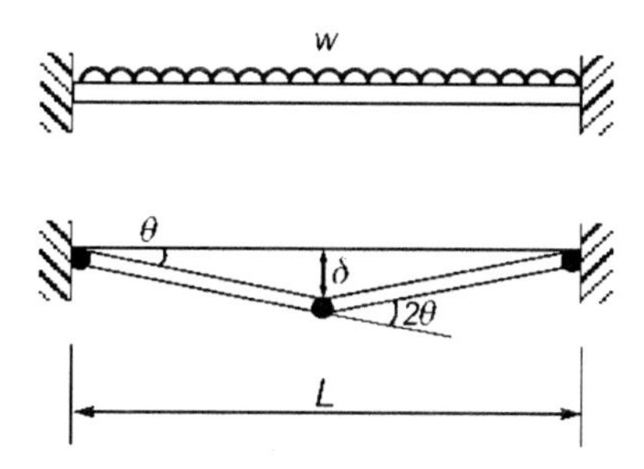

Figure 7.17 Plastic deformation of a fixed-end beam

$$wL^2/8 = M_p^- + M_p^+ \tag{7.30}$$

If the positive and negative plastic moments are equal, $M_p = M_p^- = M_p^+ = wL^2/16$. If they are different, either one can be determined if the other is known.

This approach can be used for any system of continuous beams, for any ductile material. If the plastic failure mechanism is not known exactly, several alternative mechanisms can be tried, and the one giving the least amount of internal work will be the correct answer. Calculus can be used to determine the exact solution if necessary, but it is usually sufficiently accurate to assume that all the positive plastic hinges are at mid-span, making the solution very easy. An exception is the end span of a continuous beam or propped cantilever where the positive plastic hinge may be nearer the pinned support (as shown in Figure 7.13(c)) in which case equations (7.27) and (7.28) can be used.

As an example of a different structure Figure 7.18 shows a beam with two spans and two supports. Two separate plastic mechanisms should be considered, as shown. This approach can be used for a wide variety of problems.

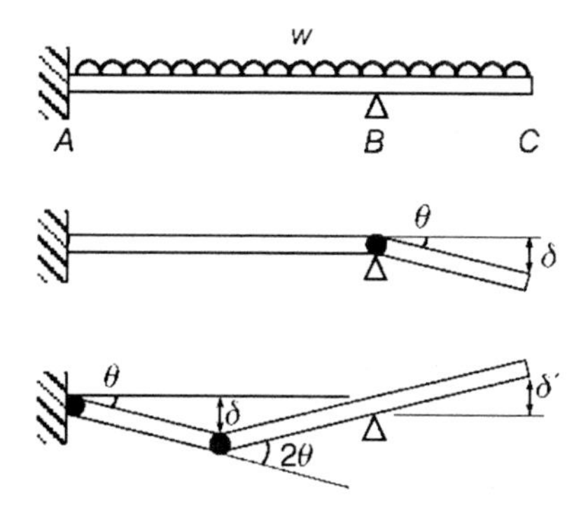

Figure 7.18 Plastic failure mechanisms for indeterminate beam

7.6.6 Axial Restraint

Another significant influence on fire performance of some structures is *axial restraint*. Axial restraint refers to the effect of axial forces which occur when a heated member is restrained from thermal expansion by a more rigid surrounding structure. Axial restraint is particularly important for reinforced or prestressed concrete slabs or beams, and for composite concrete-steel deck slabs, where the axial restraint force can partly or completely compensate for the loss in strength of steel reinforcing at elevated temperatures. The effects of axial restraint can be particularly beneficial when a fire occupies only part of a floor of a building, leaving a considerable area of surrounding structure at normal temperatures, able to resist the restraint forces. Axial restraint can be significant in concrete and steel structures, but not in timber structures because wood has a low thermal conductivity, and a low coefficient of thermal expansion. Flexural continuity is also sometimes called 'restraint', but the effects of these two phenomena are quite different, as explained previously, so the two should not be confused.

Some types of construction have two listed fire-resistance ratings, one for 'restrained' conditions and another for 'unrestrained' conditions. The difference in behaviour is most easily described with an example, as given below, considering first the heated structural member then the surrounding structure.

Effect of restraint on heated members

Figure 7.19 shows a simply supported concrete beam, located between rigid supports which permit rotation but no elongation at the ends. As the bottom of the beam heats up, it tries to expand, but is unable to do so because of the rigid supports. An axial thrust T develops in the beam, contributing to its strength. The thrust may be thought of as external prestressing. Figure 7.19 shows a situation where the elevated temperature moment capacity M_f can drop to less than the applied moment M^*_{fire} without collapse because the flexural resistance is enhanced by the moment $T.e$, where e is the eccentricity between the line of action of the thermal thrust and the centroid of the compression block near the top of the beam. The total flexural resistance $R_{fire} = M_f + T.e$ is then greater than the applied moment M^*_{fire}. The $T.e$ line is curved as shown because of the deflection of the beam.

In some situations where the surrounding structure has sufficient stiffness, the elevated temperature moment capacity M_f can drop to zero without failure, with all of the moment resisted by the $T.e$ couple. This explains the large difference between the listed 'restrained' and 'unrestrained' ratings for some assemblies which have been tested under both conditions (e.g. UL, 1996).

Figure 7.20 shows a free body diagram of a reinforced concrete beam subjected to a compressive axial restraint force T. The compression stress block must now develop a force C equal to the sum of the axial force T and the tensile force in the reinforcing T_y.

Axial restraint does not always have a beneficial effect on fire resistance. Restraint can have a negative effect if mid-span deflections become excessive, or if the axial thrust develops near the top of the cross section. In order to utilize the beneficial effects of axial restraint, it is essential that the line of thermally induced thrust is below the centroid of the compression region of the beam or slab, so that the eccentricity 'e' shown in Figures 7.19 and 7.20 has a positive value. It can sometimes be difficult to calculate the axial restraint because the position of the axial thrust can

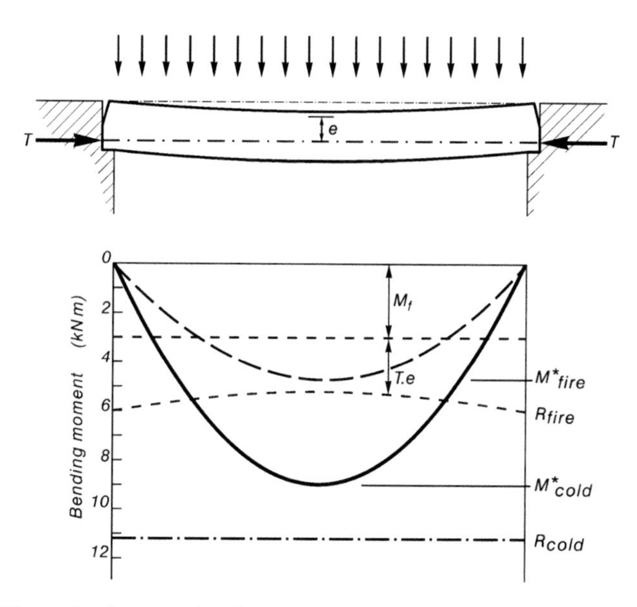

Figure 7.19 Effect of axial restraint force on bending moment diagram

change from being positive to negative and *vice versa* as deformations and rotations occur during fire exposure. Figure 7.21 (Carlson *et al.*, 1965) shows how the location of the axial restraint force depends on the support conditions of the beam or slab. An axial restraint force near the top of the beam as shown in Figure 7.21(a) would lead to premature failure of the floor system. This can be a problem with 'Double Tee' precast prestressed concrete floor panels if the webs are cut away at the ends and all the support is provided at the level of the top flange as shown in Figure 9.23. For built-in construction where the line of action of the restraint force is not known (Figure 7.21(d)) the thrust will usually be near the bottom where most of the heating and thermal expansion occurs.

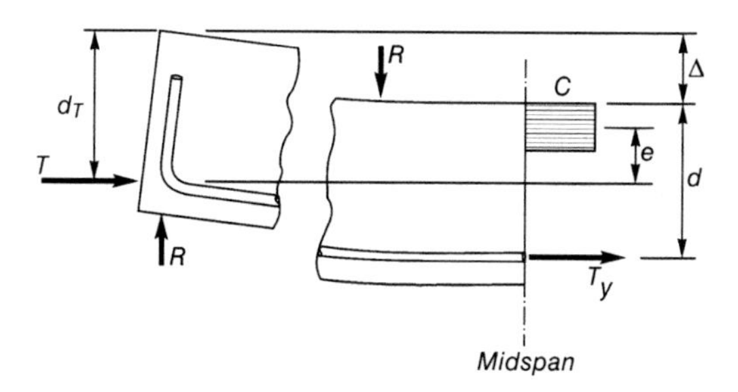

Figure 7.20 Free body diagram of beam with axial restraint force

Effect of restraint on the surrounding structure

A very stiff surrounding structure is necessary in order to develop the beneficial effects of axial restraint. Fire resistance tests can be used to assess the performance of

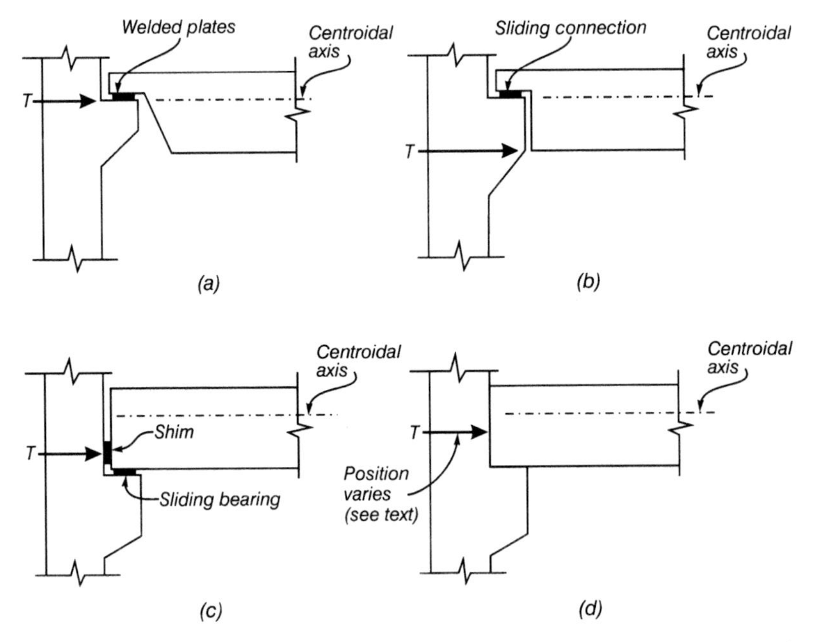

Figure 7.21 Location of axial thrust for several support conditions (Reproduced from Carlson *et al.* (1965) by permission of Portland Cement Association)

an assembly in a restrained or unrestrained condition, but cannot be used to predict behaviour in a real building. A building structure will only be able to provide axial restraint if the part heated by the fire is surrounded by cooler structural members which have sufficient strength and stiffness to restrain the thermal elongation. This is most likely when a fire occurs in a restricted area of a large building, or where a fire exposed concrete slab is surrounded by massive concrete beams. Rotter *et al.* (1999) have shown how restraint can occur in fire-protected composite edge beams, even when the fire occurs near the corner of a building. A structural engineering assessment is required on a case-by-case basis, considering the design of the actual building, but it may be difficult to assess how the surrounding structure can provide the necessary resistance to the axial restraint forces.

A negative effect of axial restraint may be serious damage to the surrounding structure, caused by large forces resulting from the thermal expansion. Such damage is more likely in concrete or masonry buildings rather than in steel buildings, because of the inability of a stiff and brittle surrounding structure to absorb the imposed thermal deformations.

Code requirements for restraint

This section has shown that consideration of axial restraint in fire-resistance assessment is difficult because it requires information about the tested assembly and also about the structure in which the assembly is to be used. Codes do not handle axial restraint very well. Recognizing the difficulty of assessing a structure to resist axial restraint forces, an appendix to ASTM E119 gives 'interim' guidance for determining conditions of restraint for floor and roof assemblies and individual beams. The following clauses and Table 7.2 are extracted directly from ASTM E119 (ASTM, 1988(a)).

> 'X3.4 For the purposes of this guide, restraint in buildings is defined as follows: 'Floor and roof assemblies and individual beams in buildings shall be considered restrained when the surrounding or supporting structure is capable of resisting substantial thermal expansion throughout the range of anticipated elevated temperatures. Constructions not complying with this definition are assumed to be free to rotate and expand and shall therefore be considered as unrestrained.'

> 'X3.5 This definition requires the exercise of engineering judgement to determine what constitutes restraint to 'substantial thermal expansion'. Restraint may be provided by the lateral stiffness of supports for floor and roof assemblies and intermediate beams forming part of the assembly. In order to develop restraint, connections must adequately transfer thermal thrusts to such supports. The rigidity of adjoining panels or structures should be considered in assessing the capability of a structure to resist thermal expansion. Continuity, such as that occurring in beams acting continuously over two or more supports, will induce rotational restraint which will usually add to the fire resistance of structural members.'

These extracts from ASTM E119 are useful but confusing because flexural continuity and axial restraint are mixed up, and the table mentions only the type of construction with no reference to the extent of the fire or the stiffness of the building. If structural calculations show that the effects of axial restraint are essential to ensure structural stability in fire conditions, then the stiffness of the structure must be calculated. Methods of calculating the effects of axial restraint and the required stiffness of the surrounding structure for reinforced and prestressed concrete buildings are described in Section 9.6.5. For wood structures, Table 7.2 recommends that all types of wood construction should be considered to be unrestrained, which ignores the beneficial effect of continuity in glulam beams or timber decking, for example.

The three issues of continuity, redundancy and restraint all require structural engineering assessment on a case-by-case basis because they cannot be tested in the traditional fire-resistance tests.

Large displacements

This discussion has been based on the assumption that the restrained slab or beam does not buckle, which is usually a good assumption for reinforced concrete structures. Rotter *et al.* (1999) have shown that fire-exposed steel beams with

Table 7.2 ASTM E119 classification for restrained and unrestrained construction (ASTM, 1988a)

Construction	Classification
I. Wall bearing:	
Single span and simply supported end spans of multiple bays[A]	
(1) Open-web steel joists or steel beams, supporting concrete slab, precast units, or metal decking	unrestrained
(2) Concrete slabs, precast units, or metal decking	unrestrained
Interior spans of multiple bays:	
(1) Open-web steel joists, steel beams or metal decking, supporting continuous concrete slab	restrained
(2) Open-web steel joists, or steel beams, supporting precast units or metal decking	unrestrained
(3) Cast-in-place concrete slab systems	restrained
(4) Precast concrete where the potential thermal expansion is resisted by adjacent construction[B]	restrained
II. Steel framing:	
(1) Steel beams welded, riveted, or bolted to the framing members	restrained
(2) All types of cast-in-place floor and roof systems (such as beams-and-slabs, pan joists, and waffle slabs) where the floor or roof system is secured to the framing members	restrained
(3) All types of prefabricated floor or roof systems where the structural members are secured to the framing members and the potential thermal expansion of the floor or roof system is resisted by the framing system or the adjoining floor or roof construction[B]	restrained
III. Concrete framing:	
(1) Beams securely fastened to the framing members	restrained
(2) All types of cast-in-place or roof systems (such as beam-and-slabs, flat slabs, pan joists, and waffle slabs) where the floor system in cast with the framing members	restrained
(3) Interior and exterior spans of precast systems with cast-in-place joints resulting in restraint equivalent to that which would exist in condition III (1)	restrained
(4) All types of prefabricated floor or roof systems where the structural members are secured to such systems and the potential thermal expansion of the floor or roof systems is resisted by the framing system or the adjoining floor or roof construction[B]	restrained
IV. Wood construction:	
All types	unrestrained

[A]Floor and roof systems can be considered restrained when they are tied into walls with or without tie beams, the walls being designed and detailed to resist thermal thrust from the floor or roof system.
[B]For example, resistance to potential thermal expansion is considered to be achieved when:
(1) continuous structural concrete topping is used,
(2) the space between the ends of precast units or between the ends of units and the vertical face or supports is filled with concrete or mortar, or
(3) the space between the ends of solid or hollow core slab units does not exceed 0.15% of the length for normal weight concrete members or 0.1% of the length for structural lightweight concrete members.

composite concrete slabs may buckle as a result of the large axial forces induced when they try to expand axially against a stiff and strong surrounding structure. This buckling results in large downwards deflections of the beam and slab which can be beneficial because the large deflections reduce the horizontal restraint forces on the surrounding structure. There is a complex interaction between the axial forces, downwards deflections and stiffness of the structure. Once buckling occurs, resulting in large deformations, the analysis presented above for single members does not apply. Later in the fire the slab or slab-beam assembly may lose flexural strength and deform into a catenary. The slab now develops internal tensile forces which pull inwards on the surrounding structure with tensile membrane action. This behaviour is described with reference to the Cardington tests in Section 8.8.1.

7.6.7 After-fire Stability

The above discussion describes the design process for structures during a fire, without addressing the performance of the structure after the fire. After a fire the structure may be partly damaged, with some members missing or having reduced strength.

For example, many single-storey industrial buildings have unprotected roof structures, so that the roof can collapse in a fire, leaving the boundary walls cantilevering from the foundations after a fire, in danger of falling in wind or earthquake. This situation can be designed for if the design requirements are explicitly stated. Unfortunately, this is a common problem which most codes do not address properly.

The New Zealand loadings code (SNZ, 1992) requires that the residual structure be able to resist a face load of 0.5 kN/m^2 on the exposed surfaces after the fire. This is intended to allow for wind or earthquake loads on the residual fire-damaged structure after the fire has been extinguished or has burned itself out. This load also provides some un-specified resistance to failure during the fire, possibly due to wind loads or catenary loads from collapse of a fire damaged roof.

A common older form of construction for multi-storey buildings consists of unreinforced masonry perimeter walls with timber floors supported on heavy timber beams and columns. The timber floors provide lateral stability to the walls. A severe fire can damage the timber floors and beams, hence removing lateral support, possibly leading to serious collapse of the exterior walls, endangering the lives of firefighters and others.

WORKED EXAMPLE 7.1

Calculate the factored bending moment at the centre of a simply supported floor slab, for normal temperature design and for fire design. Use these bending moments to calculate the load ratio for fire design.

The slab has a span of 3.6 m. The characteristic dead load is 2.0 kN/m^2 and the characteristic live load is also 2.0 kN/m^2. Make all calculations for a strip of slab one metre wide. Refer to Figure 7.9.

Calculate load combinations:

Characteristic dead load $G_k = 2.0$ kN/m^2

Characteristic live load $Q_k = 2.0$ kN/m^2

Factored load for cold design $1.4\ G_k = 2.8$ kN/m^2

$1.2\ G_k + 1.6\ Qk = 5.6$ kN/m^2 (governs)

Factored load for fire design $G_k + 0.4\ Q_k = 2.8$ kN/m^2

Convert to uniformly distributed loads on a strip 1 m wide:

Factored load for cold design $w_c = 5.6 \times 1 = 5.6$ kN/m

Factored load for fire design $w_f = 2.8 \times 1 = 2.8$ kN/m

Calculate bending moments in centre of 3.6 m span:

Design bending moment for cold design $M^*_{cold} = w_c L^2/8 = 5.6 \times 3.6^2/8 = 9.0$ kNm

Design bending moment for fire design $M^*_{fire} = w_f L^2/8 = 2.8 \times 3.6^2/8 = 4.5$ kNm

Calculate the required flexural design capacity of the slab, using a strength reduction factor $\Phi = 0.8$, and the load ratio for fire design, assuming that the slab is provided with exactly the required strength.

Flexural design capacity $R_{cold} \geqslant M^*_{cold}/\Phi$

$\geqslant 9.0/0.8 = 11.25$ kNm

Load ratio for fire design $r_{load} = M^*_{fire}/R_{cold}$

$= 4.5/11.25 = 0.40$

This shows that the slab would not be expected to fail in a fire until its strength drops to 40% of its strength at normal temperatures.

WORKED EXAMPLE 7.2

Re-calculate the bending moments from Worked Example 7.1 for a slab which is continuous over several supports. Assuming that the slab has equal positive and negative flexural capacity, calculate the load ratio considering re-distribution to equal positive and negative moments. Refer to Figure 7.14.

Calculate the design bending moments for cold conditions

Design bending moment at support (negative moment)

$M^*_{cold} = w_c\ L^2/12 = 5.6 \times 3.6^2/12 = 6.0$ kNm

Design bending moment at mid-span (positive moment)

$M^*_{cold} = w_c\ L^2/24 = 5.6 \times 3.6^2/24 = 3.0$ kNm

These bending moments are shown by the solid curve in Figure 7.14. The sum of the negative and positive moments is $6.0 + 3.0 = 9.0$ kNm, the same as M^*_{cold} in Figure 7.9. The minimum flexural capacity that must be provided to resist the negative moment is now $R_{cold} = M_{cold}/\Phi = 6.0/0.8 = 7.5$ kNm, so a weaker slab can be used than in the simply supported case. The flexural capacity provided is shown by the line R_{cold} for equal positive and negative moments. With the loads expected in fire conditions, the negative and positive bending moments reduce to 3.0 and 1.5 kNm respectively, shown by the dashed curve M^*_{fire}. The sum of the negative and positive moments is $3.0 + 1.5 = 4.5$ kNm, the same as M^*_{fire} in Figure 7.9.

Consider the effect of a fire which causes the flexural capacity R_{cold} to drop at the same rate for both positive and negative bending. If there is no redistribution of moments, the slab would fail at the supports when the flexural capacity drops to 3.0 kN.m. The load ratio is $3.0/7.5 = 0.40$. However, with redistribution, the bending moment diagram can take any position provided that the sum of positive and negative moments remains at $w_f L^2/8 = 4.5$ kNm. If positive and negative flexural strengths are equal, the optimum location of the bending moment diagram is the shape shown by the dotted curve in Figure 7.14, with positive and negative moments both $M^*_{fire,red} = 2.25$ kNm (which could also have been obtained from Equation 7.30). This corresponds to

$$\text{Load ratio for fire design } r_{load} = M^*_{fire,red}/R_{cold} = 2.25/7.5 = 0.30$$

Final failure of the slab will not be expected until its strength in fire conditions has reduced to 30% of its strength at normal temperatures.

WORKED EXAMPLE 7.3

Reconsider the slab from Worked Example 7.2, assuming that the slab can have different positive and negative flexural capacities. Calculate the load ratio assuming that the flexural capacity at mid-span (positive moment) drops to zero during fire exposure. Refer to Figure 7.15.

Before moment redistribution, the bending moments shown by the solid curve (M^*_{cold}) and the dashed curve (M^*_{fire}) are the same as worked in Worked Example 7.2. The maximum flexural capacities are shown by the two lines marked R_{cold} for unequal positive and negative values.

With moment redistribution, the bending moment diagram M^*_{fire} can again take any position, provided that it remains the same shape. The dotted line marked $M^*_{fire,red}$ in Figure 7.15 shows the bending moment diagram lifted to give a maximum value at the supports and a zero value at mid-span.

In this case the slab could survive a fire even though the positive flexural capacity at mid-span drops to zero during the fire. The slab must be able to resist a negative bending moment at the supports of $M^*_{fire,red} = 4.5$ kNm, which corresponds to:

$$\text{Load ratio for fire design } r_{load} = M^*_{fire,red}/R_{cold} = 4.5/7.5 = 0.60$$

Failure of the slab will not be expected until its cantilever strength in fire conditions has reduced to 60% of its strength at normal temperatures and the mid-span strength has dropped to zero.

WORKED EXAMPLE 7.4

Consider the continuous beam shown in Figure 7.18. The span AB is 6.0 m and BC is 2.0 m. The uniformly distributed load during the fire conditions is $w_f = 22.0\,\text{kN/m}$.

Considering the two separate failure mechanisms shown in Figure 7.18,
(a) Calculate the minimum flexural capacity of the positive and negative plastic hinges.

Span BC

By the principle of conservation of energy, the magnitude of the external virtual work is equal to the internal virtual work.

$$\text{external work} = \text{internal work}$$
$$w_f\, L_{BC}\, \delta/2 = M_p^- \theta$$
$$\text{Substituting } \delta = \theta L_{BC}$$
$$w_f\, L_{BC}^2\, \theta/2 = M_p^- \theta$$
$$\text{Substituting } w_f = 22\,\text{kN m and } L_{BC} = 2\,\text{m gives:}$$
$$M_p^- \geqslant 22 \times 2^2/2 = 44\,\text{kNm} \tag{7.31}$$

Span AB

Assume that the negative hinge occurs in the centre of span AB.

$$\text{external work} = \text{internal work}$$
$$w_f\, L_{AB}\, \delta/2 - w_f\, L_{BC}\, \delta'/2 = M_p^- \theta + 2M_p^+ \theta$$
$$\text{from geometry, } L_{AB} = 6\,\text{m},\ L_{BC} = 2\,\text{m}$$
$$\delta' = 2\delta/3,$$
$$\delta = \theta L_{AB}/2$$
$$\text{Making these substitutions with } w_f = 22\ \text{kN/m gives}$$
$$M_p^- + 2M_p^+ \geqslant 154\ \text{kNm} \tag{7.32}$$

Combining Equations (7.31) and (7.32) gives

$$M_p^- \geqslant 44\ \text{kNm}$$
$$M_p^+ \geqslant 55\ \text{kNm}$$

(b) Calculate the required flexural capacity of the negative plastic hinge if the strength of the positive plastic hinge decreases to 30 kNm under fire exposure.

If M_p^+ is reduced to 30 kNm, substituting into (7.32) gives

$$M_p^- \geqslant 94\ \text{kNm}$$

8

Steel Structures

8.1 OVERVIEW

This chapter provides the information needed for calculating the performance of steel buildings exposed to fires. Simple methods are described for designing individual steel members to resist fire exposure, including calculations of elevated temperatures, methods of fire protection, and information on the thermal and mechanical properties of steel at elevated temperatures. Fire behaviour of large steel buildings is also discussed.

This chapter draws information from Eurocode 3: Design of Steel Structures (EC3, 1995), which with the other structural Eurocodes, summarizes the results of a large international co-operative research programme over the past decade.

8.2 BEHAVIOUR OF STEEL STRUCTURES IN FIRE

When a steel structure is exposed to a fire, the steel temperatures increase and the strength and stiffness of the steel are reduced, leading to possible deformation and failure, depending on the applied loads and the support conditions (Figure 8.1). The increase in steel temperatures depends on the severity of the fire, the area of steel exposed to the fire and the amount of applied fire protection. There are many methods of protecting steel members from the effects of fire, so that structural steel buildings with applied fire protection can be designed to have excellent fire resistance.

Unprotected steel structures tend to perform poorly in fires compared with reinforced concrete or heavy timber structures, because the steel members are usually much thinner (Figure 8.2(a)). Steel also has a higher thermal conductivity than most other materials. Unprotected steel structures can survive some fires if the severity is low and the steel does not get too hot (Figure 8.2(b)). Full-scale tests and some real fires in large steel buildings have shown that well-designed structures can resist severe fires without collapse, even if some of the main load-bearing members are unprotected. Thermal expansion of steel members can cause damage elsewhere in the building (Figure 8.2(c)). A review of steel behaviour in many fire tests is given by Cooke (1996).

The main factors affecting the behaviour of steel structures in fire, as discussed further in this chapter, are as follows:

Figure 8.1 Typical fire damage to unprotected steel frames in an industrial building

- the elevated temperatures in the steel members,
- the applied loads on the structure,
- the mechanical properties of the steel, and
- the geometry and design of the structure.

8.3 FIRE-RESISTANCE RATINGS

8.3.1 Verification Methods

The design process for fire resistance requires verification that the provided fire resistance exceeds the design fire severity. Using the terminology from Chapter 5, verification may be in the *time domain*, the *temperature domain* or the *strength domain*. All three domains are often used for assessing the fire resistance of steel structures. The traditional method of using fire-resistance ratings in the time domain is described first.

In the time domain, the required fire resistance may be prescribed by a code or calculated from a time equivalent formula if the fire load and ventilation are known. The required fire resistance can then be compared with the fire-resistance rating of the selected assembly. The fire-resistance rating can be obtained from listings of generic ratings, proprietary ratings or expert opinion ratings all based on the standard fire test, or from calculations of the time to reach the limiting temperature during exposure to the standard fire.

Figure 8.2 (a) Severe fire in a movie theatre, showing collapsed steel roof trusses in the foreground; the gallery seating which did not collapse is visible in the upper background. (b) Buckling of a compression member in the heavy steel truss supporting the gallery seating; this truss is close to failure, but did not collapse. (c) Holes punched through a reinforced concrete wall by thermal expansion of the heavy steel truss supporting the gallery seating

In the temperature domain, the limiting steel temperature is compared with the maximum temperature reached in the design fire exposure. The limiting temperature is the steel temperature at which the load-bearing capacity of the member would just equal the design loads, or the steel temperature above which the member would be expected to fail. Eurocode 3 (EC3, 1995) and the British Standard (BSI, 1990(a)) give a limiting temperature option for fire design of single members. This type of calculation is most suitable when the steel cross section is assumed to be at a uniform temperature.

In the strength domain, the load-bearing capacity is compared with the expected loads on the member at the time of the fire. This is the method recommended in this book. Calculations in the strength domain must be used for accurate assessment of structural behaviour if there are temperature gradients across the steel cross sections, and for assessing fire behaviour of whole structures.

8.3.2 Generic Ratings

Generic ratings, or 'tabulated ratings' are those which assign a time of fire resistance to materials with no reference to individual manufacturers or to detailed specifications. Many national codes and some trade organizations provide lists of generic ratings for fire protection of structural steel members. The most common ratings are for encasement in concrete or some other generic material, with a table of the minimum thickness of material needed to provide certain ratings. Table 8.1 shows a typical example of generic ratings for solid concrete protection to steel columns, from the National Building Code of Canada (NBCC, 1995), which also requires the concrete to be nominally reinforced with wire or wire mesh. Tables such as this, which is typical of many similar codes around the world, are of limited use because they are empirically derived, apply only to standard fire exposure, and make no allowance for the size and shape of the fire exposed member or the level of load. The Uniform Building Code (UBC, 1997) gives a slightly more sophisticated table where the thickness depends on the weight of the steel member, but not the applied load, as shown with an example in Table 8.2. Some tabular ratings have been shown to be unsafe.

8.3.3 Proprietary Ratings

Many manufacturers of passive fire protection products provide listings of approved ratings. These are similar to generic ratings in that they generally provide ratings for exposure to the standard fire, but they may be less conservative because they relate to more closely defined products. Proprietary ratings usually make no allowance for the level of load, but they often include reference to the size and shape of the member using the section factor. An example is given in Table 8.3 which gives the required thickness of a particular proprietary spray-on protection (as shown in Figure 8.3) to provide fire resistance to a steel beam or column depending on the *section factor* F/V (or the *effective thickness* V/F). In comparison with Table 8.2, the section factor method shown in Table 8.3 allows a large number of steel sections to be compressed into one table rather than having a separate table for each steel size. Linear interpolation can be used within these tables. Section factors are described more fully in Section 8.4.3. Each proprietary system must have at least one fire-resistance rating obtained from a full scale load-bearing test to demonstrate that the fire protection material can remain in place for the duration of the expected fire as the member deforms.

Table 8.1 Minimum thickness of solid concrete protection to steel columns to provide fire resistance (NBCC, 1995)

Time (hours)	$\frac{1}{2}$	3/4	1	1.5	2	3	4
Thickness (mm)	25	25	25	25	39	64	89

Table 8.2 Thickness of solid concrete protection (mm) required to provide fire resistance to a W14X steel column (UBC, 1997)

<table>
<tr><th colspan="3">Steel size</th><th colspan="4">Fire resistance</th></tr>
<tr><th>Designation</th><th>lb/ft</th><th>kg/m</th><th>1 hour</th><th>2 hours</th><th>3 hours</th><th>4 hours</th></tr>
<tr><td rowspan="7">W14X
(W360)</td><td>233</td><td>347</td><td rowspan="7">25</td><td rowspan="3">25</td><td rowspan="2">38</td><td>51</td></tr>
<tr><td>176</td><td>262</td><td rowspan="3">64</td></tr>
<tr><td>132</td><td>196</td><td rowspan="3">51</td></tr>
<tr><td>90</td><td>134</td><td rowspan="4">38</td></tr>
<tr><td>61</td><td>91</td><td rowspan="3">76</td></tr>
<tr><td>48</td><td>72</td><td rowspan="2">64</td></tr>
<tr><td>43</td><td>64</td></tr>
</table>

Table 8.3 Thickness of proprietary spray-on protection required to provide fire resistance to a steel beam or column (ASFPCM, 1988)

Section size		Fire resistance			
F/*V* (m^{-1})	*V*/*F* (mm)	1 hour	2 hours	3 hours	4 hours
70	14.3	10	22	36	50
110	9.1	10	28	47	65
150	6.7	12	33	54	75
190	5.3	13	37	60	83
230	4.3	14	39	64	89
270	3.7	15	41	68	94

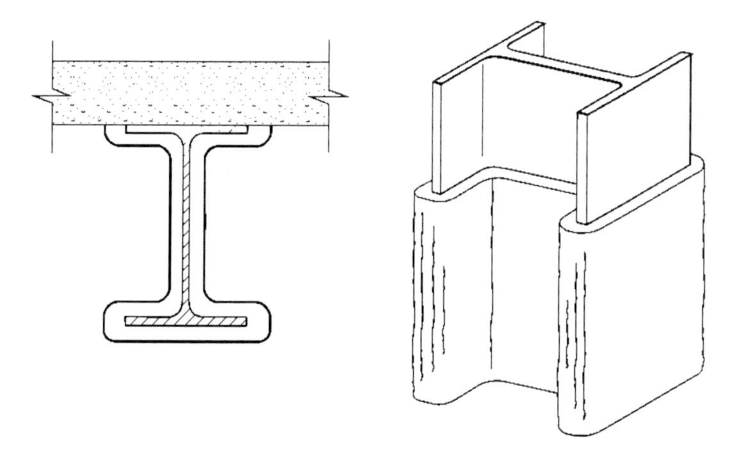

Figure 8.3 Steel beam and column protected with spray-on material

8.3.4 Calculated Ratings

In addition to summarizing various tabulated ratings, this chapter also describes calculation methods for steel structures exposed to fires. Most calculations will compare loads with load capacity in the strength domain. Using the terminology of the Eurocodes, there are two main types of calculation, namely the simple calculation model and the general calculation model. The simple model is used for single members in uncomplicated structures, using the equations presented below. The general method requires the use of a computer program for analysis of complex structures using the material properties from this chapter and as described in Eurocode 3 (EC3,1995).

8.4 STEEL TEMPERATURES

8.4.1 Fire Exposure

In any design of steel structures to resist fires, it is essential to know the temperature of the steel. The fire exposure may be the standard time–temperature curve or a more realistic fire curve, depending on the design philosophy. Generic and proprietary protection methods are all based on standard fire exposure, but calculations are often based on simulated real fires.

8.4.2 Calculation Methods

The most simple hand calculation method is to use a best-fit empirical formula to obtain the temperature of steel members exposed to the standard fire, assuming that the steel temperature is uniform over the cross section.

A computer can be used with a step-by-step calculation technique, assuming a lumped mass of steel at uniform temperature, hence constant temperature over the cross section. This method can be used with any design fire curve as input. Design charts are also available, providing graphical results of these calculations (Pettersson *et al.*, 1976, Malhotra, 1982, ECCS, 1995). Figures 8.8 and 8.9 are such charts, produced for standard fire exposure.

More sophisticated computer-based methods can calculate temperatures within the cross section, for any combination of materials or shapes, for exposure to any desired fire. Thermal properties can be temperature-dependent. A two-dimensional calculation is suitable for most situations, based on an assumption of the same temperatures at each point along the member, which is reasonable if the fire temperature is assumed to be the same throughout the fire compartment. Three-dimensional heat transfer calculations may be useful at member junctions or other special situations. Figure 8.4 shows the temperature contours in an unprotected heavy steel section (Universal Column, $356 \times 406 \times 634$ kg/m) after 30 minutes exposure to the standard fire curve. These contours have been calculated by the SAFIR program (Franssen *et al.*, 2000). It can be seen that there are temperature differences of over 100°C within the cross section, the largest difference being

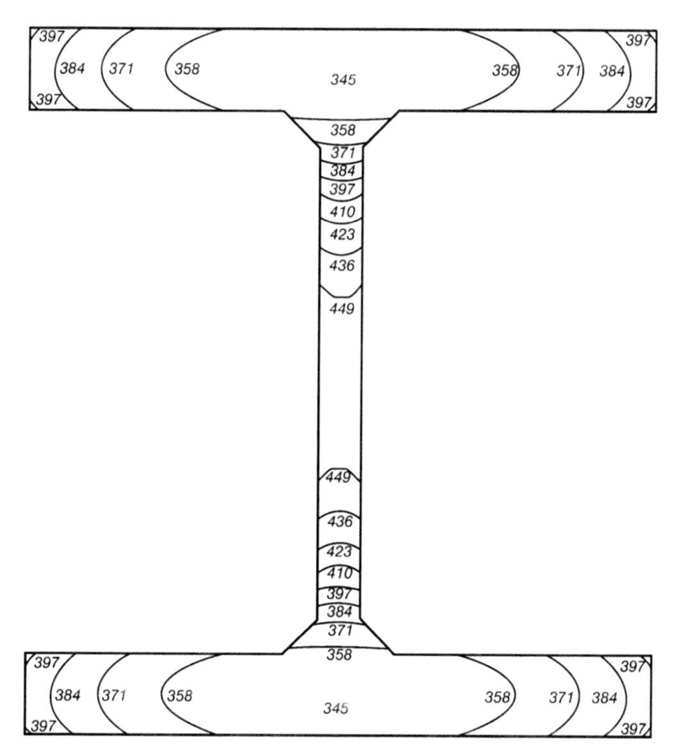

Figure 8.4 Temperature contours (°C) in a heavy steel beam exposed to fire

between the high temperatures in the thin web and the lower temperatures in the much thicker flanges.

8.4.3 Section Factor

The rate of temperature rise of a protected or unprotected structural steel member exposed to fire depends on the *section factor*, or *massivity factor*, which is a measure of the ratio of the heated perimeter to the area or mass of the cross section. The section factor is important because the rate of heat input is directly proportional to the area exposed to the fire environment, and the subsequent rate of temperature increase is inversely proportional to the heat capacity of the member (equal to the product of the specific heat, the density and the volume of the steel segment).

The section factor can be expressed in one of four different ways, as:

- the ratio of heated surface area to volume, both per unit length, F/V (m^{-1}),
- the ratio of heated perimeter to cross section area, H_p/A (m^{-1}),
- the ratio of heated surface area to mass, both per unit length, F/M (m^2/tonne), or
- the effective thickness, V/F or A/H_p (m or mm),

where F is the surface area of unit length of the member (m^2), V is the volume of steel in unit length of the member (m^3), H_p is the heated perimeter of the cross section (m), A is the cross-sectional area of the section (m^2), and M is the mass per unit length of the member (tonne).

The first two ratios are identical and can be easily converted to the third using the density of steel (7850 kg/m^3 or 7.85 tonne/m^3). The heated surface is the actual surface area of unprotected members or members with sprayed-on fire protection, and the area of the equivalent rectangle for box protection, with allowance for any unexposed surfaces, as shown in Table 8.4.

The fourth ratio listed above is V/F or A/H_p with units in metres (or millimetres). This gives much better physical understanding because it is an effective thickness of the cross section. Calculated in this way, the section factor for a steel plate exposed to a fire on both sides is $V/F = t/2$ where t is the thickness of the plate. For a hollow tube of thickness t, the section factor becomes $V/F = t$. For an I-beam, the section factor V/F is one half of the average thickness of the different parts. Mistakes in the calculation of the section factor are much less likely when it is defined in this way.

Tables of section factors for common structural steel shapes are available from distributors of steel products. Section factors for steel members are listed by UBC (1997), ASFPCM (1988) and HERA (1996) for American, British and New Zealand steel sections, respectively. Some section factors are listed in Appendix C. For protected steel members, it may be accurate to use the external surface area exposed to the fire but most references use the surface area of the steel, inside the protective material, as in Appendix C, because this dimension does not change if the insulation thickness changes.

8.4.4 Thermal Properties

In order to make calculations of temperatures in fire-exposed structures, it is necessary to know the thermal properties of the materials. The density of steel is 7850 kg/m^3, remaining essentially constant with temperature. The specific heat of steel varies according to temperature as shown in Figure 8.5 (EC3, 1995) where the peak results from a metallurgical change at about 730°C. For simple calculations the specific heat c_p (J/kg K) can be taken as 600 J/kg K, but it is more accurate to use the equations given below:

$$
\begin{aligned}
c_p &= 425 + 0.773\,T - 1.69 \times 10^{-3} T^2 + 2.22 \times 10^{-6}\,T^3 && 20^\circ\text{C} \leqslant T < 600^\circ\text{C} \\
&= 666 + 13002/(738 - T) && 600^\circ\text{C} \leqslant T < 735^\circ\text{C} \\
&= 545 + 17820/(T - 731) && 735^\circ\text{C} \leqslant T < 900^\circ\text{C} \\
&= 650 && 900^\circ\text{C} \leqslant T \leqslant 1200^\circ\text{C}
\end{aligned}
\quad (8.1)
$$

where T is the steel temperature (°C).

The thermal conductivity of steel varies according to temperature as shown in Figure 8.6, reducing linearly from 54 W/mK at 0°C to 27.3 W/mK at 800°C (EC3, 1995). For simple calculations the thermal conductivity k (W/mK) can be taken as 45 W/mK but it is more accurate to use the equations given below:

Table 8.4 Definition of section factor in the Eurocode (Reproduced from (EC3, 1995) by permission of CEN

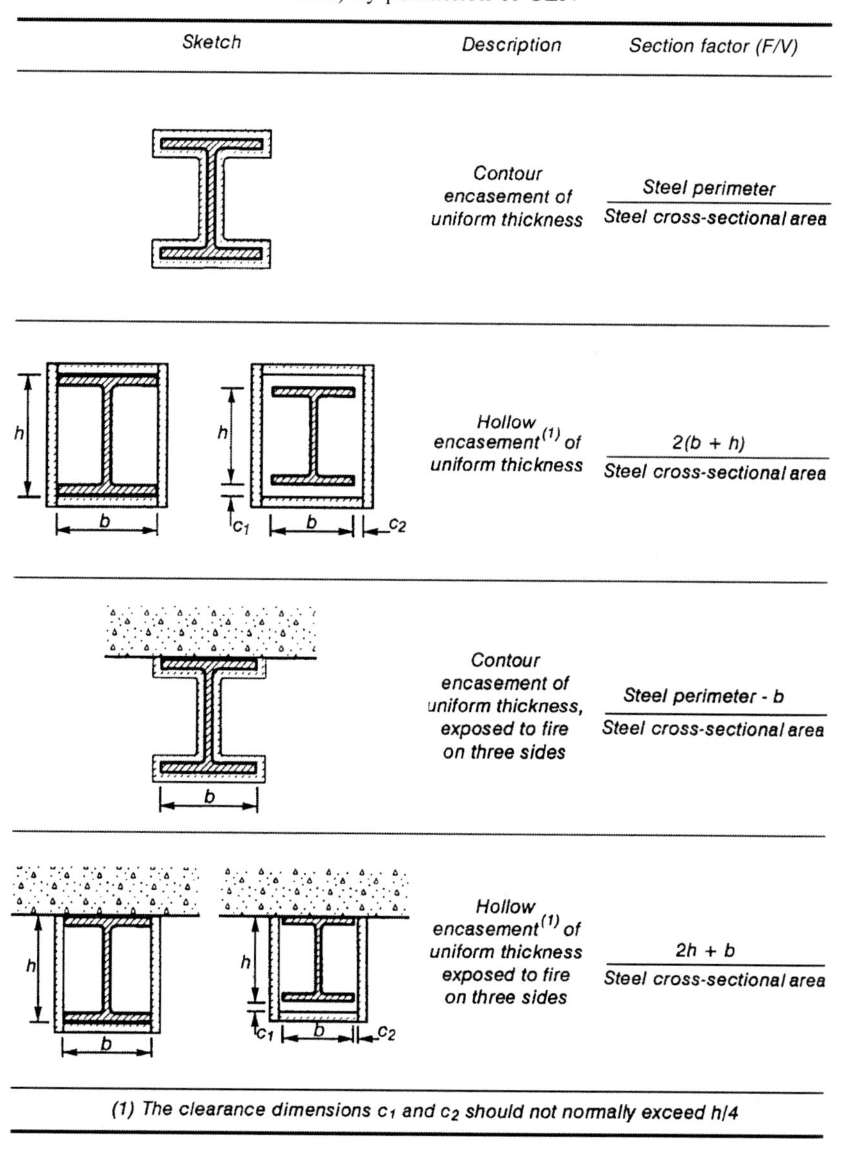

Sketch	Description	Section factor (F/V)
	Contour encasement of uniform thickness	Steel perimeter / Steel cross-sectional area
	Hollow encasement(1) of uniform thickness	2(b + h) / Steel cross-sectional area
	Contour encasement of uniform thickness, exposed to fire on three sides	Steel perimeter - b / Steel cross-sectional area
	Hollow encasement(1) of uniform thickness exposed to fire on three sides	2h + b / Steel cross-sectional area

(1) The clearance dimensions c_1 and c_2 should not normally exceed $h/4$

$$
\begin{aligned}
k &= 54 - 0.0333\,T \qquad && 20^\circ\mathrm{C} \leqslant T < 800^\circ\mathrm{C} \\
&= 27.3 && 800^\circ\mathrm{C} \leqslant T \leqslant 1200^\circ\mathrm{C}
\end{aligned}
\tag{8.2}
$$

8.4.5 Temperature Calculation for Unprotected Steelwork

Unprotected steel members can heat up quickly in fires, especially if they are thin and have a large surface area exposed to the fire. Two 'lumped mass' methods of

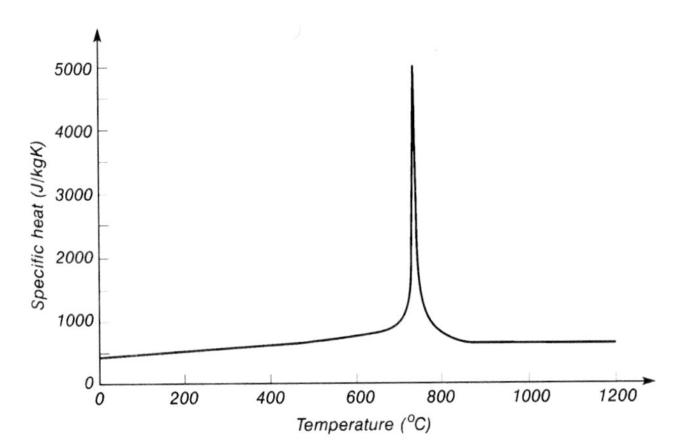

Figure 8.5 Specific heat of steel as a function of temperature. (Reproduced from (EC3, 1995) by permission of CEN)

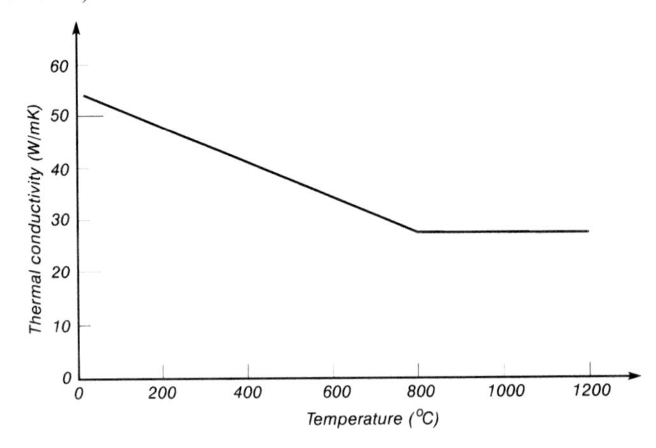

Figure 8.6 Thermal conductivity of steel as a function of temperature. (Reproduced from (EC3, 1995) by permission of CEN)

calculating temperature are given below. In many cases, such as an unprotected beam with a concrete slab on the top flange, there will be significant temperature gradients within the cross section, and these temperatures cannot be calculated with a lumped-mass method, so a finite-element method is necessary (see Chapter 3).

Best-fit method

An approximate empirical expression for predicting the time t in minutes for an unprotected steel member to reach a limiting temperature $T_{\lim}$ (°C) when exposed to the standard fire is given by ECCS (1985) as

$$t = 0.54(T_{\lim} - 50)/(F/V)^{0.6} \tag{8.3}$$

where F is the surface area of unit length of the member (m^2), and V is the volume of steel in unit length of the member (m^3).

This is stated to be valid for F/V in the range 10 to 300 m^{-1} (V/F in the range 3.3 to 100 mm) and T_{lim} in the range 400°C to 600°C, for times between 10 and 80 minutes. This formula gives results reasonably close to finite-element predictions, but the range of applicability can be extended to 100°C to 800°C for common section sizes (Lewis, 2000). The equation can be inverted to give the expected temperature at a particular time. A similar equation given in AS 4100 (SAA, 1990(a)) and NZS 3404 (SNZ, 1996) is considerably less accurate and should not be used.

This type of best-fit equation is an empirical approximation and the range of validity has not been clearly established for all section sizes. For more accurate design, it is recommended that temperatures should be calculated using a finite-element method or the step-by-step method described below.

Step-by-step method

The step-by-step calculation method for unprotected steelwork is based on the principle that the heat entering the steel over the exposed surface area in a small time step Δt (s) is equal to the heat required to raise the temperature of the steel by ΔT_s (°C) assuming that the steel section is a lumped mass at uniform temperature, so that

heat entering = heat to raise temperature

$$\dot{q}'' F \Delta t = \rho_s c_s V \Delta T_s \tag{8.4}$$

where ρ_s is the density of steel (kg/m^3), c_s is the specific heat of steel (J/kg K), ΔT_s is the change in steel temperature in the time step (°C or K), and $\dot{q}''$ is the heat transfer at the surface (W/m^2), given by

$$\dot{q}'' = h_c(T_f - T_s) + \sigma\varepsilon(T_f^4 - T_s^4) \tag{8.5}$$

where h_c is the convective heat transfer coefficient (W/m^{2}K), σ is the Stefan–Boltzmann constant (56.7×10^{-12} kW/m^{2}K^4), ε is the resultant emissivity, T_f is the temperature in the fire environment (K), and T_s is the temperature of the steel (K).

These equations can be re-arranged to give:

$$\Delta T_s = \frac{F}{V}\frac{1}{\rho_s c_s}\{h_c(T_f - T_s) + \sigma\varepsilon(T_f^4 - T_s^4)\}\Delta t \tag{8.6}$$

The convective heat transfer coefficient is recommended to have a value of 25 W/m^{2}K. The Eurocode (EC1, 1994) recommends 25 W/m^{2}K for the standard fire and 50 W/m^{2}K for the hydrocarbon fire. Heat transfer in typical fires is not very sensitive to this value because radiative heat transfer dominates at typical fire temperatures (Thomas, 1997).

A value of resultant emissivity of 0.50 is recommended by EC1 (1994). A value of resultant emissivity of 0.50 is consistent with an emissivity of 0.67 for the transmitting hot gases and the receiving surface, using Equation (3.20).

A spreadsheet for calculating steel temperatures using this method is shown in Table 8.5 (from Milke and Hill, 1996, based on Gamble, 1989). EC3 (1995) suggests

Table 8.5 Spreadsheet calculation for temperatures of unprotected steel sections

Time	Steel temperature T_s	Fire temperature T_f	Difference in temperature	Change in steel temperature ΔT_s
$t_1 = \Delta t$	Initial steel temperature T_{so}	Fire temperature halfway through time step (at $\Delta t/2$)	$T_f - T_{so}$	Calculate from Equation (8.6) with values of T_f and T_{so} from this row
$t_2 = t_1 + \Delta t$	T_s from previous time step + ΔT_s from previous row	Fire temperature half way through time step (at $t_1 + \Delta t/2$)	$T_f - T_s$	Calculate from Equation (8.6) with values of T_f and T_s from this row

a time step of no more than 30 s, and a minimum value of the section factor F/V of $10\,\text{m}^{-1}$ (maximum value of V/F of 100 mm). Kay *et al.* (1996) have shown that this type of calculation can give a very good prediction of unprotected steel beam temperatures in standard fire-resistance tests.

8.4.6 Temperature Calculation for Protected Steelwork

Protected steel members heat up much more slowly than unprotected members because of the applied thermal insulation which protects the steel from rapid absorption of heat. Two approximate calculation methods are given below. When using these methods with thick insulation, the section factor F/V should strictly be calculated using the fire-exposed perimeter rather than the inside face of the insulating material, but the inside perimeter is more often used because it is published in tables such as those in Appendix C. For steel members protected with heavy insulating materials or those with temperature-dependent thermal properties, a finite-element computer program should be used for calculating the temperatures.

Best-fit method

ECCS (1985) gives the following approximate formula for predicting the time t in minutes for a steel member protected with light, dry insulation to reach a limiting temperature T_{lim} (°C) when exposed to the standard fire:

$$t = 40(T_{lim} - 140)\left[\frac{d_i/k_i}{F/V}\right]^{0.77} \tag{8.7}$$

where k_i is the thermal conductivity of the insulation (W/m K), and d_i is the thickness of the insulation (m).

This equation is valid in the range of t from 30 to 240 minutes, T_{lim} from 400°C to 600°C, F/V from 10 to 300 m^{-1} (V/F in the range 3.3 to 100 mm) and d_i/k_i from 0.1 to 0.3 m^{2}K/W. Lewis (2000) shows that the range of applicability can be extended to 800°C for common section sizes.

For insulation containing moisture, a time delay t_v in minutes can be added to the time t calculated from the above equation, using

$$t_v = m\, \rho_i\, d_i^2/(5\, k_i) \tag{8.8}$$

where ρ_i is the density of the insulation (kg/m^3), and m is the moisture content of the insulation (%). Again it should be noted that this is an empirical approximation. It is recommended that the step-by-step method or a finite-element method be used for design.

Step-by-step method

The iterative calculation method for protected steelwork is similar to that for unprotected steel. The equation is slightly different and does not require heat transfer coefficients because it is assumed that the external surface of the insulation is at the same temperature as the fire gases. It is also assumed that the internal surface of the insulation is at the same temperature as the steel. The equation is

$$\begin{aligned}\Delta T_s = {} & (F/V)(k_i/d_i\, \rho_s\, c_s) \\ & \{\rho_s\, c_s/(\rho_s\, c_s + (F/V) d_i\, \rho_i\, c_i/2)\} \\ & (T_f - T_s)\Delta t\end{aligned} \tag{8.9}$$

where c_i is the specific heat of the insulation (J/kg K).

The spreadsheet calculation is similar to that shown in Table 8.5 except that Equation (8.9) is used instead of Equation (8.6). EC3 (1995) suggests a time step of 30 s, but Gamble (1989) shows that much longer steps can be used. EC3 (1995) has a slightly different expression with a '3' rather than '2' in the heavy insulation term to allow for the temperature gradient across the insulating material, and an extra term accounting for the increase in the fire gas temperature in the time Δt.

If the insulation is of low mass and specific heat such that the heat capacity of the insulation will not significantly slow the temperature increase of the steel, then Equation (8.9) can be simplified by omitting the central term in { } brackets. ECCS (1985) suggests ignoring the heat capacity of the insulation if it is less than half of that of the steel section, such that

$$\rho_s\, c_s\, A/2 > \rho_i\, c_i\, A_i \tag{8.10}$$

where A_i is the cross-sectional area of the insulating material (m^2), and A is the cross sectional area of the steel (m^2).

The effect of the time delay for moist materials can be incorporated into the step-by-step calculation method by modifying the specific heat of the insulating material to include a local increase of specific heat at 100°C. Typical values of

Table 8.6 Thermal properties of insulation materials

Material	Density ρ_i (kg/m^3)	Thermal conductivity k_i (W/mK)	Specific heat c_i (J/kg K)	Equilibrium moisture content %
Sprays:				
Sprayed mineral fibre	300	0.12	1200	1
Perlite or vermiculite plaster	350	0.12	1200	15
High-density perlite or vermiculite plaster	550	0.12	1200	15
Boards:				
Fibre-silicate or fibre-calcium silicate	600	0.15	1200	3
Gypsum plaster	800	0.20	1700	20
Compressed fibre boards:				
Mineral wool, fibre silicate	150	0.20	1200	2

thermal properties of insulating materials are given in Table 8.6, from ECCS (1995).

8.4.7 Typical Steel Temperatures

Figure 8.7(a) shows steel temperatures for a beam with $F/V = 200\,m^{-1}$ ($V/F = 5$ mm) exposed to the standard fire (ISO (1975)), calculated by the step-by-step method using a spreadsheet. The top curve is the fire temperature and the second curve is the temperature of an unprotected steel beam. The lower two curves are protected with insulating material, using thicknesses of 15 and 50 mm.

Figure 8.7(b) shows steel temperatures for the same beam exposed to a parametric fire, also calculated using the step-by-step method. The top curve is the fire temperature and the second curve, following the fire closely, is the steel beam with no protection. The lower two curves are protected with insulating material, using thicknesses of 15 and 50 mm as before.

Temperatures of any steel members exposed to the standard fire can easily be obtained from Figures 8.8 and 8.9, with minimal calculation. Figure 8.8 gives the temperature of unprotected steel members after any time of exposure, as a function of the section factor F/V (m^{-1}).This can be used for any member for which the section factor is known. Figure 8.9 gives the temperature of protected steel members after any time of exposure, as a function of the modified section factor $(F/V)(k_i/d_i)$ (W/m^3K).This can be used for any member for which the section factor and the insulation are known.

8.4.8 Temperature Calculation for Composite Construction

This section describes the calculation of temperatures in composite structures. Structural design of composite structures is described in Chapter 9. Composite

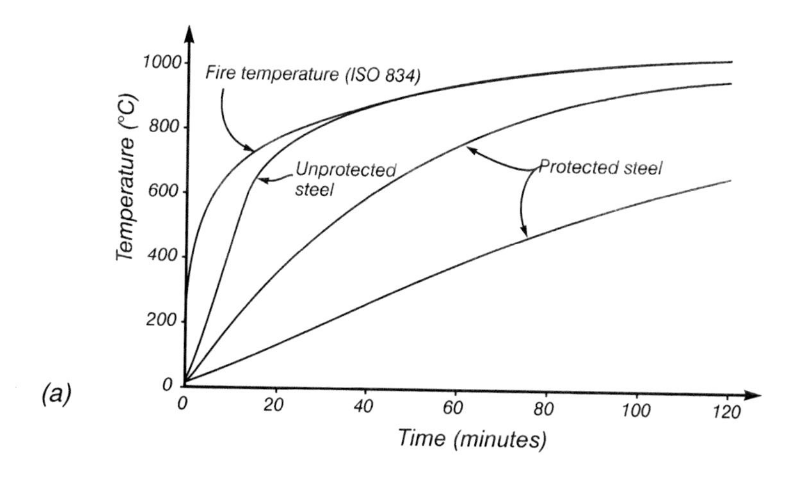

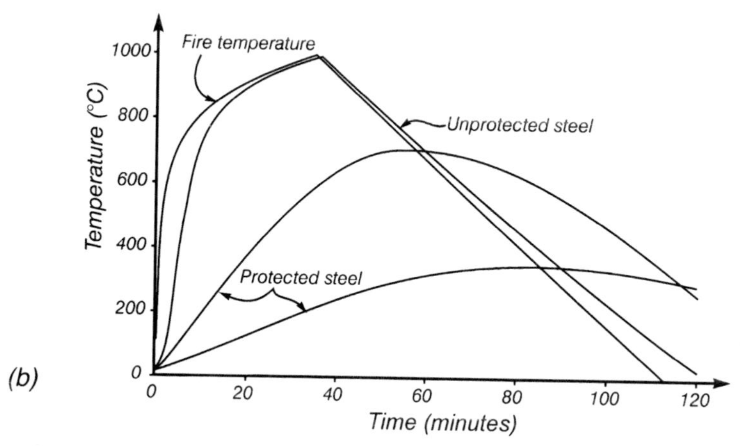

Figure 8.7 Typical steel temperatures for unprotected and protected steel beams exposed to (a) the standard fire, and (b) a parametric fire

construction refers to combined structural systems of steel and concrete, where both materials contribute to the load-bearing capacity. In many composite structures the steel member is partly or fully protected from direct fire exposure by concrete. The most common examples are where a concrete slab acts compositely with a steel deck or supporting beam as shown in Figure 8.10. Sometimes the steel member is partly or completely buried in the concrete as shown in Figure 8.11. The system with the beam completely buried in the concrete floor slab is often called 'slim-floor'.

If most of the steel beam is exposed to fire as shown in Figure 8.10, the equations above can be used to estimate the steel temperature. If a large part is buried in concrete as shown in Figure 8.11, then there will be large temperature gradients and the only accurate way to calculate the steel temperatures is to use a heat-transfer computer program. A rough calculation can be made using the step-by-

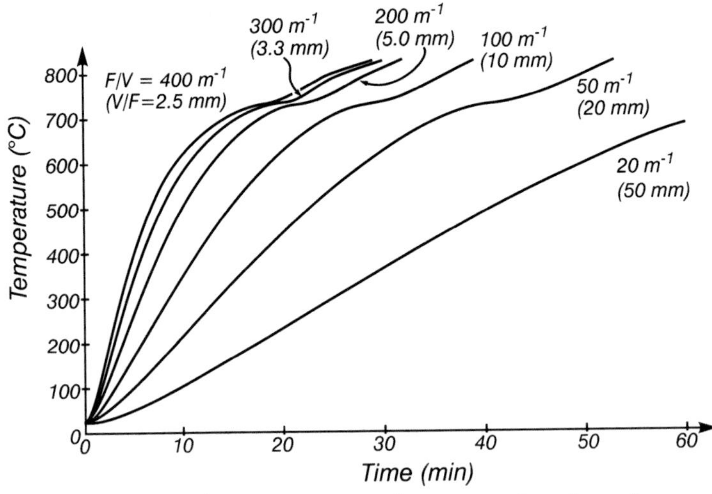

Figure 8.8 Chart for calculating temperatures of unprotected steel members exposed to the standard fire

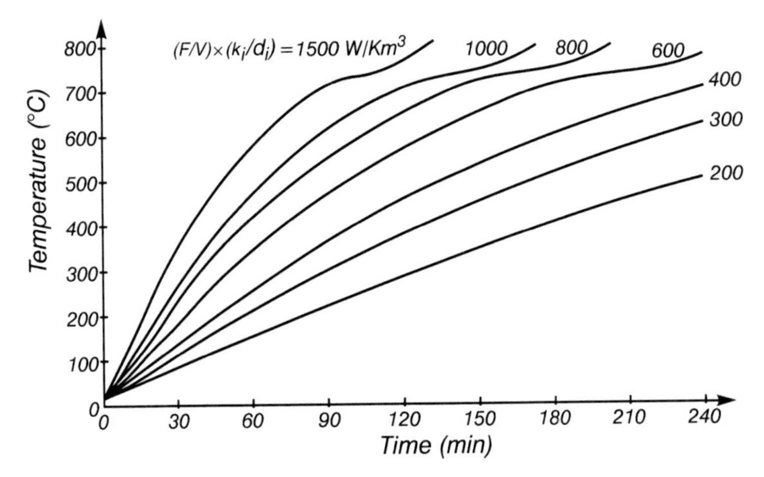

Figure 8.9 Chart for calculating temperatures of protected steel members exposed to the standard fire

step method, where the section factor should be obtained using the fire-exposed perimeter and the cross-sectional area of that portion of the steel section not buried in concrete.

For other combinations of concrete and steel. Figure 8.12 shows a steel column completely surrounded with reinforced concrete, Figure 8.13 shows a steel column with the spaces between the flanges filled with concrete, and Figure 8.14 shows a rectangular hollow steel section filled with concrete. In all these cases, the concrete may be either considered to be solely acting as fire protection to the steel, or it may be designed as a load-bearing material as part of a composite structure. In all three of these designs the concrete is increasing the thermal mass of the assembly, reducing the rate of increase of steel temperatures. In the first two designs the concrete is fully or partially reducing the area of steel surface exposed to the fire environment. A

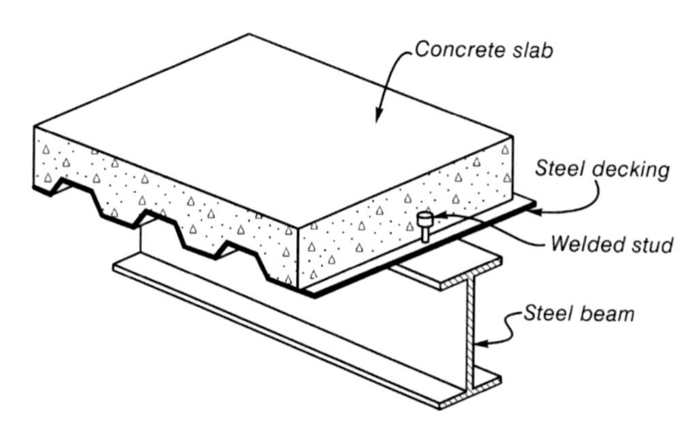

Figure 8.10 Composite construction with concrete slab on steel deck and steel beam

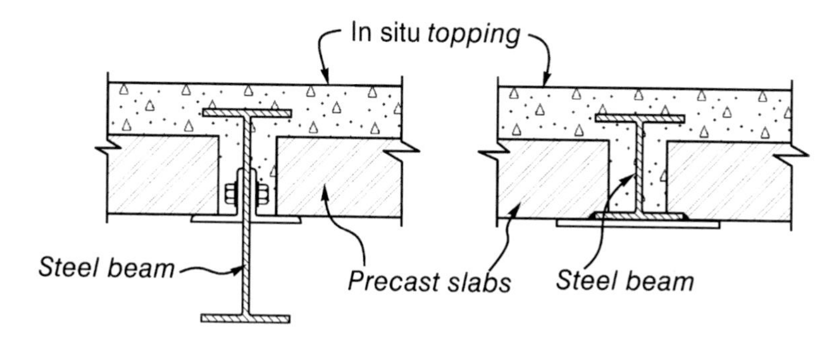

Figure 8.11 Composite construction with steel members protected by concrete

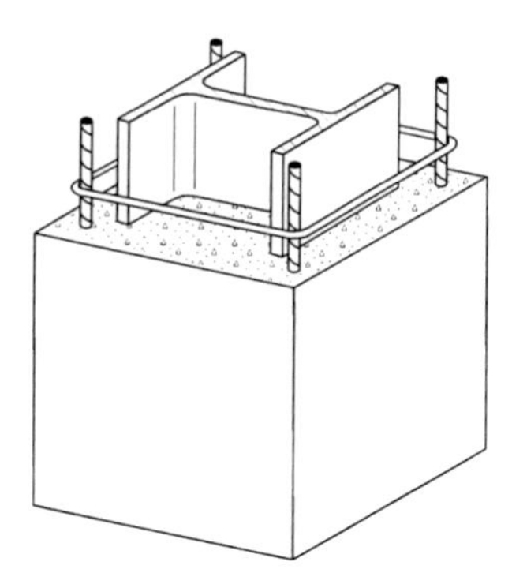

Figure 8.12 Steel column protected with concrete encasement

heat-transfer computer program is necessary if accurate predictions of temperature are to be made for these composite assemblies. Guidance is given in Eurocode 4 (EC4, 1994).

Light steel framing members are used in wall and floor assemblies where the steel is protected by the gypsum plasterboard linings as shown in Figure 11.4. There is limited proprietary information available from wallboard manufacturers, but accurate prediction of temperatures requires a heat transfer computer program. Fire resistance of light steel framing construction is described in Chapter 11.

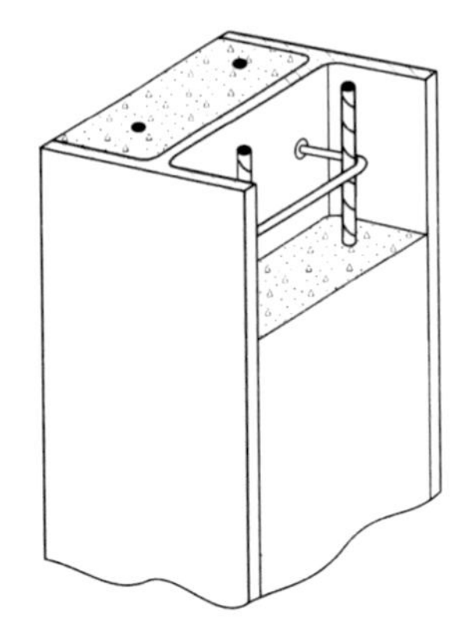

Figure 8.13 Steel column protected with concrete between the flanges

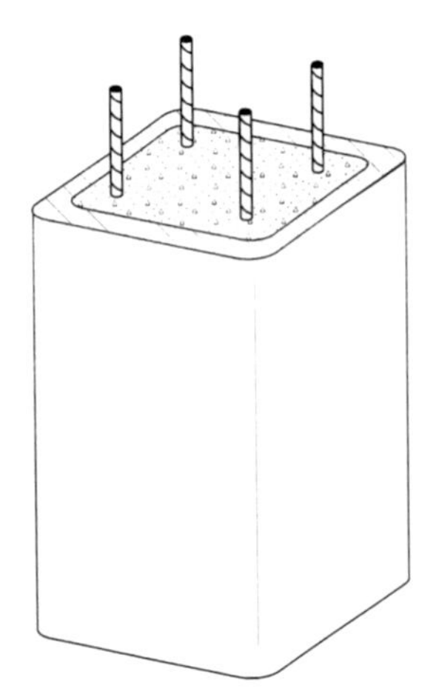

Figure 8.14 Concrete filled hollow steel column

8.4.9 Temperature Calculation for External Steelwork

Steel beams and columns outside a fire compartment may be subjected to elevated temperatures as a result of radiation from the window opening, radiation from flames, or engulfment in flames. Methods for estimating the temperature of such exposed steel members have been developed by Law and O'Brien (1989) and are incorporated into the Eurocodes. Flame sizes, temperatures and heat transfer coefficients are given in Eurocode 1 (EC1, 1994) and methods for calculating the steel temperatures are given in Eurocode 3 (EC3, 1995). The design method allows for conditions with or without a wind creating a forced draught to influence the shape of the idealized flame.

Typical flame shapes and radiation geometries are shown in Figure 8.15 for two conditions of forced draught and no forced draught, which produce different flame shapes. The design documents show many additional shapes for conditions with cross winds, flame deflectors and other variations. Figure 8.15 shows three possible column locations which require different designs. Columns at locations A and C are

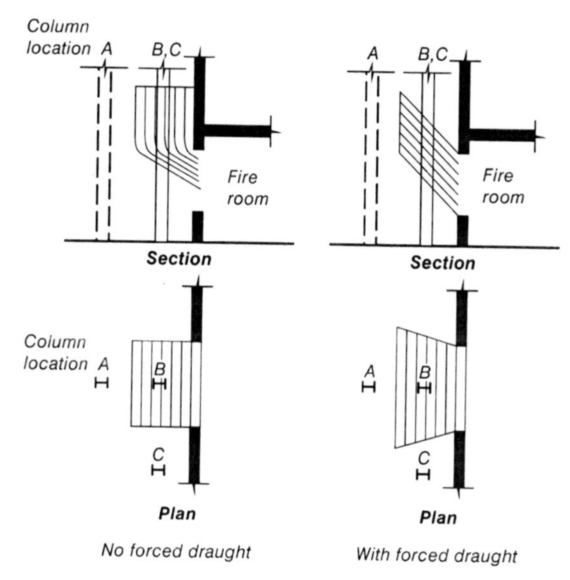

Figure 8.15 Fire exposure of external steel columns. (Reproduced from (EC3, 1995) by permission of CEN)

exposed to radiation from the flame itself and also from the window opening behind the flame, but column C has less severe exposure. The column at location B is engulfed in the flame. The documents mentioned above give design equations for all of these situations. Computer programs are available for making these calculations (IISI 1993).

8.5 PROTECTION SYSTEMS

There are many alternative passive fire protection systems available to reduce the rate of temperature increase in steel structures exposed to fire. Even if fire resistance is assessed by calculation, each type of fire protection system should have at least one fire-resistance rating obtained from full-scale load-bearing testing, to demonstrate that the insulating material has sufficient 'stickability' to remain in place for the duration of the expected fire. The toughness of passive fire protection often depends on the quality of the building materials and workmanship.

8.5.1 Concrete Encasement

A traditional method for fire protection of steelwork is encasement in poured concrete, as shown in Figure 8.12. An advantage of this system is excellent durability in corrosive environments. The required thickness of concrete to achieve standard fire-resistance ratings is given in prescriptive building codes. The reinforcing in the concrete may be nominal reinforcing simply to hold the concrete in place in the event of a fire, or it may be substantial in which case the member will be designed for composite behaviour of all three materials. This form of construction is very

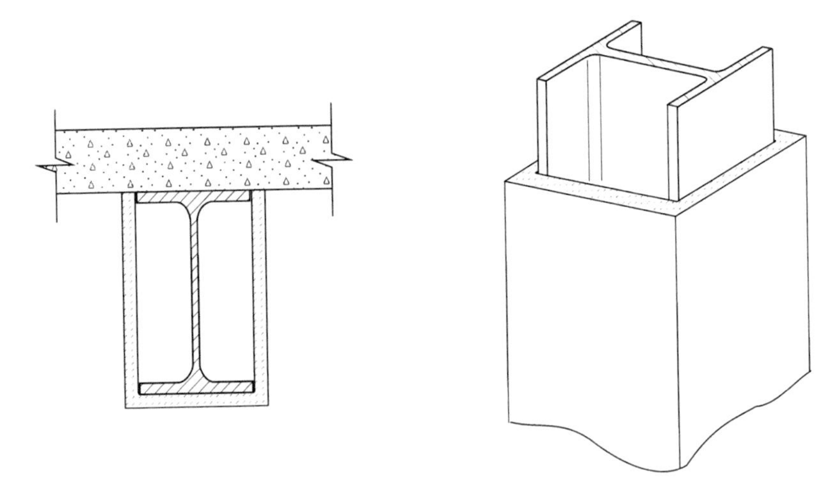

Figure 8.16 Steel beam and column protected with board materials

common in Japan where it is called *steel-reinforced-concrete*. Elsewhere, concrete encasement is not widely used because it is expensive, bulky, and time-consuming, requiring the combined cost of a steel frame plus boxing for all of the concrete.

8.5.2 Board Systems

There are many proprietary board systems for protecting structural steelwork as shown in Figure 8.16. Most of the boards are manufactured from calcium silicate or gypsum plaster. Calcium silicate board is more expensive than gypsum board in many places because it is imported from manufacturers in only a few countries. Calcium silicate boards are made of an inert material that is designed to remain in place without damage for the duration of the fire, protecting the steel by its insulating properties. Gypsum board also has good insulating properties, and its behaviour is enhanced by the water of crystallization which is driven off as the board is heated. This dehydration process gives an additional time delay at about 100°C, but it reduces the strength of the residual board after fire exposure, as described in Section 11.5.

Board systems have the advantages that they are easy to install in a dry process, and easy to finish with decorative materials. Board systems are more often used for columns than for beams because columns are more often visible in the finished building, but they are slower and more expensive than spray-on materials. The boards are usually glued or screwed to metal or wood framing which has been fastened to the steel member. The number and thickness of layers can be easily adapted to the particular application. Possible fixing arrangements are shown in Figures 8.17 and 8.18. Calcium silicate boards can be fixed to additional pieces of board wedged between the flanges of steel I-beams. Empirical formulae for calculating the required thickness of gypsum board to achieve standard fire resistance ratings are given by Milke (1995) and the Uniform Building Code (UBC, 1997).

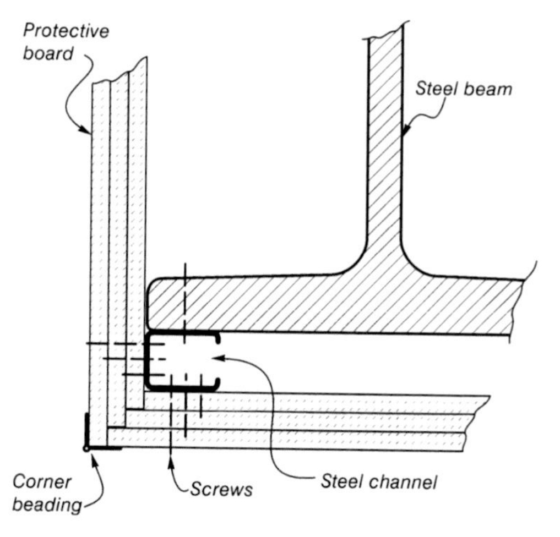

Figure 8.17 Detail of steel beam protected with board materials. (Reproduced from Milke (1995) by permission of SFPE)

Figure 8.18 Box protection being placed on a steel column using sheet material

8.5.3 Spray-on Systems

Spray-on proprietary protection as shown in Figure 8.3 is usually the cheapest form of passive fire protection for steel members. Spray-on materials are usually cement-based with some form of glass or cellulosic fibrous reinforcing to hold the material together. Earlier spray-on materials used asbestos fibres which are no longer used for health reasons. Disadvantages of spray-on protection are that the process is wet and messy, and the resulting finish is not suitable for decorative finishes. The spray-on material is often rather soft, so that it has to be protected from damage if it is in a vulnerable location. For these reasons, spray-on materials are more often used for beams than for columns (Figure 8.19). Spray-on protection is easy to apply to complicated details such as bolted connections or steel brackets. Approved spray-on systems must have proof that they have sufficient 'stickability' to remain in place during fire exposure. Test methods are available for testing the cohesion and adhesion of spray-on fire protection (UBC, 1997). The required thickness of proprietary spray-on fire protection to achieve fire-resistance ratings can be found in individual manufacturers' literature or trade publications (e.g. ASFPCM, 1988). Some generic ratings are available for spray-on systems (e.g. NBCC, 1995) but proprietary ratings from individual manufacturers are more likely to be used.

Figure 8.19 Sprayed-on fire protection to steel beams supporting precast concrete floor slabs

8.5.4 Intumescent Paint

Intumescent paint is a special paint material that swells up into a thick charry mass when it is heated. The intumescent material provides insulation to the steel member beneath. Several coats of paint may have to be applied to obtain the necessary thickness. Intumescent paints have the advantages that they do not take up much space, they can be applied quickly, and they allow the structural steel members to be seen directly, without any covering other than the paint. A disadvantage is the high cost compared to board and spray-on materials, especially for longer duration fire-resistance ratings. Many intumescent paints are not suitable for external use because of unknown durability. All intumescent paints are proprietary products, and many are under continual development. A minor disadvantage of intumescent coatings is that the protection is not obvious to casual observers, and it can be difficult to verify at a later date. Some specialist intumescent products incorporating multiple layers of fibre-glass reinforcing have been developed for high-level protection of structural steel in the offshore oil industry.

8.5.5 Protection with Timber

It is possible to provide fire protection to steel beams and columns with timber boards. Twilt and Witteveen (1974) describe fire tests and fixing details for fire exposed steel columns. Using a conservative critical steel temperature of 200°C they show that 35 mm thick softwood boards can provide 60 minutes fire resistance to a steel column with F/V 100 m^{-1}. It is essential that the timber completely encloses the steel member, and is firmly fixed in place with a thermosetting adhesive such as resorcinol. The wood must be well seasoned to prevent shrinkage cracks.

8.5.6 Concrete Filling

Hollow steel sections can be filled with concrete as shown in Figure 8.14 to improve the fire performance. A major advantage is the lack of bulky external protection, and the steel can be finished with normal paint. There are several structural possibilities. The filling concrete can either be considered simply as a heat sink to reduce the temperature increase, or as a structural material which can carry an increasing proportion of the load as the steel temperatures increase. The filling concrete can be plain concrete, or it can be reinforced with conventional bars or with fibres. The steel tube can provide excellent structural confinement to the concrete under non-fire conditions, for example during seismic loading. It is essential to provide vent holes to prevent excessive steam pressure from exploding the hollow member during heating. A variation on this theme is to fill the two spaces between the flanges of a steel I-beam with concrete as shown in Figure 8.13, with reinforcing to hold the concrete in place. The reinforcing must be welded to the web as shown and not be welded between the flanges. Structural design of concrete filled columns exposed to fire is covered in Section 9.7.3.

Figure 8.20 External steel frame building with fire resistance achieved by filling the hollow steel members with water [Hannover, Germany] (Reproduced from IISI (1993) by permission of International Iron and Steel Institute, Brussels)

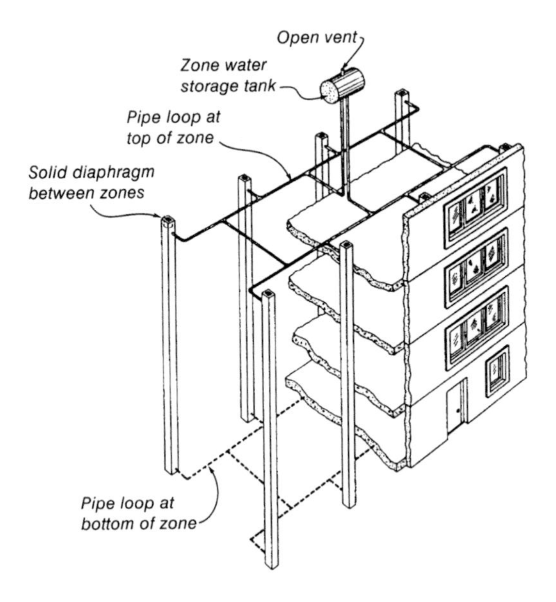

Figure 8.21 Schematic sketch of water filled column fire protection system (Reproduced from Seigel (1970) by permission of National Fire Protection Association)

8.5.7 Water Filling

A less common but effective way of preventing rapid heating of hollow steel sections is to fill them with water (Figure 8.20). A plumbing system is necessary to ensure that the water can flow by convection from member to member and to avoid excessive pressures when the water is heated. A schematic of a possible system is shown in Figure 8.21 (Seigel, 1970).This will require imaginative detailing of the connections between individual elements. Additives may be necessary to prevent corrosion, and to prevent freezing in cold climates. This method of protection is expensive and is only used for special structures. Design information is given by Bond (1975).

8.5.8 Flame Shields

In some situations it is possible to use flame shields to protect external structural steelwork from radiation or direct impingement by flames coming out of window openings. In these cases the temperatures of the steel exposed to flame contact or radiation can be calculated using the methods referred to above for external steelwork. An example of a flame shield protecting the flanges of a deep steel beam in a 54-storey building in New York is shown in Figure 8.22 (Seigel, 1970).

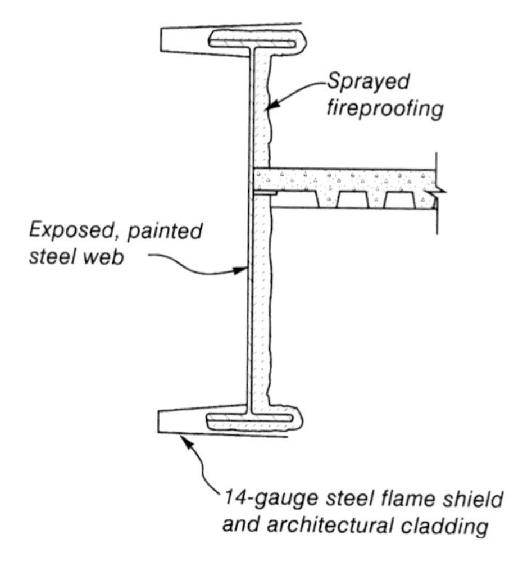

Figure 8.22 Flame protection of exterior steel beam (Reproduced from Seigel (1970) by permission of National Fire Protection Association)

8.6 MECHANICAL PROPERTIES OF STEEL AT ELEVATED TEMPERATURE

This section reviews the effects of the mechanical properties of the steel on the behaviour of steel structures in fire.

8.6.1 Components of Strain

The deformation of steel at elevated temperatures is usually described by assuming that the change in strain $\Delta\varepsilon$ consists of three components, as described in Section 7.4.2.

$$\Delta\varepsilon = \varepsilon - \varepsilon_i = \varepsilon_{th}(T) + \varepsilon_\sigma(\sigma,T) + \varepsilon_{cr}(\sigma,T,t) \tag{8.11}$$

where ε is the total strain at time t, ε_i is the initial strain at time $t=0$, $\varepsilon_{th}(T)$ is the thermal strain being a function only of temperature, T, $\varepsilon_\sigma(\sigma,T)$ is the stress-related strain, being a function of both the applied stress, σ, and the temperature T, and $\varepsilon_{cr}(\sigma,T,t)$ is the creep strain, being a function of stress, temperature and time.

These three components of strain are discussed in more detail below. For simple structural members such as simply supported beams, only the stress-related strain needs to be considered, allowing the reduced strength at elevated temperatures to be calculated without reference to the deformations. For more complex structural systems, especially where members are restrained by other parts of the structure, the thermal strain and the creep strain must also be considered, using a computer model for the structural analysis.

8.6.2 Thermal Strain

The thermal strain is the well-known thermal expansion that occurs when most materials are heated. Anderberg (1988) reports four studies which obtained very similar linear relationships for the thermal expansion of steel. At room temperatures, the coefficient of thermal expansion is usually taken to be 11.7×10^{-6}/°C. At higher temperatures such as those experienced in fires, the coefficient increases, and a discontinuity occurs between 700 and 800°C. On the basis of extensive testing, Poh (1996) has proposed an equation which includes all of these effects. For normal design purposes, Eurocode 3 (EC3, 1995) recommends a linear coefficient of 14.0×10^{-6}/°C, hence the thermal elongation of steel $\Delta L/L$ can be approximated by a linear function of temperature T (°C) given by

$$\Delta L/L = 14 \times 10^{-6}(T - 20) \tag{8.12}$$

For the design of simple members such as single beams and columns it is not usually necessary to calculate and include the effects of thermal strains. If thermal restraint forces develop in beams they are usually beneficial to the fire performance, although they may increase the axial load in columns.

8.6.3 Creep Strain

Creep is relatively insignificant in structural steel at normal temperatures. However, it becomes very significant at temperatures over 400 or 500°C. Poh (1996) has carried out many experiments on the creep behaviour of steel at elevated temperatures. Figure 8.23 shows the results of transient tests (regime (6) in Figure 7.6) by Kirby

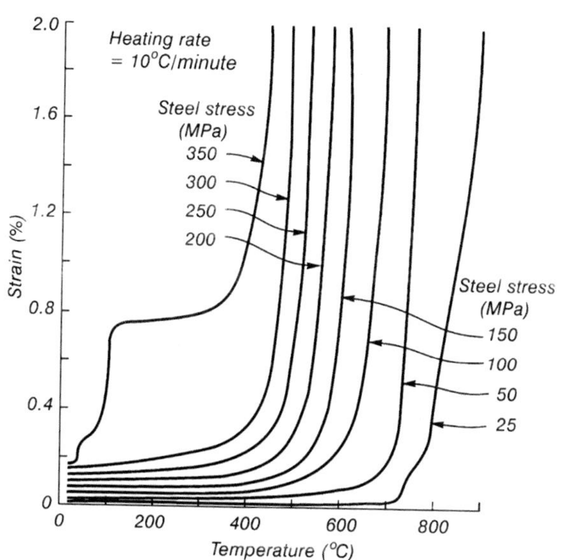

Figure 8.23 Creep of steel tested in tension (Reproduced from Kirby and Preston (1988) with permission from Elsevier Science)

and Preston (1988) where it can be seen that the creep is highly dependent on temperature and stress level. The creep deformations accelerate rapidly where the creep strain curve becomes nearly vertical.

Despite the great importance of creep deformations in fire exposed steel structures which are approaching their collapse loads, creep is not usually included explicitly in the computer-based fire design process because of lack of data and the difficulty of the calculations. The usual assumption is that the stress–strain relationships used for design are 'effective' relationships which implicitly include the likely deformations due to creep during the time of exposure to the fire (EC3, 1995). On the other hand, Anderberg (1986), Srpcic (1995) and Poh and Bennetts (1995) have shown how creep deformations can be explicitly included in a computer model, finding that the effects of creep and the nature of the strain hardening of the material can have a significant influence on predicted behaviour.

8.6.4 Stress-related Strain

Stress–strain relationships at elevated temperatures can be obtained directly from steady-state tests at certain elevated temperatures or they can be derived from the results of transient tests such as those shown in Figure 8.23. Typical stress-strain relationships for structural steel at elevated temperatures are shown in Figure 8.24, where it can be seen that yield strength and modulus of elasticity both decrease with increasing temperature, but the ultimate tensile strength increases slightly at moderate temperatures before decreasing at higher temperatures. Similar curves for a cold-drawn prestressing steel are shown in Figure 8.25, where there is a less well-defined yield point and slightly different behaviour at elevated temperatures.

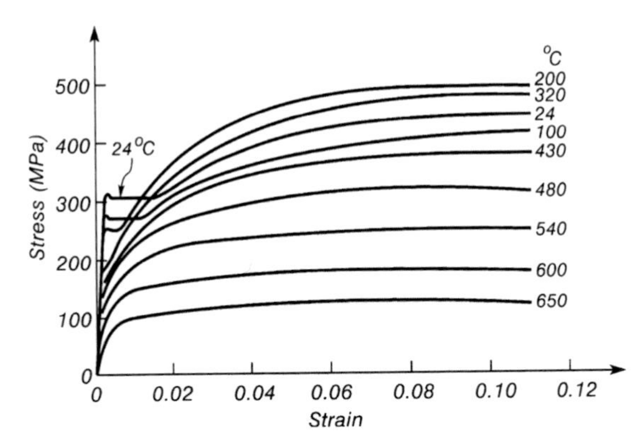

Figure 8.24 Stress–strain curves for typical-hot rolled steel at elevated temperature (Reproduced from (Harmathy, 1993) by permission of Pearson Education Ltd)

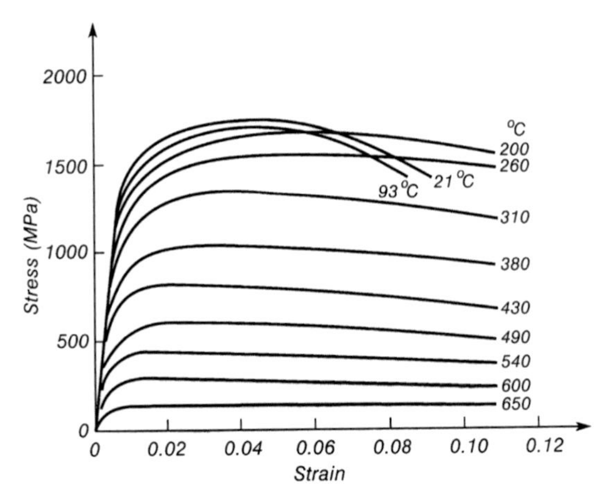

Figure 8.25 Stress–strain curves for prestressing steel at elevated temperature (Reproduced from (Harmathy, 1993) by permission of Pearson Education Ltd)

Proof strength and yield strength

Design of structural steel members at normal temperatures requires knowledge of the yield strength of the steel. Most normal construction steels have a very well-defined yield strength at normal temperatures, but this disappears at elevated temperatures, as shown in Figure 8.24. Figure 8.26 is a sketch of stress–strain relationships for a typical steel, showing a well-defined yield strength at normal temperatures and a much softer curve at elevated temperatures. A value of yield strength is required for design at elevated temperatures. Kirby and Preston (1988) recommend using the 1% proof strength as the effective yield strength in fire engineering calculations. In Figure 8.26 the line AB has been constructed so that it passes through 1% strain on the *x*-axis and is parallel to the linear elastic portion of the 400°C curve. The vertical value of point B is defined as the 1% proof strain. This could be done for any level of proof strain at any steel temperature.

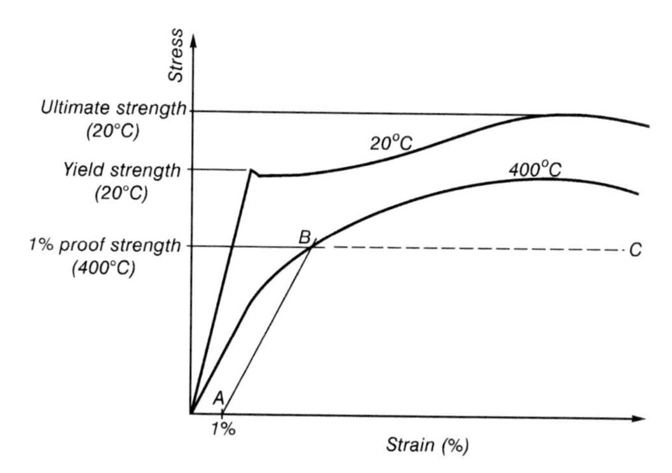

Figure 8.26 Stress–strain curves for steel illustrating yield strength and proof strength

Design values

Test reports of steel properties at elevated temperatures show considerable scatter. Harmathy (1993) reviewed a large amount of literature, resulting in Figure 8.27 and Figure 8.28 which show the scatter in published data for hot-rolled steel and cold-worked steel, respectively. Some of this scatter may be due to lack of a clear

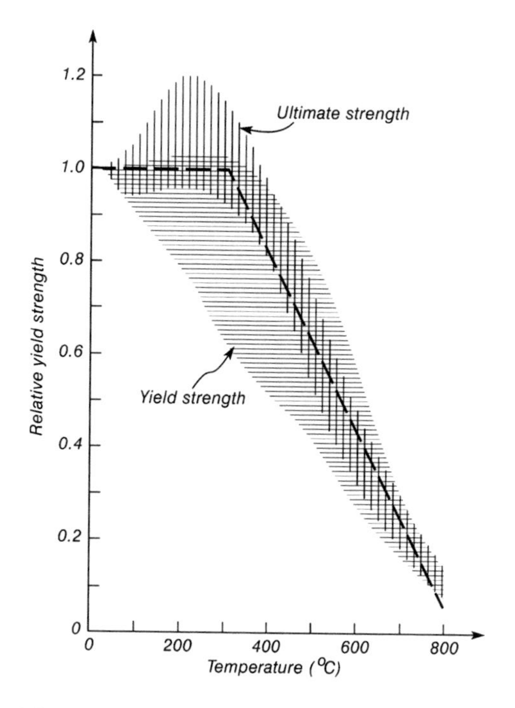

Figure 8.27 Scatter in published results of hot-rolled steel (Reproduced from (Harmathy, 1993) by permission of Pearson Education Ltd)

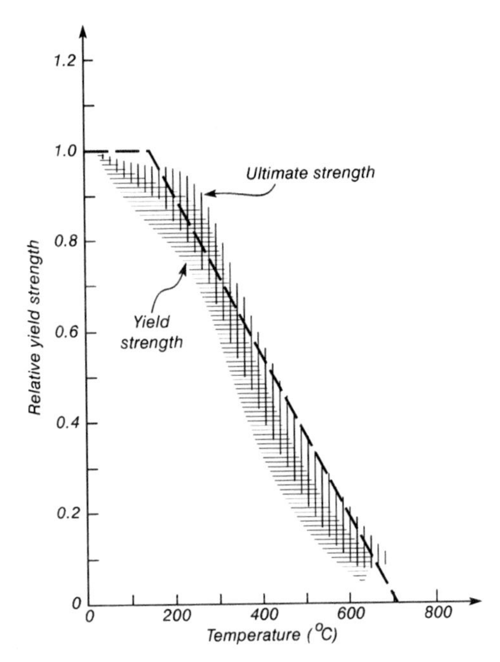

Figure 8.28 Scatter in published results of cold-worked steel (Reproduced from (Harmathy, 1993) by permission of Pearson Education Ltd)

definition of yield strength, as discussed above. The dotted straight lines in these figures show suggested values for design (ISE, 1978).

For design purposes, many national codes have proposed slightly different approximations to the published test data. Typical relationships are shown in Figure 8.29, where the line for structural steel is from AS 4100 and NZS 3404, and the lines for reinforcing steel and prestressing steel are from BS 8110, AS 3600 and NZS 3101. The equations of the lines (below $k_{y,T} = 1.0$) are

$$\begin{aligned} k_{y,T} &= (905 - T)/690 && \text{structural steel} \\ k_{y,T} &= (720 - T)/470 && \text{reinforcing steel} \\ k_{y,T} &= (700 - T)/550 && \text{prestressing steel} \end{aligned} \tag{8.13}$$

where $k_{y,T}$ = is the ratio of $f_{y,T}$ (the yield strength at elevated temperature) to f_y (the yield strength at 20°C).

The relationships in Equation (8.13) can be reversed to give the limiting temperature for a given load ratio r_{load}. The limiting temperature is that at which an individual steel member is expected to fail, assuming no load sharing or redundant behaviour. The limiting temperatures are given by

$$\begin{aligned} T_{lim} &= 905 - 690\, r_{load} && \text{structural steel} \\ T_{lim} &= 720 - 470\, r_{load} && \text{reinforcing steel} \\ T_{lim} &= 700 - 550\, r_{load} && \text{prestressing steel} \end{aligned} \tag{8.14}$$

Similar curves from Eurocode 3 are shown in Figure 8.30. The reduction in yield strength, proportional limit and modulus of elasticity are defined by a number of

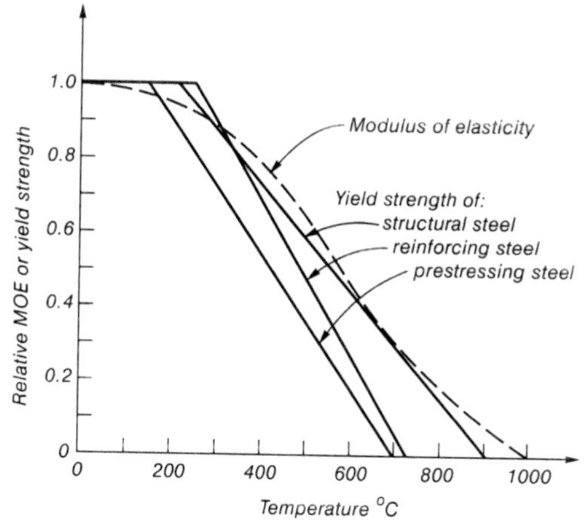

Figure 8.29 Design curves for reduction in yield strength and modulus of elasticity of steel with temperature

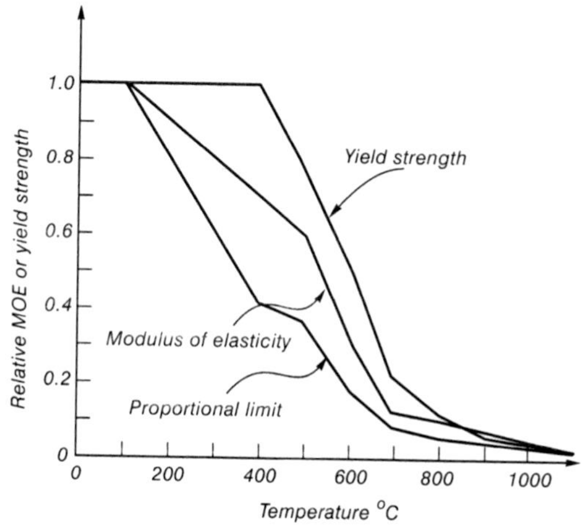

Figure 8.30 Reduction in yield strength and modulus of elasticity with temperature (Reproduced from (EC3, 1995) with permission of CEN)

points. Eurocode 3 gives an expression for critical temperature, from which an approximate curve for the reduction in yield strength is given by:

$$k_{y,T} = [0.9674(1 + \exp[(T - 482)/39.19])]^{-1/3.833} \tag{8.15}$$

The structural Eurocodes for steel (EC3, 1995) and concrete (EC2, 1993) have more detailed expressions, with equations for the stress–strain relationship of various steels, both with and without strain hardening included. These have not been quoted here in the interests of providing a consistently simple document but they can be consulted as necessary. It is interesting to note that the curves used for reduction in yield strength in various countries are quite different, as seen by a comparison of Figures 8.29 and 8.30, even though the materials are very similar. This may be more to do with different definitions of yield strength than differences in materials.

Modulus of elasticity

The modulus of elasticity is needed for buckling calculations. The modulus of elasticity would also be required for elastic deflection calculations, but these are rarely attempted under fire conditions because elevated temperatures lead rapidly to plastic deformations. The reduction in modulus of elasticity shows the same trend as the reduction in yield strength. There can be obvious numerical difficulties if both properties do not reach zero at the same temperature (as shown in Figure 8.29). Minor changes such as shown by the dotted line in Figure 9.16 have been proposed to overcome this problem. The Eurocode 3 reduction in modulus of elasticity with temperature is shown in Figure 8.30, with some nominal strength and stiffness up to 1200°C. In AS 4100 and NZS 3404 the relationship for modulus of elasticity is given by the curve shown in Figure 8.29, with equation given by:

$$\begin{aligned} k_{E,T} &= 1.0 + T/[2000\ln(T/1100)] && 0 < T \leqslant 600^{\circ}\mathrm{C} \\ &= 690(1 - T/1000)/(T - 53.5) && 600 < T \leqslant 1000^{\circ}\mathrm{C} \end{aligned} \qquad (8.16)$$

where $k_{E,T}$ = is the ratio of E_T (the modulus of elasticity at elevated temperature) to E (the modulus of elasticity at 20°C)

Residual stresses

Residual stresses are internal stresses which exist in unloaded steel members, resulting from the hot-rolled manufacturing process. Residual stresses do not usually have a significant effect on the ultimate load capacity of steel structures at normal temperatures, and even less during fire exposure. Sophisticated computer-based structural analysis models used for fire design of large structures can include the effects of residual stresses, and can accurately assess their effects on structural performance if necessary.

8.7 DESIGN OF STEEL MEMBERS EXPOSED TO FIRE

8.7.1 Design Methods

There are two main methods for structural design of steel structures exposed to fire; the *simplified method* for single elements and the *general method* for restrained members, more complex assemblies, or large frames (EC3, 1995). The simplified method is described in this section.

Verification

As for members of any material, verification in the strength domain requires that

$$U^*_{\mathrm{fire}} \leqslant R_{\mathrm{fire}} \qquad (8.17)$$

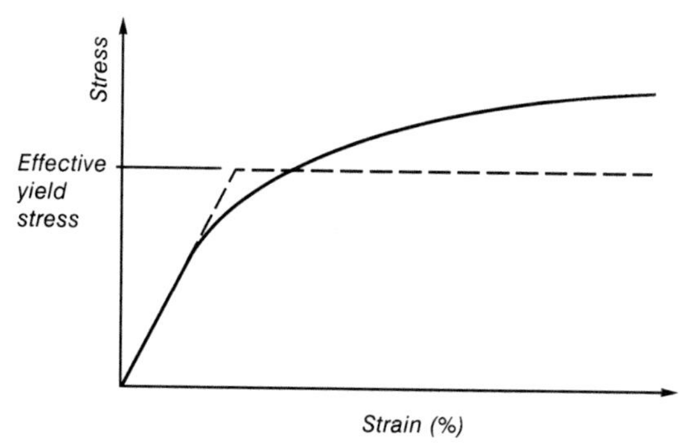

Figure 8.31 Stress–strain curve with elasto-plastic approximation

where U^*_{fire} is the design force resulting from the applied load at the time of the fire, and R_{fire} is the load-bearing capacity in the fire situation.

Applied loads have been described in Chapter 7. Design forces are obtained from the applied loads by conventional structural analysis. Calculations of the load capacity are described below, based on the mechanical properties of steel at elevated temperatures. The design force U^*_{fire} may be axial force N^*_{fire}, bending moment M^*_{fire}, or shear force V^*_{fire} occurring singly or in combination, with the load capacity calculated accordingly as axial force N_f, bending moment M_f or shear force V_f in the same combination.

Note that Equation (8.17) does not include a partial safety factor for mechanical properties γ_M (or a strength reduction factor Φ) because both have a value of 1.0 in fire conditions, as described in Chapter 7.

The recommendations for member design presented below are based on the simplified method of Eurocode 3 (EC3, 1995) presented in a form which allows adaptation to national steel design codes in any country. The simplified method follows the ultimate strength design method as for normal temperatures, except that there are reduced loads for the fire condition and reduced values of modulus of elasticity and yield strength of steel at elevated temperatures. The effects of restraint caused by thermal deformations are not included.

Flexural continuity can be included by ensuring that the collapse mechanism, formed after plastic hinges occur, has sufficient strength to resist the applied loads. Design of steel structures is based on the assumption that steel is ductile, with a long flat yield plateau, so that under fire conditions the stress–strain relationship follows the dotted line in Figure 8.31, rather the actual solid curve. Such design will be conservative if the actual curve rises much above the assumed straight line, depending on the criteria for deriving the 'effective yield strength' from the actual stress–strain relationship (Figure 8.26).

Structural design at normal temperatures requires prevention of collapse (meeting the strength limit state) and preventing excessive deformations (the serviceability limit state). Much of the effort in the normal temperature design process is to ensure that excessive deformations do not occur. Design for fire resistance is mainly concerned with preventing collapse. Large deformations are expected under severe fire exposure, so they are not normally calculated unless they are going to affect the structural performance.

8.7.2 Design of Individual Members

Tension members

Single tension members are relatively simple elements to design because there is no possibility of buckling and the stresses are often uniform over the cross section. The design equation is

$$N^*_{\mathrm{fire}} \leqslant N_{\mathrm{f}} \tag{8.18}$$

Design for fire depends on whether the temperature is uniform over the cross section. If the temperature is uniform, the tensile load-bearing capacity is obtained from

$$N_{\mathrm{f}} = A\, k_{\mathrm{y,T}}\, f_{\mathrm{y}} \tag{8.19}$$

where A is the area of the cross section (mm^2), $k_{\mathrm{y,T}}$ is the reduction factor for yield strength of the steel at temperature T, and f_{y} is the yield strength of the steel at 20°C (MPa).

In the unlikely event that there is a temperature gradient over the cross section, the strength of the member can be obtained by summing the contributions of the respective parts, considering the temperature-reduced yield strength of each part. This equation, and others to follow, is based on the assumption that steel is a ductile material, so that sufficient elongation can occur for each elemental area to develop its yield strength. The equation is

$$N_{\mathrm{f}} = \sum_{i=1,n} A_i\, k_{\mathrm{y,Ti}}\, f_{\mathrm{y}} \tag{8.20}$$

where A_i is an elemental area of the cross section with a temperature T_i, and $k_{\mathrm{y,Ti}}$ is the reduction factor for yield strength of the steel at temperature T_i.

If there is a temperature gradient over the cross section it is conservative to assume that the whole of the cross section is at the maximum temperature. Figure 8.32 shows the distribution of internal forces at ultimate load for an idealized rectangular steel tension member with a uniform and non-uniform temperature gradient.

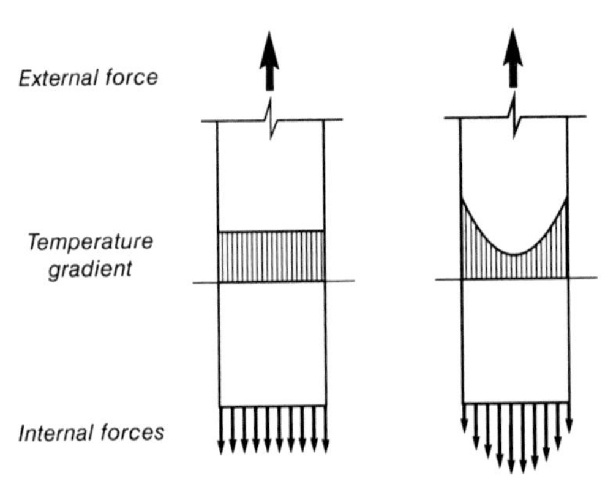

Figure 8.32 Internal forces in a steel tension member

Simply supported beams

The design equation for flexure is given by

$$M^*_{\text{fire}} \leqslant M_{\text{f}} \tag{8.21}$$

As with tension members, the strength of bending members in fire depends on whether the temperature is uniform over the cross section. An additional consideration for beams is the susceptibility of the cross section to local buckling. If the temperature is uniform, the design load-bearing capacity is obtained from

$$\begin{aligned} M_{\text{f}} &= S\,k_{\text{y,T}}\,f_{\text{y}} \quad \text{(plastic design), or} \\ M_{\text{f}} &= Z\,k_{\text{y,T}}\,f_{\text{y}} \quad \text{(elastic design)} \end{aligned} \tag{8.22}$$

where S is the plastic section modulus (mm^3), Z is the elastic section modulus (mm^3), $k_{\text{y,T}}$ is the reduction factor for yield strength of the steel at temperature T, and f_{y} is the yield strength of the steel at 20°C (MPa).

The decision whether to use the elastic or plastic design equation depends on the compactness of the selected cross section. Figure 8.33 shows a plot of mid-span moment *versus* deflection for a simply supported steel beam (Kulak *et al.*, 1995), showing how excellent plastic behaviour can be achieved for compact sections, but not for others.

The equation for plastic design applies if the shape of the steel section is such that full plastic moment can be achieved without local buckling occurring (Class 1 or Class 2 section in Canada or Eurocodes, or 'compact' section in Australian codes). The equation for elastic design should be used for steel sections where only the elastic moment can be achieved without local buckling occurring (Class 3 or 'non-compact' section). For light cold-rolled sections vulnerable to local buckling (Class 4), a simple design approach is to ensure that the steel temperature does not exceed 350°C or 400°C as described in Chapter 11.

If there is a temperature gradient over the cross section, there are several options for design. The most accurate method is to calculate the temperature of each part, so that the strength of the member can be obtained by summing the contributions of the

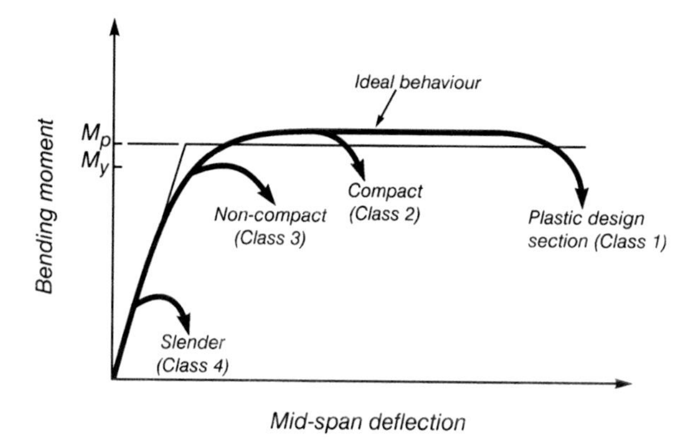

Figure 8.33 Moment–deflection relationship for a steel beam

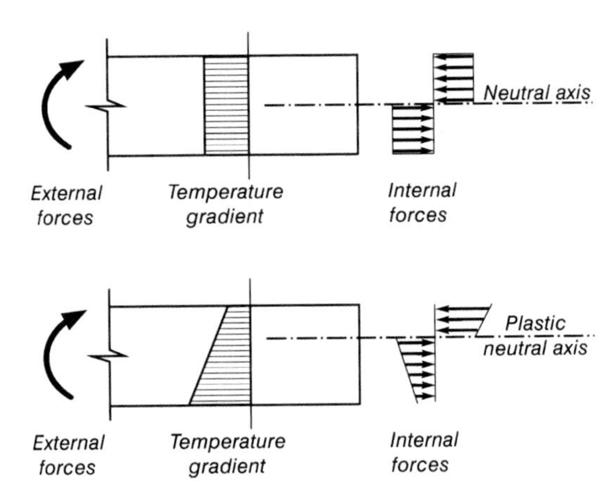

Figure 8.34 Internal forces in a steel flexural member

respective parts, considering the temperature-reduced yield strength of each part, to give

$$M_{\mathrm{f}} = \sum_{i=1,n} A_i z_i k_{\mathrm{y,Ti}} f_{\mathrm{y}} \tag{8.23}$$

where z_i is the distance from the plastic neutral axis to the centroid of the elemental area, and A_i and the other terms are as defined above.

The plastic neutral axis of a flexural section with a non-uniform temperature distribution is the axis perpendicular to the plane of bending such that the elemental areas yielding in tension and compression on either side of the axis are in equilibrium. The axis is then located such that

$$\sum_{i=1,n} A_i k_{\mathrm{y,Ti}} f_{\mathrm{y}} = 0 \tag{8.24}$$

Figure 8.34 shows the distribution of internal forces in a simple rectangular flexural member with a uniform and non-uniform temperature gradient.

If the temperature gradient over the cross section is known, it is conservative to assume that the whole of the cross section is at the maximum temperature. If the top surface of the beam is protected from fire by a concrete slab, Eurocode 3 (EC3, 1995) allows the strength calculated using the maximum temperature to be increased by a factor 1/0.7. This increase does not apply if the temperature of the steel is calculated using a lumped-mass method.

If the beam is statically indeterminate, with continuity at the supports, Eurocode 3 (EC3, 1995) allows the calculated strength to be increased by another empirical factor to allow for the steel temperature being lower in the support region than in the span of the beam, resulting from heat conduction into the columns or other supports. The factor is 1/0.85 for four-sided exposure or 1/0.60 if the top surface of the beam is protected from fire by a concrete slab.

Lateral torsional buckling Lateral torsional buckling must be considered for beams. Slender beams with no lateral restraint to the compression edge can fail by buckling before the flexural capacity of the cross section is reached. Lateral torsional buckling does not occur if the compression edge is restrained against lateral movement, or if the cross section is reasonably compact and the slenderness is not too large. In typical design of steel beams, buckling is allowed for with a strength reduction factor, or buckling factor, which reduces the design strength by an amount depending on the unrestrained length of the beam and the compactness of the cross section.

The Eurocode provisions (EC3, 1995) permit buckling to be ignored for well restrained fire-exposed beams of Class 1 or Class 2 cross sections. For beams with larger distances between locations of lateral restraint, the flexural capacity, allowing for buckling, is reduced by a factor

$$\chi_{\mathrm{LT,fi}}/1.2$$

where $\chi_{\mathrm{LT,fi}}$ is the normal temperature reduction factor for lateral-torsional buckling, but calculated using the effective length which applies in the fire design situation. The constant 1.2 is an empirical correction factor which allows for a number of effects including the strain at failure being greater than the yield strain. This method can be used with any national code for steel design. In the Australian and New Zealand steel design codes (SAA, 1990(a), SNZ, 1996) the term χ_{LT} is the beam slenderness factor, denoted by α_s.

For laterally unrestrained beams, British Standard 5950 (BSI, 1990(a)) gives a method of calculating the limiting temperature which is about 65°C lower than for restrained beams. Bailey *et al.* (1996(b)) used a finite-element model to predict that theoretical failure would occur at an even lower temperature than the limiting temperatures from BS5950 or Eurocode 3, but they also point out that real beams will often have support conditions which provide considerably more continuity and restraint than assumed in the computer model.

Shear The design equation to resist a shear force V^*_{fire} during fire is

$$V^*_{\mathrm{fire}} \leqslant V_{\mathrm{f}} \tag{8.25}$$

The design shear resistance V_{f} under fire conditions is calculated from

$$V_{\mathrm{f}} = k_{\mathrm{y,T}}\, V_{\mathrm{c}} \tag{8.26}$$

where V_{c} is the design shear resistance of the cross section for normal temperature design.

If there is a temperature gradient over the cross section, Equation (8.26) should be based on the maximum temperature in the cross section. As for bending, Eurocode 3 allows the shear strength calculated in this way to be increased by a factor 1/0.7 if the top surface of the beam is protected from fire by a concrete slab.

Continuous beams

Beams which are continuous over several supports or form part of a moment-resisting frame are different from simply supported beams in several ways. The main

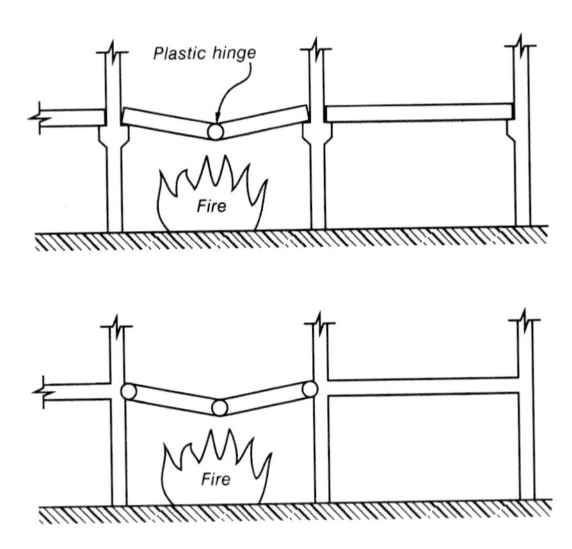

Figure 8.35 Failure mechanisms for simply supported and continuous beams

advantage of continuity in fire design is the possibility for considerable moment redistribution during the fire, which can lead to a considerable increase in fire resistance as described in Chapter 7. A possible negative aspect of flexural continuity for steel beams is the lack of lateral restraint to the lower flange of the beam where it is in compression in the negative moment regions near the supports.

With reference to Figure 8.35, a simply supported beam will fail as soon as one plastic hinge forms at the centre, when the flexural strength becomes equal to the applied bending moment. A continuous beam will not fail until three plastic hinges form, which can give greatly increased fire resistance in many cases. The end span of a continuous beam is intermediate between these two cases, as shown in Figure 7.13. Design of continuous beams is essentially the same as for simply supported beams, but including the redistribution of moments using plastic analysis as described in Chapter 7.

Columns

The design of columns is often more difficult than the design of beams because lateral buckling must usually be considered and the prediction of behaviour is less reliable. If there is a temperature gradient over the cross section, it is not possible to consider accurately the variation of strength segment by segment without a computer program because thermal bowing and instability considerations dominate the behaviour. Following EC3 (1995), an approximate design method is based on the assumption that the whole cross section is at the maximum temperature T_m. It is important to note that the steel temperature for the calculation must be the maximum temperature, not the average temperature obtained from a lumped-mass calculation. This approximate method is not always conservative if the thermal gradient causes significant bowing. Columns with thermal gradients across the section should preferably be analysed with a specialist computer program. The design equations for a column subjected to an axial load N^*_{fire} is

$$N^*_{\text{fire}} \leqslant N_{\text{f}} \tag{8.27}$$

In the approximate method, the compressive load-bearing capacity is obtained from

$$N_{\text{f}} = (\chi_{\text{fi}}/1.2) A\, k_{\text{y,Tm}}\, f_{\text{y}} \tag{8.28}$$

where χ_{fi} is the normal temperature buckling factor, calculated using the effective length for the fire design situation, A is the area of the cross section (mm^2), $k_{\text{y,Tm}}$ is the reduction factor for yield strength of steel which is at the maximum temperature T_{m}, and f_{y} is the yield strength of the steel at 20°C (MPa)

The constant 1.2 is an empirical correction factor similar to that used above for beams. Hence it is recommended (EC3, 1995) that design of columns in fire conditions should follow the same principles as for normal temperature design, but in addition to reducing the steel strength for elevated temperatures, there should be a 17% reduction (1/1.2) in strength to allow for other effects. In the Australian and New Zealand steel design codes (SAA, 1990(a), SNZ, 1996) the term χ is the column buckling factor, denoted by α_{c}. The Eurocode 3 method has been critically reviewed by Franssen *et al.* (1995, 1996) who propose an alternative formulation, not yet adopted by any codes.

The buckling length of a column should usually be calculated in the same way as for normal temperature design. However, in a braced frame, the buckling length may be determined by considering it to have fixity to the columns above and below, as shown in Figure 8.36, provided that the building design is such that the fire is not able to spread to an upper floor.

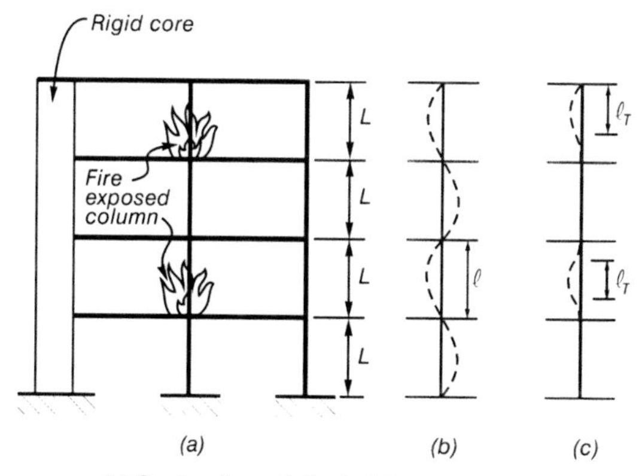

Figure 8.36 Effective lengths of fire exposed columns in a multi-storey frame (EC3, 1995)

8.7.3 Bolted and Welded Connections

Welded connections perform as well or better than the connected members in fire conditions. Bolted connections generally behave well despite the drop in bolt strength with elevated temperatures. There has not been much research into the fire performance of bolted steel connections. Lawson (1990) tested eight beam-to-column connections, with some of the beams supporting composite concrete slabs. All of the connections were exposed to the ISO 843 standard time–temperature curve. Bolt temperatures were found to be lower than those of the exposed flange and the grade 8.8 bolts behaved very well. There were no failures of bolts or welds; the rotations at the connections resulted from local flexural deformation of the end-plates welded to the ends of the beams. In composite beams, the mesh reinforcing in the slab was found to provide a significant contribution to flexural capacity. More recently Al-Jabri *et al.* (1998) tested a series of bolted connections, also finding that typical connections which are considered to be pinned at normal temperature are capable of resisting considerable bending moments at elevated temperatures. In all tests the deformation was small up to about 400°C, beyond which there was a progressive increase in rotation. Failure modes were similar to those observed at normal temperatures, generally localized bending or fracture of the end-plates. In the composite slabs there was some failure of the reinforcing mesh and the shear studs connecting the slab to the beam. Finite element modelling of bolted connections in steel and steel composite construction is described by Liu (1999).

In support of the above observations, Eurocode 3 suggests that bolted and welded connections in steel structures should be designed for normal temperature conditions and be protected in the same manner as the main members. However, if the members are over-designed for any reason, the connections should be over-designed by at least the same amount because the over-design results in additional fire resistance, which must be matched by the connections.

The design of bolted beam-to-column connections should allow for very high tensile forces which can occur during the decay stage of the fire, especially in buildings with unprotected steel beams acting compositely with concrete floor slabs (Rotter *et al.*, 1999).

8.7.4 Cast-iron Members

Cast-iron was manufactured and used widely in buildings throughout the 19th century, being replaced by rolled-steel sections in the early years of the 20th century. Fire engineering assessment of historical buildings sometimes requires calculation of the fire resistance of cast-iron columns or beams. Barnfield and Porter (1984) have confirmed earlier reports that the behaviour of cast-iron under fire conditions is difficult to predict accurately because brittle fracture can occur in some circumstances, usually associated with distortion resulting from applied loads, casting defects or thermal movements of adjacent elements. They state that cast-iron elements are unlikely to fail as long as the cast-iron temperature remains below a limiting temperature. The suggested limiting temperature is 300°C for cast-iron members attached to iron or steel elements, and 550°C for members attached to timber elements which are likely to impose much lower thermal deformations in a fire. Cast-iron has similar thermal properties to structural steel, so the same methods of thermal analysis can be used. Intumescent paint is the best method of protecting cast-iron members which are intended to remain visible in the finished building.

8.8 DESIGN OF STEEL BUILDINGS EXPOSED TO FIRE

Steel buildings or significant assemblies in steel buildings cannot be designed economically by the simple methods described above. It becomes necessary to use a specialist computer program for analysis of the fire-exposed structure (the 'general' method in Eurocode 3). Such a program will impose deformations on the structure and calculate the total strain in each member resulting from those deformations. The stress-related strain will be calculated from Equation (8.11), leading to calculation of the internal forces in each member for comparison with the applied loads. The general method is essential for any structures with large displacements. The calculated fire resistance of a structural steel member is different if it is considered to be part of a frame rather than a single member. Fire resistance is usually enhanced if the member is part of a frame.

8.8.1 Multi-storey Steel Framed Buildings

In recent years, a number of large fires in steel framed buildings have demonstrated that the fire performance of large steel frame structures is often much better than can be predicted by consideration of the fire resistance of the individual structural elements (Moore and Lennon, 1997). This excellent behaviour results from the ductility of steel, allowing large rotations and deflections without significant loss of strength. These observations have been supported by extensive computer analyses, including Franssen *et al.* (1995) who showed that when axial restraint from thermal expansion of the members is included in the analysis of a portal frame building, the behaviour is completely different from of that of the column and beam analysed separately.

An often quoted example is the severe fire which occurred in the Broadgate complex in London in 1990 (Lawson *et al.*, 1991). The fire occurred in a contractor's hut on the second storey of a 14 storey building nearing completion, before most of the columns had been protected with fire resisting materials. It is estimated that many of the structural members in the fire area reached temperatures of 650°C. The fire caused severe distortion of trusses and beams supporting the floor slabs, and axial shortening of five columns, but there was no structural collapse and no loss of integrity of the floor slabs (Figure 8.37). The building was repaired with no serious difficulties.

A large series of full-scale fire tests was carried out between 1994 and 1996 in the Cardington Laboratory of the Building Research Establishment in England (Figure 8.38). A full-size eight-storey steel building was constructed with composite reinforced concrete slabs on exposed metal decking, supported on steel beams with no applied fire protection other than a suspended ceiling in some tests. The steel

Figure 8.37 Local buckling of an unprotected steel column during a fire in a building under construction [Broadgate, London, 1990] (Lawson *et al.*, 1991)

columns were fire-protected. A number of fire tests were carried out on parts of one floor of the building, resulting in steel beam temperatures up to 1000°C, leading to deflections up to 600 mm, but no collapse and generally no integrity failures (Armer and O'Dell, 1996, Martin and Moore, 1997).

The good performance of the floor/beam systems in such buildings has been attributed to a complex interrelated sequence of events, described rather simply as follows.

(1) The fire causes heating of the beams and the underside of the slab.

(2) The slab and beam deform downwards as a result of thermal bowing.

(3) Thermal expansion causes compressive axial restraint forces to develop in the beams.

(4) The reaction from the stiff surrounding structure causes the axial restraint forces to become large.

(5) The yield strength and modulus of elasticity of the steel reduce steadily.

(6) The downward deflections increase rapidly due to the combined effects of the applied loads, thermal bowing and the high axial compressive forces.

(7) The axial restraint forces reduce due to the increased deflections and the reduced modulus of elasticity, limiting the horizontal forces on the surrounding structure.

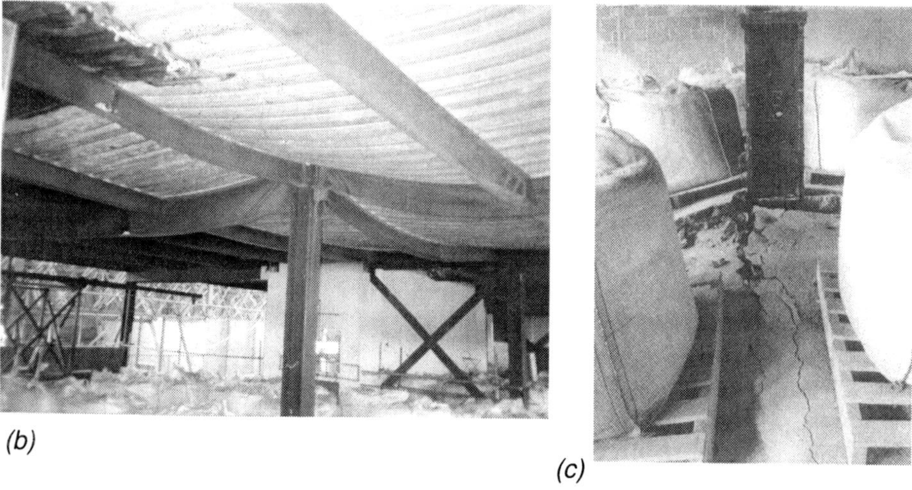

Figure 8.38 (a) Flames coming from the window during a post-flashover fire in the Cardington fire test building (Reproduced from Kirby (1999) by permission of Corus UK Ltd). (b) Large vertical deflections of unprotected steel beams supporting a composite steel-concrete floor slab; the column was protected during the fire. (c) Cracking in the top surface of the composite slab around the column; integrity was maintained (Reproduced from Kirby (1999) by permission of Corus UK Ltd)

(8) Higher temperatures lead to a further reduction of flexural and axial strength and stiffness.

(9) The slab beam system deforms into a catenary, resisting the applied loads with tensile membrane forces.

(10) As the fire decays, the structural members cool down and attempt to shorten in length.

(11) High tensile axial forces are induced in the slab, the beam and the beam connections.

These actions can take place in two dimensions or three dimensions, depending on the geometry of the building and the layout of the structure. The large deformations are often accompanied by local buckling of the steel members. The high axial tensile forces can result in fractures of buckled beams after the fire (Tide, 1998).

Modern computing power has recently made it possible to model the structural response of steel frame buildings exposed to fires. Computer modelling has been used to help interpret the behaviour of the Cardington building (including Wang *et al.*, 1995, Rose *et al.*, 1998, O'Connor and Martin, 1998, and Rotter *et al.*, 1999). Some of the studies have found that the building can be modelled using two-dimensional sub-frames rather than the complete three-dimensional frame, but others have emphasized the three-dimensional behaviour. Other studies have found that column yielding causes beams to behave as if they are on pinned supports, and beam behaviour is significantly influenced by web buckling. The development of tensile membrane action in the composite steel/concrete floors is described by Wang (1996).

8.8.2 Car-Parking Buildings

Fires in car-parking buildings are less severe than fires in many other occupancies. Many studies have shown that the fire load is low, and fires do not often spread from car to car because each car body acts like a form of enclosure. Schleich *et al.* (1999) suggest that the fire should be considered to spread from car to car every 12 minutes in an unsprinklered building. Even if there is no structural problem, burning cars can produce large volumes of toxic smoke which is a major hazard to life.

The required level of fire protection depends on whether the car-parking building is open to the outside air, or enclosed. A burning car in an enclosed car-parking building can result in high temperatures in the structural members, so that installation of sprinklers or applied fire protection is necessary to ensure no collapse of steel members. Maximum temperatures in structural members may be much more localized in car-parking buildings than in other occupancies, so local temperatures should be checked if flashover does not occur.

Tests in Australia (Bennetts *et al.*, 1999) have shown that fires are much less severe in car-parking buildings which are open to the air on at least two opposite sides. Much of the heat and smoke from burning cars in open car-parking buildings is carried directly to the outside, so that hot gas temperatures remain low. Maximum temperatures measured in exposed unprotected steel beams do not exceed 260°C, so that unprotected steel can be used for the beams and columns of such buildings, provided that the steel members are not too light. Bennetts *et al.* (1999) recommend a maximum section factor F/V of $230\,m^{-1}$ (minimum effective thickness 4.3 mm) provided that the beam is in continuous contact with a reinforced concrete slab designed for composite action. The same recommendations apply to closed car-parking buildings which have sprinklers installed. Schleich *et al.* (1999) also recommend the use of unprotected steel for columns and composite beams in closed car-parking buildings which are protected with sprinklers.

8.8.3 Single-storey Portal-frame Buildings

A very common form of steel construction is single-storey portal-frames for industrial buildings as shown in Figure 8.39. There is considerable debate about the objectives and strategies for fire-resisting design of such buildings. Typical buildings have portal frames 5 to 10 m apart, spanning from 20 to 50 m, or more with internal columns. The roof usually has a slope between 2° and 15°, consisting of thin steel sheeting and translucent plastic sheeting supported on timber or steel purlins up to 1 m apart, spanning between the portal frames. Other types of roofing include deep trough steel sheeting or sandwich panels, which can span much larger distances. Walls of single-storey portal-frame buildings are often brick, concrete masonry or concrete 'tilt-panels'. Concrete tilt-panels are precast concrete panels, cast on the floor slab on the site before lifting into place. In some buildings the side walls are load-bearing, such that there is no steel column and the concrete wall panel provides vertical and flexural support to the steel rafters.

For an uncontrolled fire in a single-storey industrial building, the possible heat release rate is shown in Figure 8.40 (Cosgrove, 1996), where the fire is initially limited by available ventilation, but becomes fuel controlled after the skylights melt and the roof eventually collapses. The steel roof structure is seldom fire-rated, so it will usually collapse in a fully developed fire. The area of fully developed fire and hence

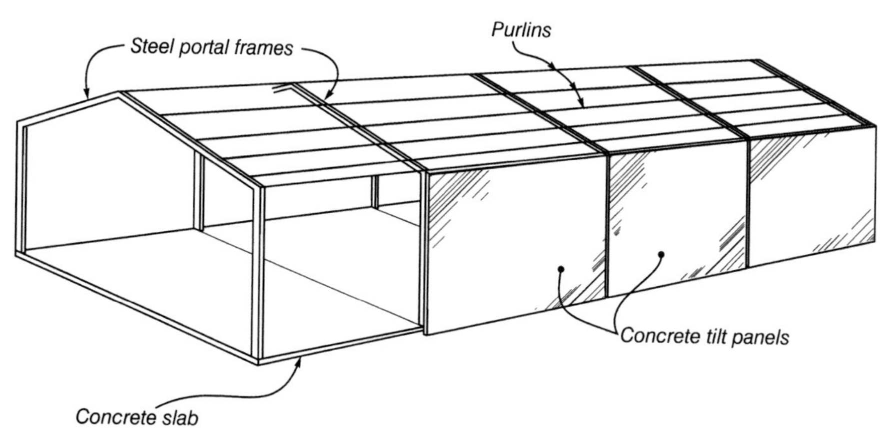

Figure 8.39 Single-storey portal-frame industrial building

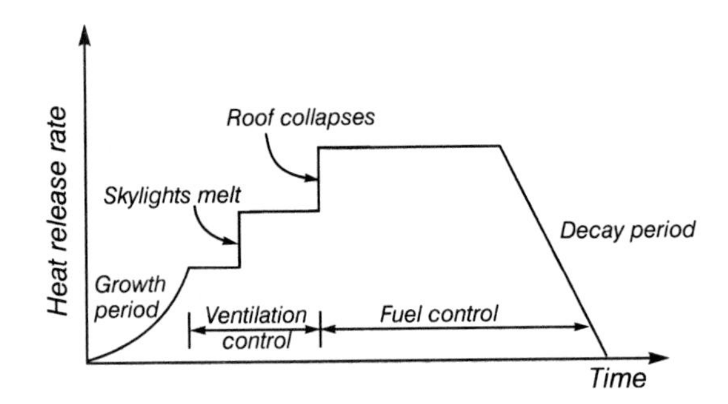

Figure 8.40 Heat release rate for fire in an industrial building

the area of collapsing roof will move around the building as the fire grows and spreads. If the purlins supporting the roofing are of timber, the roofing will probably fall into the fire before the main beams collapse. Steel purlins will tend to deform into a tensile catenary shape, holding the roofing in place between the main beams until they collapse.

Large amounts of roof venting, caused by melting of large plastic skylights or aluminium roof cladding, result in lower temperatures because much of the heat from the fire is released directly to the atmosphere without heating the steel structure.

Portal-frame buildings are often constructed near property boundaries, so one of the main fire safety objectives may be prevention of fire spread to neighbouring properties. Fire spread is controlled with the use of fire-resisting boundary walls. Special design for the fire situation is necessary because under normal temperature conditions the wall panels would be supported by non-fire-rated steel frames. There are alternative strategies for preventing fire spread: cantilever construction to ensure that the exterior walls remain in place for the duration of the fire, or pinned column bases which allow the wall panels to fall inwards towards the fire.

The traditional approach is to provide the steel portal-frames with a fixed (or partially fixed) base, and to apply passive fire protection to the portal-frame columns. In this case the fire-resisting walls are supported before the fire by portal-frame action, then during and after the fire by the cantilevered steel columns with moment-resisting base connections. With fixed or partially fixed column bases, Newman (1990) has shown how a fire initially causes an outward thrust on the columns as the rafters expand and deflect, followed by in inwards tensile force as the rafters droop into a catenary (Figure 8.43). The initial outwards movement can cause masonry or precast concrete walls to collapse outwards as shown in Figures 8.41 and 8.42. Newman describes how to calculate the overturning moment at the column base, and shows that design for full fixity is not required.

A common variation is to rely on reinforced concrete columns instead of steel columns for the portal frames, in which case the after-fire condition is of less concern. Yet another approach is to provide the concrete wall panels with a cantilever base connection to a strong foundation, on the assumption that the panels will remain free-standing after the steel roof structure collapses,

Figure 8.41 Failure of an unreinforced brick masonry wall of an industrial building; the wall was pushed outwards by thermal expansion of the steel portal frames, which later collapsed inwards. (Reproduced by permission of Cement and Concrete Association of Australia)

Figure 8.42 Outward collapse of a precast concrete wall panel during a fire in an industrial building; such a collapse would endanger fire-fighters and allow fire spread (Reproduced by permission of Cement and Concrete Association of Australia)

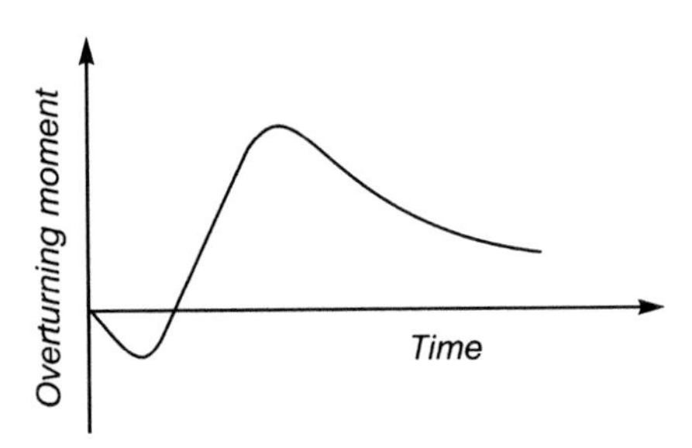

Figure 8.43 Axial thrust in rafter of portal frame during fire (Reproduced from 'The behaviour of steel portal frames in boundary conditions' by permission of the Director, The Steel Construction Institute)

depending on the connection between the panels and the steel roof framing. A fire inside the building will cause cantilevered walls to deform outwards due to the temperature gradient through the wall (Cooke, 1988, Lim, 2000) possibly leading to undesirable outwards collapse if the separate panels are not tied together.

An alternative approach, promoted in Australia by O'Meagher *et al.* (1992) is to design the portal frames with pinned bases and no applied fire protection, supporting concrete wall panels which have pinned connections at top and bottom. In this case it is essential that the concrete wall panels are tied together at their tops, the roofing has some diaphragm stiffness, and the connections between the top of the panels and the steel frames be provided with adequate fire resistance. O'Meagher *et al.* (1992) have used computer-based structural analysis to demonstrate that the walls of such a building will collapse inwards, not outwards, and there will be good protection of adjacent properties if the walls remain tied together as they collapse or partially collapse inwards, as shown in Figure 8.44.

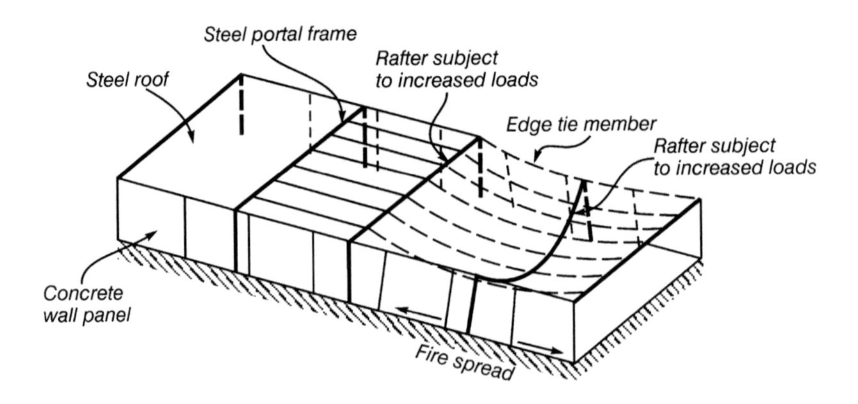

Figure 8.44 Failure mechanism for single-storey industrial building in fire (Reproduced from (O'Meagher *et al.*, 1992) by permission of Australian Institute of Steel Construction)

WORKED EXAMPLE 8.1

Calculate the section factor for a steel H-section column, of dimensions 300×300 mm. The column is exposed to fire on all four sides.

Make calculations for (a) box-type protection, and (b) spray-on protection.

GIVEN:

Height of section	$h = 300$ mm
Width of section	$b = 300$ mm
Flange thickness	$T = 20$ mm
Web thickness	$t = 8$ mm

CALCULATION:

(a) **Box-type protection**

Area of cross section	$A = 2\ (b \times T) + t\ (h - 2T)$
	$= 14\,080\ \text{mm}^2 = 0.01408\ \text{m}^2$
Volume of 1 m length	$V = A \times 1.0\ \text{m} = 0.01408\ \text{m}^3$
Perimeter of section	$H_P = 2\ (b + h) = 1200\ \text{mm} = 1.2\ \text{m}$
Surface area of 1 m length	$F = H_P \times 1.0\ \text{m} = 1.2\ \text{m}^2$
Section factor	$H_P/A = H_P/A = 1.2/0.01408 = 85.2\ \text{m}^{-1}$
Section factor	$F/V = F/V = 1.2/0.01408 = 85.2\ \text{m}^{-1}$
Effective thickness	$V/F = 1000/(F/V) = 11.7$ mm

(b) **Spray-on protection**

Perimeter of section	$H_P = 2\ (b + h + (b - t)) = 1784\ \text{mm} = 1.78\ \text{m}$
Surface area of 1m length	$F = H_P \times 1.0\ \text{m} = 1.78\ \text{m}^2$
Section factor	$H_P/A = 1.78/0.01408 = 126\ \text{m}^{-1}$
Section factor	$F/V = 1.78/0.01408 = 126\ \text{m}^{-1}$
Effective thickness	$V/F = 1000/(F/V) = 7.9$ mm

WORKED EXAMPLE 8.2

Use the step-by-step method shown in Table 8.5 to calculate the steel temperature of an unprotected beam exposed to the ISO 834 standard fire.

The beam section factor F/V is $200\ \text{m}^{-1}$. Use a convective heat transfer coefficient h_c of $25\ \text{W/m}^2\text{K}$ and emissivity 0.6. The density of steel is $7850\ \text{kg/m}^3$ and the specific heat is 600 J/kgK. Use a time step of 0.5 minutes.

The first two minutes of the solution are shown in the table below. The results are plotted in Figure 8.7.

Time (minutes)	Time at half step	Steel temperature T_s	ISO fire temperature at half step T_f	Difference in temperature	Change in steel temperature
0.0	0.25	20.0	184.6	164.6	6.8
0.5	0.75	26.8	311.6	284.7	13.8
1.0	1.25	40.6	379.3	338.7	18.2
1.5	1.75	58.8	425.8	366.9	21.5
2.0	2.25	80.3	461.2	380.9	24.0
2.5					
3.0					

WORKED EXAMPLE 8.3

Use the step-by-step method shown in Table 8.5 to calculate the steel temperature of a protected beam exposed to the ISO 834 standard fire. The beam is the same as in Worked Example 8.2. The beam is protected with 50 mm of lightweight insulating material which has thermal conductivity of 0.2 W/mK, specific heat 1100 J/kg K and density 300 kg/m^3.

The first two minutes of the solution are shown in the table below. The results are plotted in Figure 8.7(a), with another curve for 15 mm of insulation.

Time (minutes)	Time at half step	Steel temperature T_s	ISO fire temperature at half step T_f	Difference in temperature	Change in steel temperature
0.0	0.25	20.0	184.6	164.6	0.62
0.5	0.75	20.6	311.6	290.9	1.10
1.0	1.25	21.7	379.3	357.6	1.35
1.5	1.75	23.1	425.8	402.7	1.52
2.0	2.25	24.6	461.2	436.6	1.65
2.5					
3.0					

WORKED EXAMPLE 8.4

Use the step-by-step method shown in Table 8.5 to calculate the steel temperature of an unprotected beam exposed to a parametric fire. The beam is the same as in Worked Example 8.2. The fire compartment is made from lightweight concrete with density 2000 kg/m^2, specific heat 840 J/kgK and thermal conductivity 0.8 W/mK. The room is 5 m square and 3 m high with one window 2.4 m wide and 1.5 m high. The fuel load is 800MJ/m^2 floor area.

Length of room	$l_1 = 5.0\,\text{m}$
Width of room	$l_2 = 5.0\,\text{m}$
Height of room	$H_r = 3.0\,\text{m}$
Area of internal surfaces	$A_t = 2\,(l_1\, l_2 + l_1\, H_r + l_2\, H_r)$
	$= 2\,(5 \times 5 + 5 \times 3 + 5 \times 3) = 110\,\text{m}^2$
Height of window	$H_v = 1.5\,\text{m}$
Width of window	$B = 2.4\,\text{m}$
Area of window	$A_v = BH_v = 2.4 \times 1.5 = 3.6\,\text{m}^2$
Ventilation factor	$F_v = A_v\sqrt{H_v}/A_t = 3.6 \times \sqrt{1.5}/110 = 0.04\,\text{m}^{-1/2}$
Fuel load (floor area)	$e_f = 800\,\text{MJ/m}^2$
Fuel load (total area)	$e_t = e_f\, A_f/A_t = 800 \times 25.0/110 = 182\,\text{MJ/m}^2$
Thermal conductivity	$k = 0.8\,\text{W/mK}$
Density	$\rho = 2000\,\text{kg/m}^3$
Specific heat	$c_p = 840\,\text{J/kg K}$
Thermal inertia	$b = \sqrt{k\rho c_p} = 1160\,\text{W s}^{0.5}/\text{m}^2\,\text{K}$ (medium)
Gamma factor	$\Gamma = (F_v/0.04)^2/(b/1900)^2 = 2.69$

The parametric fire can be calculated using this value of gamma.

The first two minutes of the solution are shown in the table below. The results are plotted in Figure 8.7(b).

Time (minutes)	Parametric fire temperature	Steel temperature T_s	Fire temperature (average of this and next step) T_f	Difference in temperature	Change in steel temperature
0.0	0.0	20.0	114.6	94.6	3.7
0.5	229.2	23.7	305.9	282.2	13.5
1.0	382.6	37.2	434.5	397.3	23.1
1.5	486.4	60.3	522.1	461.8	31.5
2.0	557.8	91.8	582.8	491.0	38.1
2.5	607.8				
3.0					

WORKED EXAMPLE 8.5

Repeat Worked Example 8.4 with the beam protected with 50 mm of the same insulation as in Worked Example 8.3.

The first two minutes of the solution are shown in the table below. The results are plotted in Figure 8.7(b).

Time (minutes)	Parametric fire temperature	Steel temperature T_s	Fire temperature (average of this and next step) T_f	Difference in temperature	Change in steel temperature
0.0	0.0	20.0	114.6	94.6	0.36
0.5	229.2	20.4	305.9	285.6	1.08
1.0	382.6	21.4	434.5	413.1	1.56
1.5	486.4	23.0	522.1	499.1	1.88
2.0	557.8	24.9	582.8	557.9	2.11
2.5	607.8				
3.0					

WORKED EXAMPLE 8.6

Fire calculation in strength domain

For a simply supported steel beam of known span, load, yield strength, and section properties, calculate the flexural strength after 15 minutes exposure to the standard fire. The beam has no applied fire protection, and is exposed on three sides.

GIVEN

Dead load	$G_k = 8.0$ kN/m (including self weight)
Live load	$Q_k = 15.0$ kN/m
Beam span	$L = 8.0$ m
Beam size	410 UB 54 (410 mm deep Universal Beam, 54 kg/m)
	This is a 'compact' section (type 1)
Plastic section modulus	$S = 1060 \times 10^3$ mm^3
Section factors:	
Area to volume ratio	$F/V = 190$ m^{-1}
(Effective thickness	$V/F = 5.3$ mm)

COLD CALCULATIONS

Strength reduction factor	$\Phi = 0.9$
Yield strength	$f_y = 300$ MPa
Design load (cold)	$w_c = 1.2G_k + 1.6Q_k = 33.6$ kN/m
Bending moment	$M^*_{cold} = w_c L^2/8 = 269$ kNm
Bending strength	$M_n = S f_y = 318$ kNm (assume adequate lateral restraint)
Design flexural strength	$\Phi M_n = 286$ kNm

Design is OK ($M^*_{cold} < \Phi M$)

FIRE CALCULATIONS

Strength reduction factor	$\Phi = 1.0$ (hence not used in the calculations)
Design load (fire)	$w_f = G_k + 0.4Q_k = 14.0$ kN/m
Bending moment	$M^*_{fire} = w_f L^2/8 = 112$ kNm
Temperature after time t	$T = 1.85\, t\, (F/V)^{0.6} + 50$ (from Equation (8.3))
Temperature after 15 minutes	$T = 1.85 \times 15 \times 190^{0.6} + 50 = 696°\text{C}$
Yield strength reduction	$k_{y,T} = (905 - \text{T})/690 = 0.30$
Flexural capacity	$M_f = S\, k_{y,T}\, f_y$ (assume adequate lateral restraint) $= 1060 \times 10^3 \times 0.30 \times 300/10^6 = 95$ kNm

Design fails ($M^*_{fire} > M_f$)

(Note: For more accurate calculations, the maximum temperature should be calculated by the step-by-step method. The flexural calculation method would be identical.)

WORKED EXAMPLE 8.7

Fire calculation in time domain

For a simply supported steel beam of known span, load, yield strength, and section properties, calculate the time to failure when exposed on three sides to the standard fire: (a) unprotected, and (b) protected with insulation of known thickness and properties.

GIVEN

Dead load	$G_k = 6.0$ kN/m (including self weight)
Live load	$Q_k = 12.5$ kN/m
Beam span	$L = 15.0$ m
Beam size	760UB147 (760 mm deep Universal Beam, 147 kg/m)
	This is a 'compact' section (type 1)

Plastic section modulus	$S = 4480 \times 10^3 \, mm^3$
Section factors:	
Area to volume ratio	$F/V = 119 \, m^{-1}$
(Effective thickness	$V/F = 8.4 \, mm$)

COLD CALCULATIONS

Strength reduction factor	$\Phi = 0.9$
Yield strength	$f_y = 300 \, MPa$
Design load (cold)	$w_c = 1.2 G_k + 1.6 Q_k = 27.2 \, kN/m$
Bending moment	$M^*_{cold} = w_c L^2/8 = 765 \, kNm$
Bending strength	$M_n = S \, f_y = 1344 \, kNm \; (= R_{cold})$
	(assume adequate lateral restraint)
Design flexural strength	$\Phi M_n = 1210 \, kNm$

Design is OK ($M^*_{cold} < \Phi M$)

FIRE CALCULATIONS

Design load (fire)	$w_f = G_k + 0.4 Q_k = 11 \, kN/m$
Bending moment	$M^*_{fire} = w_f L^2/8 = 309 \, kNm$
Load ratio	$r_{load} = M^*_{fire}/R_{cold} = 309/1344 = 0.23$
Limiting steel temperature	$T_{lim} = 905 - 690 \, r_{load} = 746^\circ C$

(a) **Unprotected steel** (three-sided exposure)

Time to reach limiting temperature

$$t = 0.54 \, (T_{lim} - 50)/(F/V)^{0.6}$$
$$= 0.54 \, (746 - 50)/119^{0.6}$$
$$= 21.4 \text{ minutes}$$

Design is OK if the equivalent fire severity is no more than 21 minutes.

(b) **Protected steel**

Thickness of insulation	$d_i = 0.020 \, m$ (20 mm)
Thermal conductivity	$k_i = 0.10 \, W/mK$

Time to reach limiting temperature

$$t = 40 \, (T_{lim} - 140) \, [(d_i/k_i)/(F/V)]^{0.77}$$
$$= 40 \, (746 - 140) \, [(0.02/0.10)/119]^{0.77}$$
$$= 177 \text{ minutes}$$

Moisture content of insulation	$m = 15\%$
Density of insulation	$\rho_i = 800 \, kg/m^3$
Time delay for insulation	$t_v = m \, \rho_i \, d_i^2/(5 \, k_i)$
	$= 15 \times 800 \times 0.02^2/(5 \times 0.10) = 9.6$ minutes
Total time	$t_{total} = t + t_v = 186$ minutes

Design is OK if the equivalent fire severity is no more than 186 minutes.

(Note: For more accurate calculations, the maximum temperature should be calculated by the step-by-step method. The flexural calculation method would be identical.)

9

Concrete Structures

9.1 OVERVIEW

This chapter describes simple methods of designing reinforced concrete structures to resist fires, including information on thermal and mechanical properties of concrete and structural design methods. Prestressed concrete and composite steel–concrete structures are also covered briefly.

9.2 BEHAVIOUR OF CONCRETE STRUCTURES IN FIRE

Concrete structures have a reputation for good behaviour in fires. A very large number of reinforced concrete buildings which have experienced severe fires have been repaired and put back into use (Figure 9.1). Concrete is non-combustible and has a low thermal conductivity. The cement paste in concrete undergoes an endothermic reaction when heated, which assists in reducing the temperature rise in fire-exposed concrete structures. Concrete tends to remain in place during a fire, with the cover concrete protecting the reinforcing steel, with the cooler inner core continuing to carry load (Figure 9.2).

Calculation of the behaviour of concrete structures in fire depends on many factors, the most important being the applied loads on the structure, the elevated temperatures in the concrete and reinforcing and the mechanical properties of the steel and concrete at those temperatures. When a reinforced concrete structure is exposed to a fire, the temperatures of both steel and concrete increase, leading to increased deformation and possible failure, depending on the applied loads and the support conditions. Most types of concrete behave similarly in fires. This chapter refers to the slightly different performance of concrete made with different types of aggregate, lightweight concrete and high-strength concrete.

Catastrophic failures of reinforced concrete structures in fire are rare, but some occasionally occur (e.g. Papaioannou, 1986, Berto and Tomina, 1988) (Figure 9.3). Observations have shown that when concrete buildings fail in real fires, it is seldom because of the loss of strength of the materials, but nearly always because of the inability of other parts of the structure to absorb the large imposed thermal deformations in the horizontal direction which can cause shear or buckling failures of columns or walls (A. van Acker, personal communication).

Figure 9.1 Non-structural fire damage to a typical reinforced concrete office building

(a) (b)

Figure 9.2 (a) A multi-storey office building engulfed in flames. The reinforced concrete structure did not collapse in the fire [Sao Paolo, Brazil, 1972]. (b) Severe spalling of a reinforced concrete wall in the fire

Figure 9.3 Major structural damage to a multi-storey reinforced concrete department store [Athens, 1980] Reproduced from Papaioannou (1986) by permission of John Wiley and Sons, Ltd)

9.2.1 High-strength Concrete

There has been considerable interest recently in high-strength concrete as a high performance construction material. High-strength concrete contains additives such as silica fume and water-reducing admixtures which result in compressive strength in the range 50 to 120 MPa. An extensive survey of high-strength concrete properties at elevated temperatures by Phan (1996) shows that they tend to have a higher rate of strength loss than normal concrete at temperatures up to 400°C, and explosive spalling is a problem in some cases. Fire tests on high-strength columns are reported by Aldea *et al.* (1997) and Kodur (1997). In some studies, the compressive strength at elevated temperatures is found to be higher when the concrete is heated under stress, rather than loaded after heating. Design recommendations are given by Tomasson (1998) who recommends the simplified method of Eurocode 2 (EC2, 1993), ignoring the strength contribution of concrete which is hotter than 500°C. For columns he suggests changing the limiting temperature to 400°C.

9.2.2 Lightweight Concrete

Lightweight concrete is usually made with normal cement and some form of lightweight aggregate such as pumice or expanded clay or shale. Other possible materials include perlite and vermiculite. Lightweight concrete has been shown to have excellent fire resistance, due to its low thermal conductivity compared with normal weight concrete. Many listings of generic fire-resistance ratings have separate tables for lightweight concrete. Many lightweight aggregates have been manufactured at high temperatures, so they remain very stable during fire exposure.

9.2.3 Fibre Reinforced Concrete

Steel-fibre reinforced concrete uses small steel fibres added to the concrete mix, to improve concrete toughness and strength. The fibres are typically 0.5 mm diameter, and 25 to 40 mm long with crimped or hooked ends to improve the bond. Thermal and mechanical properties of steel-fibre reinforced concrete at elevated temperatures are given by Lie and Kodur (1996). They show that the presence of steel fibres increases the ultimate strain and improves the ductility of the concrete.

9.2.4 Spalling

The design recommendations in this book for calculating thermal gradients and structural behaviour in concrete members, are based on the assumption that all the concrete remains intact for the duration of the fire. This assumption is not valid if the cover concrete spalls off a member during a fire, exposing some or all of the main reinforcing steel to the fire. Experiments and real fire experience have shown that most normal concrete members can withstand severe fires without serious spalling, but minor spalling often occurs.

Note that 'cover concrete' refers to the concrete outside the main reinforcing cage, protecting the reinforcing steel from moisture, corrosion and fire. The 'cover' is the distance from the surface of the concrete to the reinforcing steel. For durability considerations, the cover is usually measured from the concrete surface to the closest face of the main bars, but for fire engineering the cover is usually measured from the concrete surface to the centre of the main bars. Care must be taken to avoid confusing these two definitions.

The phenomenon of spalling is not well understood because it is a function of several different factors, often leading to unpredictable behaviour. In some cases spalling is related to the type of aggregate or to thermal stresses near corners, but it is more often linked to the behaviour of the cement paste. It is generally agreed that spalling most often occurs when water vapour is driven off from the cement paste during heating, with high pore pressures creating effective tensile stresses in excess of the tensile strength of the concrete. Experiments have shown that increased susceptibility to spalling results from high moisture content concrete, rapid rates of heating, slender members, and high concrete stresses at the time of the fire (Figure 9.4). Malhotra (1984) and Phan (1996) review studies of spalling. High-strength concrete tends to be more susceptible to spalling than normal concrete since it has smaller free pore volume (higher paste density), so that the pores become filled with high-pressure water vapour more quickly than in normal weight concrete and the low porosity results in slower diffusion of the water vapour through the concrete.

Figure 9.4 Local spalling at the corner of a concrete beam

Even though serious spalling of concrete is unlikely, the probability of occurrence requires consideration for critical structures or those containing high-strength concrete. The addition of an additional reinforcing cage to prevent spalling is impractical and expensive. The best economical method of preventing spalling is the addition of fine polypropylene fibres to the concrete mix (0.15 to 0.3%). These fibres reduce the likelihood of spalling because the polypropylene melts during fire exposure, increasing the porosity by leaving cavities through which the water vapour can escape, as described by Kodur (1997). Steel fibres added to the concrete mix will reduce the probability of spalling by increasing the fracture toughness of the concrete, but this is much more expensive than adding polypropylene fibres.

9.2.5 Masonry

Concrete masonry consists of hollow concrete blocks mortared together, most often used in walls. In many areas, especially seismic regions, reinforcing bars are placed in the hollow cores which are then filled with concrete to create solid reinforced concrete masonry which has essentially the same fire behaviour as reinforced concrete. Concrete masonry blocks are often manufactured from lightweight concrete, giving enhanced fire-resistant properties. Fire-resistance ratings of many different types of concrete masonry are given by Allen (1970). Unfilled unreinforced masonry has less thermal mass and potential lines of integrity failure at the mortar joints, but has demonstrated excellent fire resistance, provided that the foundations and supporting structure can keep the wall in place during the anticipated fire. Hollow-core block walls can be considered to have the equivalent thickness of a solid wall of the same volume of concrete, and to have the same generic fire-resistance rating (NBCC, 1995). All joints between blocks and shrinkage control joints must be able to provide the same fire rating as the rest of the wall. Some construction details are given in the Uniform Building Code (UBC, 1997). Some methods for fire design of masonry are given in Eurocode 6 (EC6, 1995).

Brick masonry also behaves well in fires. Ceramic bricks are made by firing clay at high temperatures, producing bricks which remain stable when exposed to fires. Brick masonry can be reinforced if it is made from hollow bricks, but most brick masonry consists of solid bricks joined only with lime or cement mortar. Thermal bowing of very tall unreinforced cantilever masonry walls can lead to collapse during a severe fire on one side of the wall (Cooke, 1988). See Figure 8.41.

9.2.6 Prestressed Concrete

The term *prestressed concrete* refers to concrete structures which are stressed prior to the application of any external loads. There are two main types of prestressing: pre-tensioned and post-tensioned. For pre-tensioned prestressed concrete, the steel tendons are stressed in tension against a reaction frame or the mould before the concrete is cast, so that the prestressing force is resisted by bond stresses between the concrete and the tendon after the tendons are cut. Post-tensioned prestressed concrete is cast with ducts for the steel tendons which are stressed with hydraulic jacks after the concrete has cured. The prestressing force is resisted by permanent anchorage points at the ends of the tendons.

Pre-tensioned prestressed concrete is most often used for precast components for flooring, including flat panels, hollow-core panels, double-tee floor units or concrete planks for supporting non-prestressed components. Post-tensioning is used within large components such as beams or slabs, or for connecting several precast concrete elements together.

Prestressing tendons are made of high-strength steel, often manufactured by pulling steel wires through a die, or otherwise cold-working the steel. Cold-worked steel suffers permanent loss of strength when subjected to elevated temperatures.

Most of this chapter refers to reinforced concrete. The same principles apply to prestressed concrete which is often more vulnerable in fires because prestressing steels are much more sensitive to elevated temperatures than mild steel reinforcing bars, because prestressed concrete is often manufactured in slender components with thin cover concrete and because some failure modes such as debonding, shear and spalling are more critical in prestressed concrete (Gustaferro and Martin, 1988).

Recent full-scale fire tests have shown that bond failures of pre-tensioned tendons have caused premature failures much before the calculated fire-resistance time. A series of tests on double-tee and hollow-core slabs, simply supported without axial restraint over a span of 6 m, showed that the ends of the tendons were pulled into the concrete due to loss of bond near the ends of the specimens (Andersen and Laurisden, 1999). In these tests, collapse resulted from shear failure in the web after the compressive stresses had been reduced near the ends of the slabs. Similar results for hollow-core slabs have been observed elsewhere (Fontana and Borgogno, 1995).

9.2.7 External Reinforcing

Various forms of external reinforcing are used in special structures. The most common is steel decking used as permanent formwork in composite construction (Figure 8.10). Design of composite steel decking construction exposed to fires is described later in this chapter.

The use of fibre-epoxy coatings is new technology, increasingly being used to improve the strength of existing reinforced concrete structures. The fibre-epoxy coatings consist of mats of glass, carbon or teflon fibres surrounded by epoxy resin. These fibre-epoxy coatings can be wrapped around columns to improve the confinement to the concrete or can be glued to the surface of beams to increase the flexural strength. External fibre-epoxy coatings have zero fire resistance because the epoxy will melt and burn away at low temperatures. However, the residual reinforced concrete structure will usually have sufficient strength to carry the applied loads, and the coating can be re-applied after a fire.

9.3 FIRE-RESISTANCE RATINGS

9.3.1 Verification Methods

The emphasis of this chapter is on using scientific knowledge to estimate the fire performance of concrete members and structures. The design process for fire resistance requires verification that the provided fire resistance exceeds the design fire severity. Using the terminology from Chapter 5, verification may be in the *time domain*, the *temperature domain* or the *strength domain*. For reinforced concrete structures, fire resistance is most often verified in the time domain, by comparing generic fire resistance ratings with code-specified levels of fire resistance. For more detailed assessment of important structures, the load-bearing capacity is compared with the expected loads on the structure at the time of the fire, in the strength domain. Calculations comparing critical temperatures (in the temperature domain) are not usually used for concrete structures.

9.3.2 Generic Ratings

In contrast to steel structures, there are very few proprietary ratings used for concrete structures, and generic ratings are most often used. Generic ratings, or 'tabulated ratings' are those which assign fire resistance to materials with no reference to individual manufacturers or detailed specifications. Many national codes and some trade organizations list generic ratings for concrete construction, usually giving minimum sizes and minimum concrete cover to the reinforcing steel. It is assumed in all these tables that the concrete has sufficient reinforcing to resist the loads at normal temperatures. Some generic ratings distinguish between normal weight and lightweight concrete, and others further differentiate normal weight concrete into siliceous aggregate and calcareous (limestone) aggregate concretes.

Table 9.1 Minimum width (mm) and minimum cover (mm) for generic fire-resistance ratings of reinforced concrete members (BSI, 1985)

		Beams	Columns	Slabs	Walls
0.5 hours	Width	80	150	75	75
	Cover	20	20	15	15
1.0 hours	Width	120	200	95	75
	Cover	30	25	20	15
1.5 hours	Width	150	250	110	100
	Cover	40	30	25	25
2 hours	Width	200	300	125	100
	Cover	50	35	35	25
3 hours	Width	240	400	150	150
	Cover	70	35	45	25
4 hours	Width	280	450	170	180
	Cover	80	35	55	25

Table 9.1 shows typical values of generic fire resistance ratings for reinforced concrete members, from BS 8110 (BSI, 1985). The table gives minimum width (or thickness) and minimum cover to the reinforcing steel for several different types of reinforced concrete members. The cover is measured to the surface of the reinforcing bar. Similar tables are available for prestressed concrete. This type of information is usually very conservative, it applies only to standard fire exposure, and it makes no allowance for the shape of the fire exposed member or the level of load.

There is no guidance for situations where the cover is different from that shown in the table. The minimum thicknesses of floor slabs and walls are mostly based on the insulation criterion of preventing a temperature increase of 140°C on the unexposed surface. A summary of generic approvals from 10 different sources around the world are given in Appendix D, compiled by Wade (1991(a)). The values in Table 9.1 and Appendix D are typical of many similar international codes.

More comprehensive generic ratings are given in the Eurocodes (EC2, 1993), those for load-bearing walls and columns including the level of applied load, and those for beams giving several optional combinations of beam width and concrete cover.

Because concrete structures usually have good fire performance, many will meet the generic approvals with no increase in cover from normal temperature design, so that detailed calculations may not be necessary for fire-resistance ratings up to about 1.5 hours. Calculations are most likely to be useful for members which are thin or slender, members with little concrete cover to the reinforcing, and members which are very highly stressed under gravity loads.

9.3.3 Protection Systems

It is possible to enhance the fire performance of a concrete member by the application of protective layers such as gypsum board or trowelled on plaster, but this is not often done because it is cheaper to simply increase the member size or move the reinforcing bars further from the surface to increase the cover. Some generic listings include extra plaster protection or overlays of various forms of lightweight concrete (e.g. UBC, 1997, NBCC, 1995).

9.3.4 Joints Between Precast Concrete Panels

It is often necessary to provide fire resistance to gaps between precast concrete panels. Gustaferro and Martin (1988) report tests on various types of protection for such gaps. Compressed ceramic fibre blanket can give excellent fire resistance. An additional water-proofing sealant can be used to seal the joint for weatherproofing and visual appearance.

9.4 CONCRETE AND REINFORCING TEMPERATURES

9.4.1 Fire Exposure

In any specific design of concrete structures exposed to fires, it is essential to know the temperatures of the concrete and the reinforcing steel. The fire exposure may be the standard time–temperature curve or a more realistic fire curve, depending on the design philosophy. Many design charts are available giving thermal gradients in beams, columns and slabs exposed to the standard fire, but not for realistic design fires. It is best to use computer-based thermal calculations to provide accurate temperature gradients in concrete members exposed to realistic fires.

9.4.2 Calculation Methods

When making thermal calculations in reinforced or prestressed concrete members, it is usual to assume that the heat transfer is a function of the thermal properties of the concrete alone, and the temperature of the reinforcing is the same as the temperature of the surrounding concrete. Steel has a much higher thermal conductivity than concrete, but most reinforcing steel is parallel to the fire-exposed surfaces, so does not have a significant influence on heat transfer perpendicular to the surfaces. Some authorities have suggested that the much higher specific heat of steel than concrete, and possible moisture condensation, may result in the reinforcing steel being cooler than the surrounding concrete, but this concept is not used in design.

Unlike steel members, the only accurate way to calculate temperatures is to use a two-dimensional finite-element computer program which gives the temperature distribution with time over the cross section. The advantage of such a program is that any combination of materials, shapes and voids can be included, for exposure to any desired fire. Most programs do not consider the mass transport of water or water vapour in fire-exposed concrete, although this has been studied by Ahmed and Hurst (1995).

For simple members of normal-weight concrete, empirical hand calculation methods are available, derived from computer-based thermal analysis (Wickström,

1986, Hertz, 1981). The simple lumped-mass approach used for steel members is not appropriate for concrete. Wickström's method of calculating the temperatures in a normal weight concrete slab in the standard fire is based on the fire-exposed surface temperature T_w being:

$$T_w = \eta_w T_f \tag{9.1}$$

$$\eta_w = 1 - 0.0616 t_h^{-0.88} \tag{9.2}$$

where T_f is the fire temperature and t_h is the time (hours).

At any depth x (m) into the slab, at time t_h, the concrete temperature T_c is a factor η_x of the surface temperature T_w with η_x given by:

$$\eta_x = 0.18 \ln(t_h / x^2) - 0.81 \tag{9.3}$$

Hence the concrete temperature T_c is given by

$$T_c = \eta_x \eta_w T_f \tag{9.4}$$

This formula generally gives similar results to those shown in Figure 9.7.

The method can be used for corners of beams where there is heat conduction in two directions, using η_y calculated in the same way as η_x so that the concrete temperature T_c is now given by

$$T_c = [\eta_w(\eta_x + \eta_y - 2\eta_x\eta_y) + \eta_x\eta_y]T_f \tag{9.5}$$

This approximate equation gives temperatures roughly similar to those shown in Figure 9.10 for 160 mm wide beams, but does not make any allowance for the different rates of temperature increase in wider or narrower beams. Wickström (1986) shows how these equations can be modified for other types of concrete, and also gives approximate methods of calculating temperatures in concrete members exposed to realistic fires with a decay period. Empirical calculations in the decay period are less accurate because the maximum concrete temperatures occur a considerable time after the fire temperature passes its peak value (as shown in Figure 9.9). A finite-element calculation method is recommended for thermal analysis of concrete structures exposed to real fires.

9.4.3 Thermal Properties

In order to calculate temperatures within structural assemblies it is necessary to know the thermal properties of the materials. These are discussed briefly below. For more detail see Schneider (1986), Bazant and Kaplan (1996), Neville (1997) or Harmathy (1993).

The density of concrete depends on the aggregate and the mix design. Typical 'dense' concrete has a density of about 2300 kg/m^3. There are many 'lightweight' concretes which use porous aggregates or air entrainment to reduce the density to half or two-thirds of this value. When heated to 100°C the density of most concretes will be reduced by up to 100 kg/m^3 due to the evaporation of free water, which has a

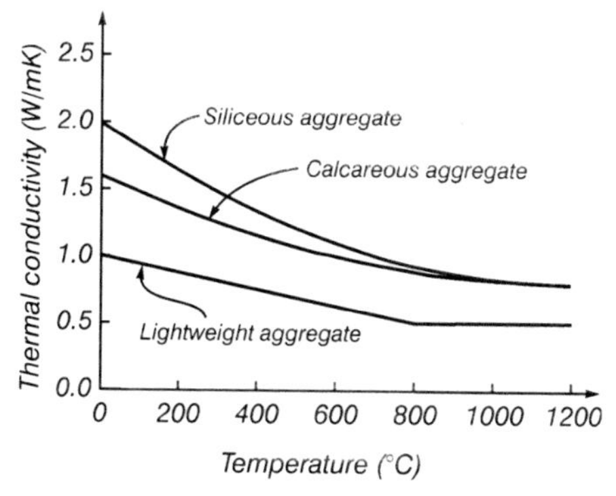

Figure 9.5 Thermal conductivity of concrete (Reproduced from (EC2, 1993) by permission of CEN)

minor effect on thermal response. Other than moisture changes, the density of concrete does not change much at elevated temperature, except for limestone (calcareous) aggregate concrete which decomposes above 800°C with a corresponding decrease in density.

The thermal conductivity of concrete is temperature dependent, and varies in a broad range, depending on the type of aggregate. Values from EC2 (1993) are shown in Figure 9.5. Approximate values for design purposes are 1.6 W/mK for siliceous concrete, 1.3 W/mK for calcareous (limestone) aggregate concrete, and 0.8 W/mK for lightweight concrete (EC2, 1993). Data for other types of concrete are given by Schneider (1988).

The specific heat of concrete also varies in a broad range, depending on the moisture content, with design values from EC2 (1993) shown in Figure 9.6. The peak between 100 and 200°C allows for water being driven off during the heating process. Approximate design values are 1000 J/kgK for siliceous and calcareous aggregate concrete, and 840 J/kgK for lightweight concrete (EC2, 1993).

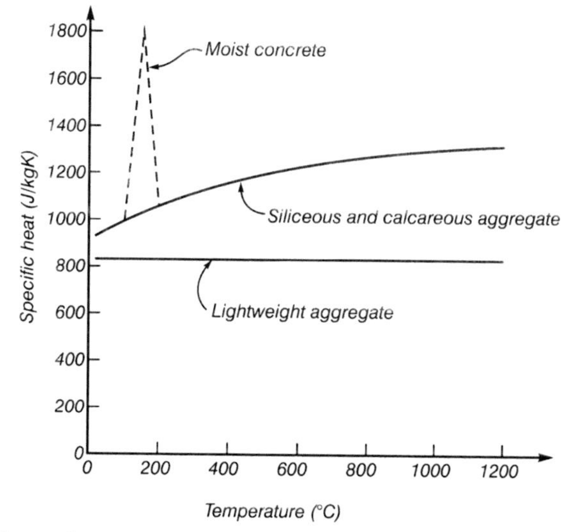

Figure 9.6 Specific heat of concrete (Reproduced from (EC2, 1993) by permission of CEN)

9.4.4 Published Temperatures

There is good published information available on temperatures within concrete members exposed to the standard fire (e.g. ACI, 1981, Fleischmann, 1995, Gustaferro and Martin, 1988 and Wade, 1991(b)). Most of these data have been derived from the work of Abrams and Gustaferro (1968). The availability of this information makes it much easier to design for standard fire exposure than for realistic fire temperatures, especially for the simple hand calculated design methods.

Slabs

Typical temperatures in concrete slabs exposed to the standard fire are shown in Figure 9.7 (from Wade, 1991(b), redrawn from ACI, 1981). Note that the distance from the fire-exposed surface is to the centre of the reinforcing steel, not the cover to the surface of the steel which is usually specified to control durability.

For exposure to typical real fires, very little published information is available on thermal gradients. Figure 9.8 shows typical peak temperatures reached at various depths in a concrete slab, calculated by Wade (1994) using the design fires proposed by Lie (1995) for a range of opening factors (ventilation factors) and a fuel load of 600 MJ/m^2 floor area. Figure 9.9 shows the progression of temperature *versus* time at various depths within the slab, for one of those real fires. It can be seen that temperatures within the slab continue to increase well beyond the time of 35 minutes when the fire reached its peak temperature. The greater the cover, the greater delay in reaching the peak temperature.

Beams

Figure 9.10 shows typical internal temperature contours at the corner of concrete beams exposed to the standard fire (from EC2, 1993).

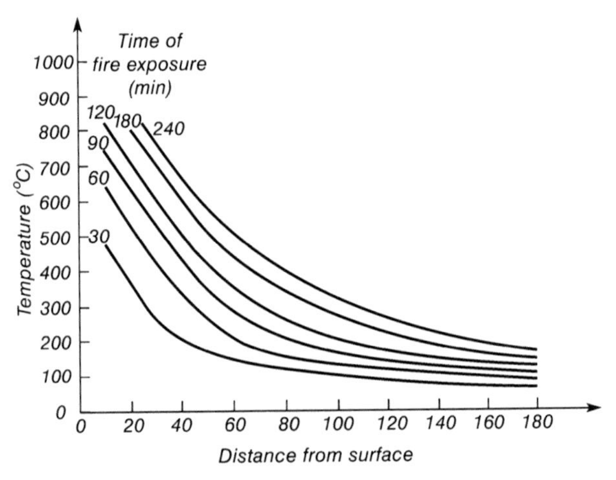

Figure 9.7 Temperatures in concrete slabs exposed to the standard fire (Reproduced from Wade (1991b) by permission of Building Research Association of New Zealand)

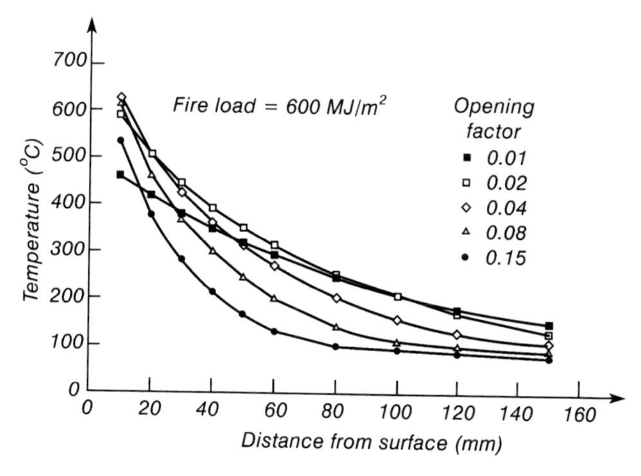

Figure 9.8 Peak temperatures in concrete slabs exposed to design fires (Reproduced from Wade (1994) by permission of Society of Fire Protection Engineers)

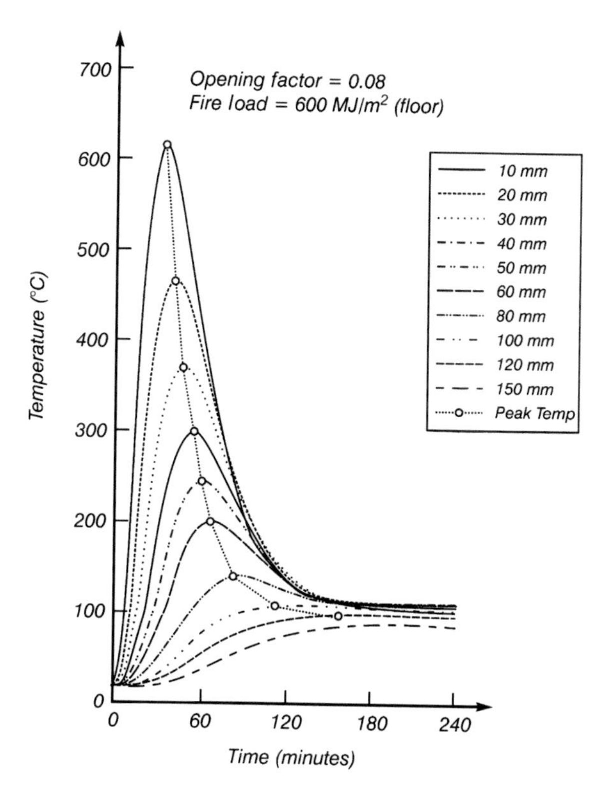

Figure 9.9 Temperature–time curves inside concrete slabs exposed to design fires (Reproduced from Wade (1994) by permission of Society of Fire Protection Engineers)

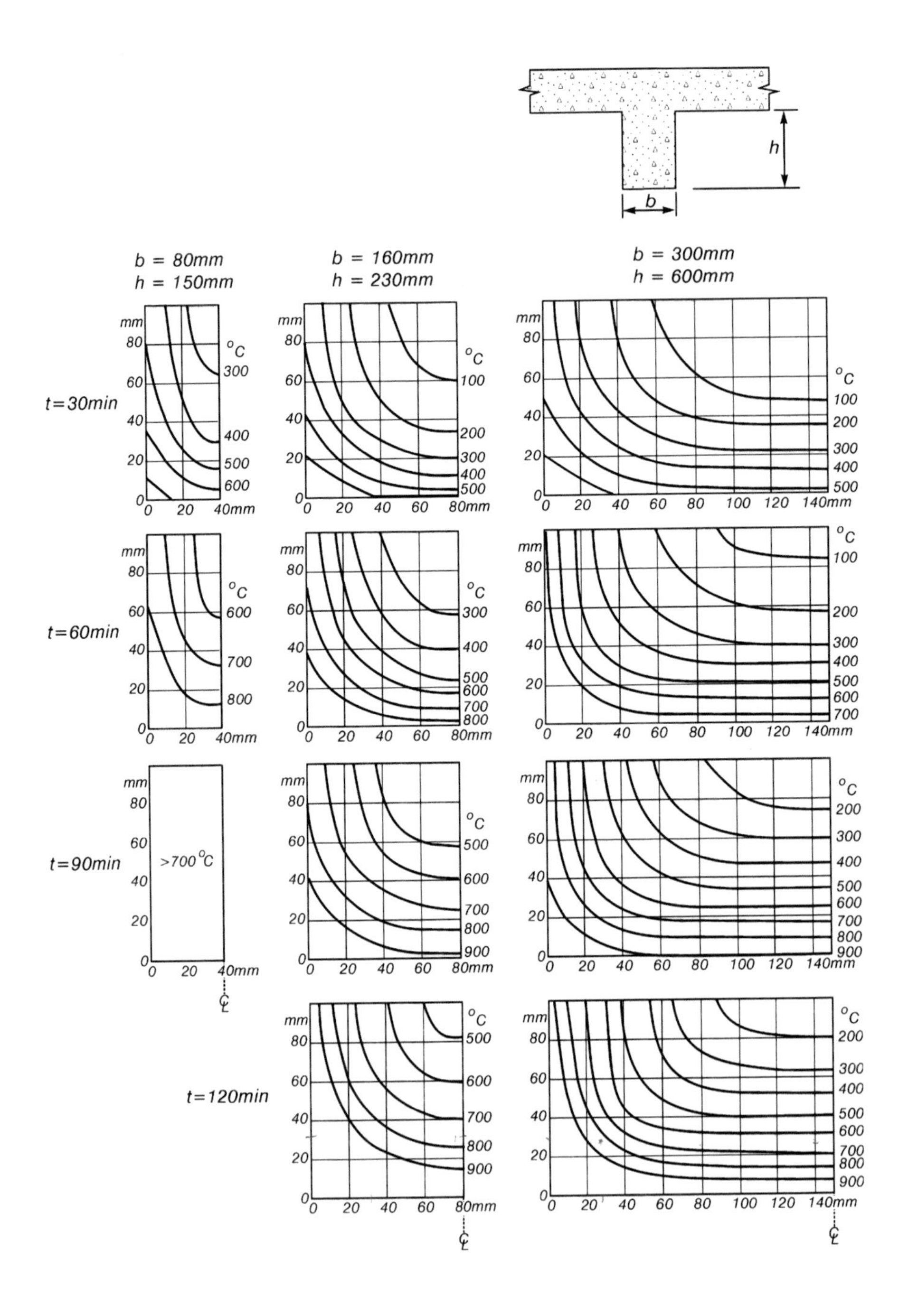

Figure 9.10 Temperature contours in concrete beams exposed to the standard fire (Reproduced from (EC2, 1993) by permission of CEN)

9.5 MECHANICAL PROPERTIES OF CONCRETE AT ELEVATED TEMPERATURES

9.5.1 Test Methods

The same test methods as described earlier for steel are applicable to concrete. Properties of concrete at elevated temperatures are described by Schneider (1986), Bazant and Kaplan (1996) and Harmathy (1993).

9.5.2 Components of Strain

The deformation of concrete at elevated temperatures is slightly more complicated than that of steel, because of an additional component of strain called transient strain. The deformation of concrete is usually described by assuming that the total strain ε consists of four components, being

$$\varepsilon = \varepsilon_{\mathrm{th}}(T) + \varepsilon_{\sigma}(\sigma,T) + \varepsilon_{\mathrm{cr}}(\sigma,T,t) + \varepsilon_{\mathrm{tr}}(\sigma,T) \tag{9.6}$$

where $\varepsilon_{\mathrm{th}}(T)$ is the thermal strain being a function only of temperature, T, $\varepsilon_{\sigma}(\sigma,T)$ is the stress related strain, being a function of both the applied stress σ and the temperature, $\varepsilon_{\mathrm{cr}}(\sigma,T,t)$ is the creep strain, being also a function of time, t, and $\varepsilon_{\mathrm{tr}}(\sigma,T)$ is the transient strain, being a function of both the applied stress and the temperature.

These components of strain are described in slightly different ways by different researchers (Anderberg, 1976, Schneider, 1988, Khoury *et al.*, 1985). Some details are given below. A slightly different strain model is given by Schneider *et al.* (1994).

For simple structures such as simply-supported beams, only the stress-related strain needs to be considered, allowing the reduced strength at elevated temperatures to be calculated without reference to the deformations. For more complex structural systems, especially where members are restrained by other parts of the structure, the thermal strain, creep strain and the transient strain must also be considered, using a computer model for the structural analysis.

9.5.3 Thermal Strain

Approximate expressions for thermal elongation $\Delta L/L$ of concrete are given by

$$\begin{aligned} \Delta L/L &= 18 \times 10^{-6} T_{\mathrm{c}} \text{ for siliceous aggregate concrete} \\ \Delta L/L &= 12 \times 10^{-6} T_{\mathrm{c}} \text{ for calcareous aggregate concrete} \\ \Delta L/L &= 8 \times 10^{-6} T_{\mathrm{c}} \text{ for lightweight concrete} \end{aligned} \tag{9.7}$$

where T_{c} is the concrete temperature. These values are from EC2 (1993) which gives more precise expressions for detailed calculations. It is difficult to separate thermal strain and shrinkage in tests, so the above expressions also include effects of shrinkage.

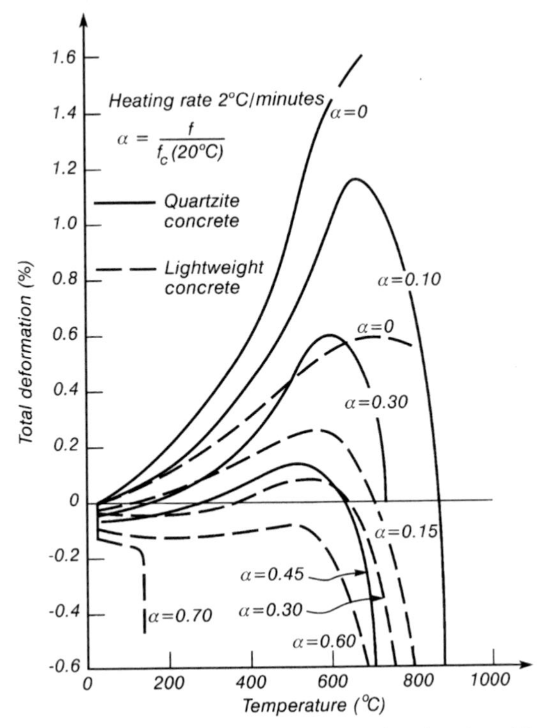

Figure 9.11 Total deformation in different concretes during heating (Reproduced from Schneider (1988) with permission from Elsevier Science)

9.5.4 Creep Strain and Transient Strain

Creep strain and transient strain are closely linked. If a concrete specimen is heated under load (Regime (5), Figure 7.6), all of the strain components described above combine to produce deformations as shown in Figure 9.11 (Schneider, 1988).

Khoury *et al.* (1985) have measured creep strains during testing under constant temperature and stress (Regime (3), Figure 7.6) producing results such as those shown in Figure 9.12. They also describe transient thermal strain, which occurs during the first time heating of concrete under load to 600°C, but not on subsequent heating. During all these processes there are complex changes in the moisture content and chemical composition of the cement paste, interacting with the aggregate which remains relatively inert (Schneider, 1988).

9.5.5 Stress-related Strain

The stress-related strain includes the elastic and plastic components of strain resulting from applied stresses. Typical stress–strain relationships for normal concrete at elevated temperatures are shown in Figure 9.13. It can be seen that the ultimate compressive strength drops, and the strain at peak stress increases with increasing temperature. Similar curves and corresponding equations are given by EC2 (1993).

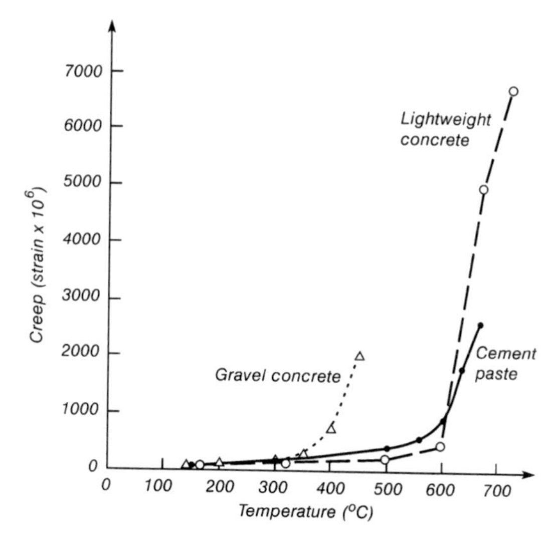

Figure 9.12 Creep in concrete one day after loading at 10% of the initial strength (Reproduced from Khoury and Sullivan (1988) with permission from Elsevier Science)

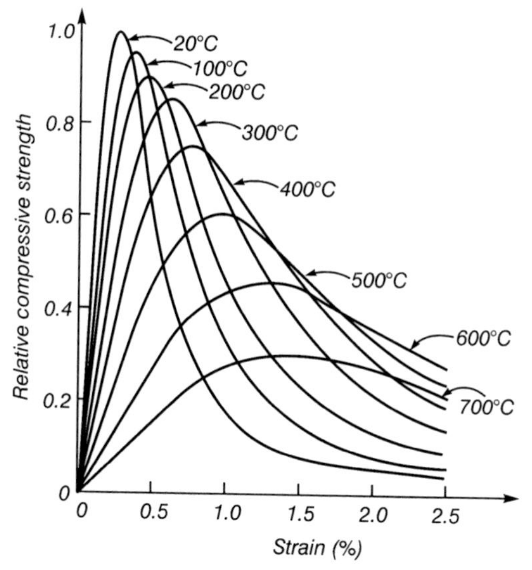

Figure 9.13 Stress–strain relationships for concrete at elevated temperatures (Reproduced from (EC2, 1993) by permission of CEN)

The reduction in ultimate compressive strength with temperature for typical structural concrete is shown in Figure 9.14 (Schneider, 1988), derived from several studies.

Confined concrete

It is well established that confinement of concrete by reinforcing such as hoops or ties gives a significant increase in ductility at normal temperatures. This is used to good

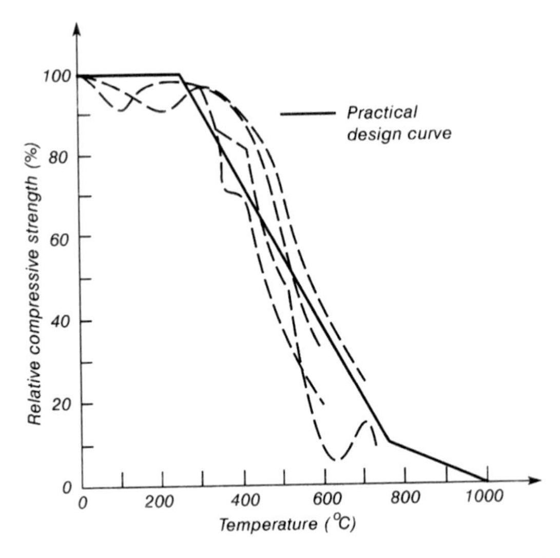

Figure 9.14 Reduction in compressive strength with temperature (Reproduced from Schneider (1988) with permission from Elsevier Science)

effect in the seismic design of concrete structures. No specific studies of such confined concrete under elevated temperature are known of, although Schneider (1988) reports studies showing that the ratio of biaxial compressive strength to uniaxial strength increases at elevated temperatures. It follows that confined concrete members designed for seismic attack probably have enhanced fire resistance as well. Franssen and Bruls (1997) describe how the flexural performance of a fire-exposed prestressed concrete tee-beam can be enhanced with confining reinforcing around the tendons.

Design values

Typical stress–strain curves for concrete at elevated temperatures have been shown above in Figure 9.13. The tensile strength of concrete is usually assumed to be zero at elevated temperatures. In similarity with steel properties, the reduction of ultimate strength with temperature is variable and a simple expression is necessary for design purposes. Figure 9.15 shows the lines used in BS 8110 (BSI, 1985) (also SAA, 1994, and SNZ, 1995 for normal weight concrete). More detailed expressions are given in EC2 (1993). The line for normal weight concrete in Figure 9.15 is given by

$$\begin{aligned} k_{c,T} &= 1.0 && \text{for } T < 350^\circ\text{C} \\ k_{c,T} &= (910 - T)/560 && \text{for } T > 350^\circ\text{C} \end{aligned} \tag{9.8}$$

The line for lightweight concrete in Figure 9.15 is given by

$$\begin{aligned} k_{c,T} &= 1.0 && \text{for } T < 500^\circ\text{C} \\ k_{c,T} &= (1000 - T)/500 && \text{for } T > 500^\circ\text{C} \end{aligned} \tag{9.9}$$

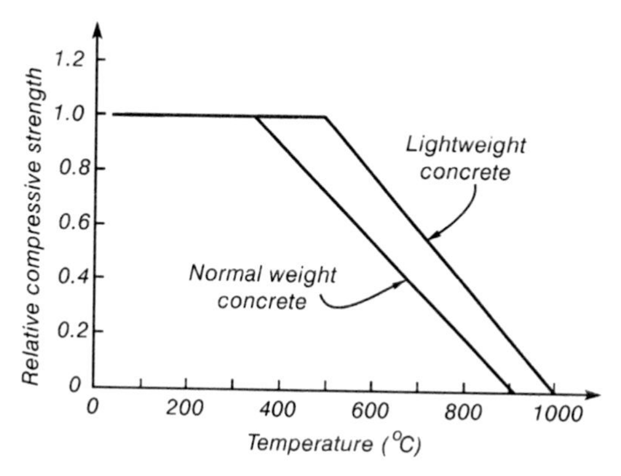

Figure 9.15 Design values for reduction of compressive strength with temperature

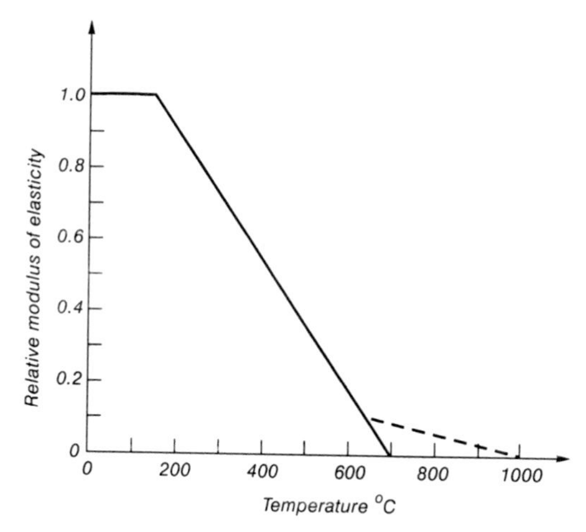

Figure 9.16 Design values for reduction of modulus of elasticity with temperature

Modulus of elasticity

The modulus of elasticity of concrete also drops with increasing temperature. Figure 9.16 shows the line used in BS 8110 (BSI, 1985). More detailed expressions are given in EC2 (1993). Lightweight and high-strength concretes behave similarly to normal weight concrete. The line in Figure 9.16 is given by

$$\begin{aligned} k_{E,T} &= 1.0 && \text{for } T < 150°\text{C} \\ k_{E,T} &= (700 - T)/550 && \text{for } T > 150°\text{C} \end{aligned} \qquad (9.10)$$

A problem occurs with the use of Figures 9.15 and 9.16 at high temperatures, because the compressive strength and the modulus of elasticity can be seen to reach zero at different temperatures. Because this is physically impossible, Inwood (1999) has proposed a minor alteration shown by the dotted line in Figure 9.16, in order to increase the temperature at which the modulus of elasticity reaches zero.

9.6 DESIGN OF CONCRETE MEMBERS EXPOSED TO FIRE

The overall strategy for the structural design of fire-exposed structures is the same as at normal temperature, but taking into account the effect of elevated temperatures on the material properties. In all cases, the design of concrete members should follow 'ultimate strength design' or 'limit states design' as used in all modern concrete design codes.

A hierarchy of design methods is as follows.

(1) For simply supported slabs or tee-beams exposed to fire from below, concrete in the compressive zone remains at normal temperatures, so the structural design need only consider the effect of elevated temperatures on the yield strength of the reinforcing steel. Simple hand calculations are possible.

(2) For continuous slabs or beams, some of the fire-exposed surfaces are in compression, so the simple hand calculation methods must consider the effects of elevated temperature on the compressive strength of the concrete.

(3) Similar methods can be applied to fire-exposed concrete walls and columns, but these methods are less accurate because of deformations caused by non-uniform heating and the possibility of instability failures.

(4) For moment-resisting frames, or structural members affected by axial restraint or non-uniform heating, it is recommended to use a special-purpose computer program for structural analysis under fire conditions.

Eurocode 2 (EC2, 1993) describes three methods of design: the generic 'tabulated' method, a 'simplified' calculation method and 'general' calculation methods. 'General' calculation methods include those which provide a realistic analysis of concrete structures exposed to fire, based on fundamental physical behaviour (design method (4) in the above list). Complex structures must be designed using general calculation methods, using a computer program for analysing the structure at elevated temperatures, including all the components of strain described above. Thermal gradients are also calculated by computer, so any type of fire exposure can be used.

The 'simplified' method is useful for single members, using the hand calculations which are used for design at normal temperatures (design methods (1) to (3) in the above list). It is essential to know the temperatures within the members. For standard fire exposure, these can be obtained from design charts, Wickström's method or by computer calculation. Computer calculation is essential for real fire exposure. For simply supported slabs or beams, only the reduction of steel strength needs to be considered because the heated concrete is all in the tensile zone. If heated concrete is in compression, in columns or continuous beams or slabs for example, the effect of temperature on concrete strength becomes important. The simplified calculation method ignores any concrete over a certain limiting temperature and may include a reduction in strength of the remaining cooler concrete core.

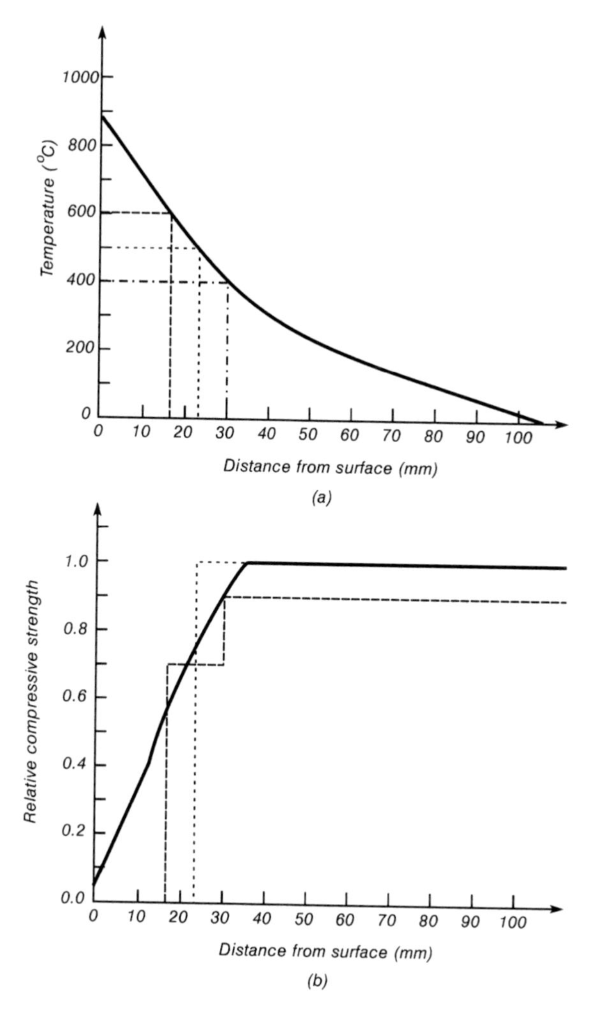

Figure 9.17 Temperature profile in a concrete slab, and the resulting reduction in compressive strength

There is no consensus on the limiting temperatures to use for the simplified method. The most simple approach, recommended here and used in the worked examples, is to assign full strength to concrete below 500°C and zero strength to concrete above 500°C. According to Anderberg (1993) the empirical charts for reduction in cross section and concrete strength with elevated temperature in Eurocode 2 (EC2, 1993) are based on this approach. This simplified method is excellent for large members, but may become inaccurate for thin concrete members where the temperature is over 500°C and the concrete has some residual strength.

As a refinement, Anderberg (1993) suggests using two layers, with 90% strength for concrete below 400°C, 70% strength for concrete between 400°C and 600°C and zero strength for concrete over 600°C. The extra computational complexity of a multi-layer method is not recommended here, except for thin members. Another approach is the method used by Gustaferro and Martin (1988) and Wade (1991(b)) which ignores concrete over 750°C and assigns a single strength to the remaining concrete core based on its average temperature. All of these simplified methods give similar results, with the more sophisticated methods requiring more time and providing slightly more accuracy. The method giving zero strength to concrete above 500°C is recommended here.

As a comparison of two simplified approaches, Figure 9.17(a) shows the temperature profile in a concrete slab after 60 minutes exposure to the standard fire calculated using Wickström's formula. The depth of the isotherms at 400°C, 500°C and 600°C are marked. Figure 9.17(b) shows the corresponding decrease in concrete compressive strength from Equation (9.8). The dotted line shows the effect of using the simple assumption of zero strength for concrete over 500°C, and the dashed line shows the effect of using two layers as recommended by Anderberg. The difference is seen to be relatively small.

9.6.1 Member Design

As for steel members, verification in the strength domain requires that

$$U^*_{\text{fire}} \leqslant R_{\text{fire}} \tag{9.11}$$

where U^*_{fire} is the design force resulting from the applied load at the time of the fire, and R_{fire} is the load-bearing capacity in the fire situation.

Applied loads have been described in Chapter 7. Design forces are obtained from conventional structural analysis. The design force U^*_{fire} may be axial force N^*_{fire}, bending moment M^*_{fire}, or shear force V^*_{fire} occurring singly or in combination, with the load capacity calculated accordingly as axial force N_{f}, bending moment M_{f} or shear force V_{f} in the same combination. Calculations of the load capacity are described below, based on the mechanical properties of concrete and reinforcing steel at elevated temperatures. Note that Equation (9.11) does not include a partial safety factor for mechanical properties γ_{M} (or a strength reduction factor Φ) because both have a value of 1.0 in fire conditions, as described in Chapter 7.

This section describes design of individual components, using the simplified calculation method with zero strength for concrete above 500°C. This design method uses the normal assumptions for reinforced concrete design, assuming that concrete has no tensile strength, and the parabolic compressive stress block in the concrete can be approximated by an equivalent rectangle. For the examples in this book, the equivalent rectangle is calculated assuming that the characteristic strength is 85% of the crushing strength of the concrete (Park and Paulay, 1975). For beams and slabs it is conservative to ignore any compression reinforcing, which simplifies the calculations.

9.6.2 Simply Supported Slabs and Beams

The most simple reinforced concrete members to design are simply supported slabs, such as shown in Figure 9.18. None of the compressive region is exposed to elevated temperatures, so the strength under fire conditions is solely a function of the

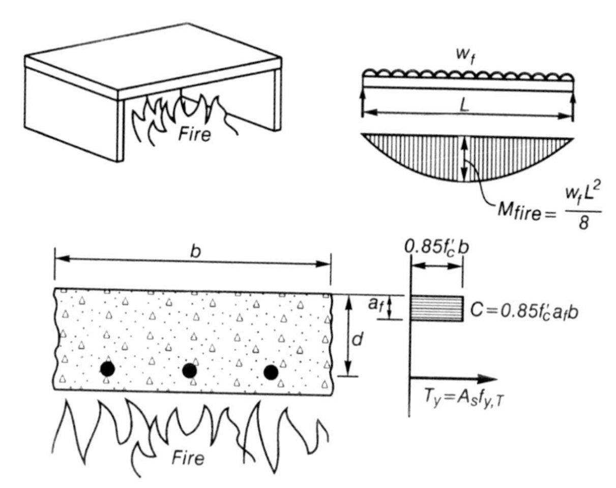

Figure 9.18 Simply supported slab exposed to fire

temperature of the reinforcing steel. There is no possibility of moment redistribution in any simply supported slabs or beams. The design equation for a member submitted to a bending moment M^*_{fire} is

$$M^*_{\text{fire}} \leqslant M_{\text{f}} \tag{9.12}$$

The flexural capacity under fire conditions M_{f} is given by

$$M_{\text{f}} = A_{\text{s}}\, f_{\text{y,T}}\, (d - a_{\text{f}}/2) \tag{9.13}$$

where A_{s} is the area of the reinforcing steel, $f_{\text{y,T}}$ is the yield stress of the reinforcing steel, reduced for temperature ($f_{\text{y,T}} = k_{\text{y,T}}\, f_{\text{y}}$), d is the effective depth of the cross section (distance from the extreme compression fibre to the centroid of the reinforcing steel), and a_{f} is the depth of the rectangular stress block, reduced by fire, given by

$$a_{\text{f}} = A_{\text{s}}\, f_{\text{y,T}}/0.85\, f'_{\text{c}}b \tag{9.14}$$

where f'_{c} is the characteristic compressive strength of the concrete, and b is the width of the beam or slab.

These calculations assume that the temperature of the concrete in the compressive zone is not high enough to cause any reduction in strength. A simply supported tee beam (Figure 9.19) has the same conditions, hence the same simple design procedure and equations.

A simply supported beam with a non-composite slab (Figure 9.20) is slightly more affected by fire because the two sides of the compressive zone of the beam are affected by elevated temperatures, so that the depth of the rectangular stress block now becomes

$$a_{\text{f}} = A_{\text{s}}\, f_{\text{y,T}}/0.85 f'_{\text{c}} b_{\text{f}} \tag{9.15}$$

where b_{f} is the fire-reduced effective width of the beam.

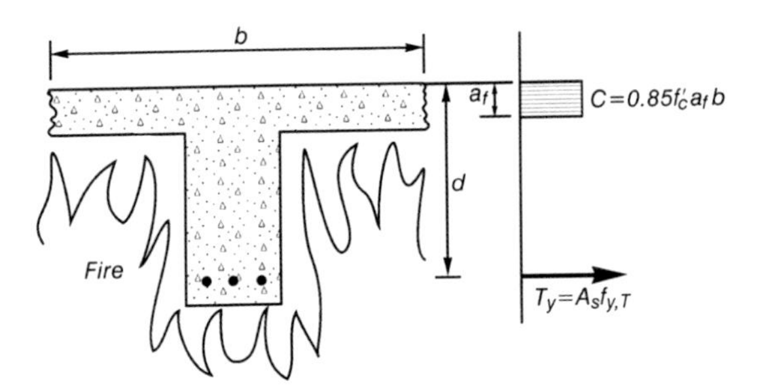

Figure 9.19 Simply supported Tee-beam exposed to fire

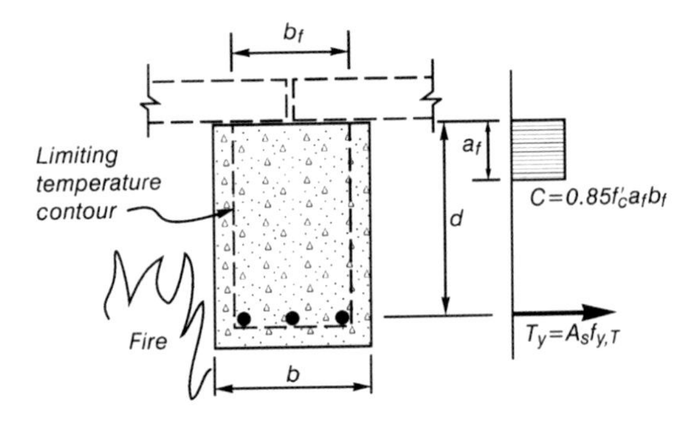

Figure 9.20 Simply supported non-composite beam exposed to fire

9.6.3 Shear Strength

Shear is not usually a problem in fire-exposed concrete structures, with the exception of precast pre-tensioned slabs with narrow webs. For shear design, Eurocode 2 (EC2, 1993) recommends using normal temperature design methods with the mechanical properties reduced for temperature and the cross section reduced to the 500°C contour. Franssen and Bruls (1997) have shown that the concrete contribution to shear strength reduces much more slowly than the contribution from the stirrup reinforcing, and any contribution from the prestressing force drops rapidly to zero as the concrete temperature increases. Shear failures have been observed in precast pre-tensioned hollow-core and double-tee slabs after loss of prestress due to bond failure near the ends of the slabs (Andersen and Laurisden, 1999).

9.6.4 Continuous Slabs and Beams

Slabs or beams which are built in to one or more supports usually have enhanced fire resistance because of the moment redistribution which must occur before a collapse mechanism can develop, as described in Chapter 7. If calculations show that the slab or beam can resist the fire as a simply supported member, no calculations for continuity are necessary.

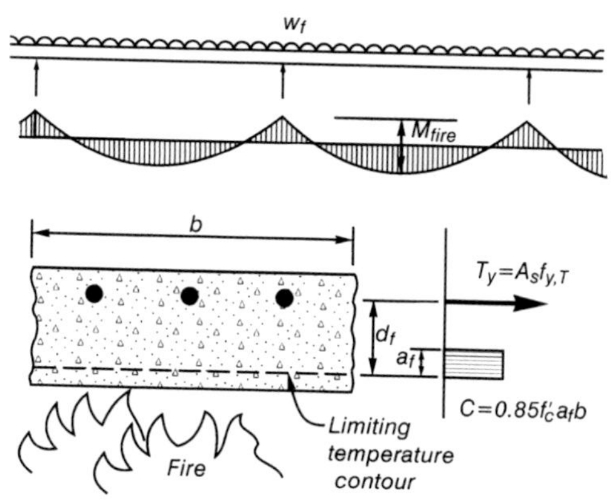

Figure 9.21 Support region of continuous slab exposed to fire

Plastic design

Reinforced concrete is very different from steel because the strength of a beam of a given size can have many possible values depending on the amount of reinforcing steel. Positive and negative flexural capacities may be very different for the same reason. The methods of moment redistribution and plastic design methods described in Chapter 7 can be used for analysis or design of reinforced concrete structures.

Negative flexural capacity

To allow for the effects of flexural continuity it is necessary to calculate the negative moment capacity at the supports during fire exposure. Part of the compressive region is now exposed to fire temperatures, which must be accounted for in the design process.

For a slab of uniform thickness, the negative flexural capacity M_f^- at the supports is given by the following equation, where the terms are illustrated in Figure 9.21.

$$M_f^- = A_s f_{y,T}(d_f - a_f/2) \tag{9.16}$$

where d_f is the effective depth of the slab, reduced to allow for the hot layer of concrete on the bottom surface, and a_f is the depth of the rectangular stress block for fire conditions, given by

$$a_f = A_s f_{y,T}/0.85 f'_c b \tag{9.17}$$

For a beam with its compression edge exposed to fire, the beam width b must be replaced by the effective width b_f of concrete below the critical temperature. The reduced cross section and reduced rectangular stress block are shown in Figure 9.22.

When the compression region of a slab or beam is exposed to fire, it becomes important to ensure that the compression capacity is not reduced so low as to cause a sudden compression failure. This can be ensured by checking that

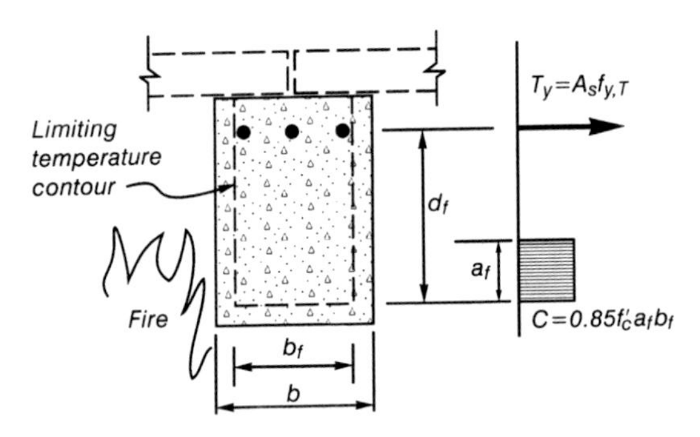

Figure 9.22 Support region of continuous beam exposed to fire

$$A_s\, f_{y,T}/b\, d_f\, f'_{c,T} < 0.30 \tag{9.18}$$

It can be shown that Equation (9.18) checks that the depth of the compression zone a_f is not more than 35% (0.3/0.85) of the effective depth of the cross section d_f. This check is not necessary if there is significant longitudinal reinforcing in the compression zone.

Curtailment of reinforcing bars

For continuous beams and slabs in reinforced concrete structures, major benefits can often be obtained by redistributing bending moments to achieve optimum behaviour, as described in Chapter 7. Redistribution of moments under fire conditions may change the location of the points of inflection, so the cut-off locations for reinforcing must allow the sections to develop the required flexural strength, with allowance for anchorage as required by national codes. Any redistribution of bending moments must be followed by a calculation check of the curtailment locations of the reinforcing bars, to avoid the possibility of structural failure due to top bars being terminated in regions of high tensile stress. Several sources recommend that the lengths of negative moment reinforcing bars should be increased by 15% of the span to avoid this problem. If adjacent spans are of different lengths, the bars on both sides of the support should be increased in length by 15% of the length of the longer of the two spans. It is also widely recommended that at least 20% of the negative moment reinforcing should be extended throughout the span of all beams.

Another related issue is the bond strength of deformed reinforcing bars in fire conditions, which has been investigated by Hertz (1982) and Schneider (1986) who show that bond strength drops by about half at 500°C, which can be a problem in some cases.

9.6.5 Axial Restraint

Axial restraint can have a significant influence on the fire resistance of reinforced concrete slabs and beams as described in Chapter 7. Axial restraint often has a more

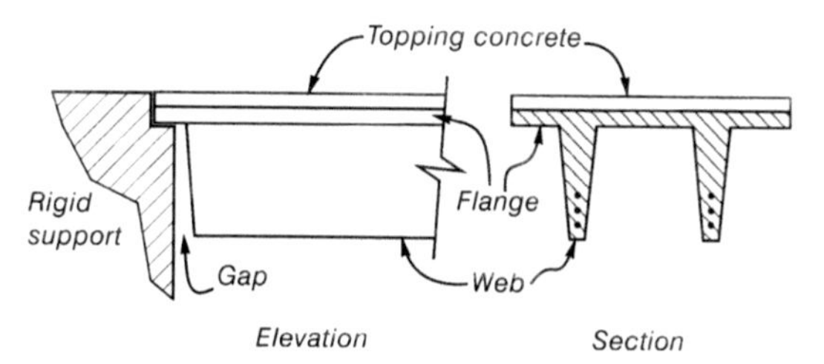

Figure 9.23 Unsatisfactory axial restraint in flange-supported double-Tee floor

profound influence on concrete structures than on steel structures because the more rapid heating and more ductile behaviour of steel structures can result in large vertical deflections which reduce the horizontal axial restraint forces. Compared with steel structures, concrete members tend to heat up more slowly, but when they become hot they tend to undergo larger horizontal displacements with lower vertical deflections, resulting in larger horizontal restraint forces if the building offers sufficient resistance to the thermal expansion.

Axial restraint is beneficial for reinforced or prestressed concrete slabs or beams, and for composite concrete and metal deck slabs, where the axial restraint force can partly or completely compensate for the loss in strength of steel reinforcing at elevated temperatures. As pointed out in Chapter 7, it is essential for the line of thrust to be below the compressive stress block if the beneficial effects of axial restraint are to be utilized. Figure 9.23 shows a double-tee concrete floor unit with cut-away webs, supported on the flange, where the line of action of the axial thrust is so high that premature failure could occur during fire exposure.

PCA method for calculating restraint

A semi-empirical method of calculating the required strength and stiffness of the surrounding structure is given by Gustaferro and Martin (1988) based on a large series of tests by the Portland Cement Association which showed that the thermal thrust for a given expansion varied directly with the heated perimeter of the member and the modulus of elasticity of the concrete. The applicability of this approach to slabs and beams other than those tested has not been demonstrated. Anderberg and Forsen (1982) have shown that this method does not give accurate results in many cases because it over-predicts the thermal strains. The PCA method is included here because it is the only method of assessing restraint without a comprehensive computer-based analysis package.

A step-by-step guide to the PCA procedure is as follows.

(1) Calculate the bending moment at midspan under fire-reduced loads M^*_{fire} (kNm) assuming simply supported behaviour.

(2) Calculate the flexural capacity at midspan during the fire M_f^+. If $M_f^+ > M^*_{fire}$ no continuity or restraint is necessary.

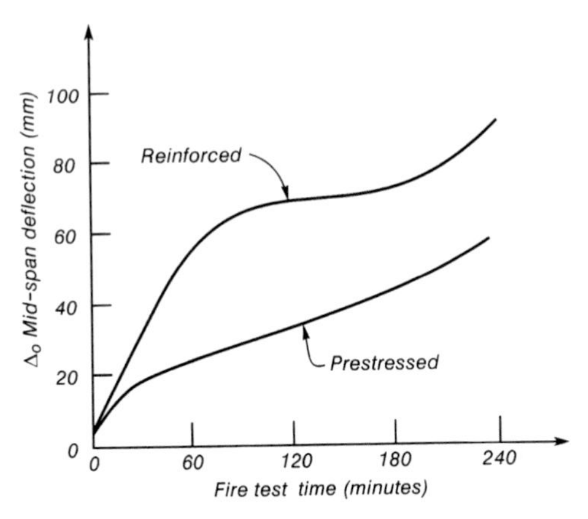

Figure 9.24 Mid-span deflection of reference specimens (Reproduced from (Gustaferro and Martin, 1988) by permission of Prestressed Concrete Institute)

(3) Calculate the flexural capacity at the supports during the fire M_f^-. If $M_f^+ + M_f^- > M^*_{fire}$ continuity is sufficient and no restraint is necessary. If the member is not symmetrical, the flexural capacity M_f^- should be the average of the two ends.

(4) Estimate the mid-span deflection Δ using

$$\Delta = L^2 \Delta_0 / 89\,000\ y_b \tag{9.19}$$

where Δ_0 is the mid-span deflection of the reference specimen, from Figure 9.24 (mm), L is the heated length of the member (mm), and y_b is the distance from the neutral axis of the member to the extreme bottom fibre (mm).

(5) Estimate the distance of the line of thrust from the top of the member d_T (mm) at the supports. For built-in construction assume that the line of thrust is 0.1 h above the bottom of the member where h is the overall depth of the member. For other support conditions an independent estimate may be necessary.

(6) Calculate the magnitude of the required axial thrust T (kN) to prevent collapse, using

$$T = 1000 \frac{M^*_{fire} - (M_f^+ + M_f^-)}{d_T - a_f/2 - \Delta} \tag{9.20}$$

where a_f (mm) is the height of the internal rectangular compression stress block in the member, approximated by $a_f \approx a M^*_{fire}/M_f^+$ where a is a_f from Equation (9.14).

(7) Calculate a_f more accurately, using

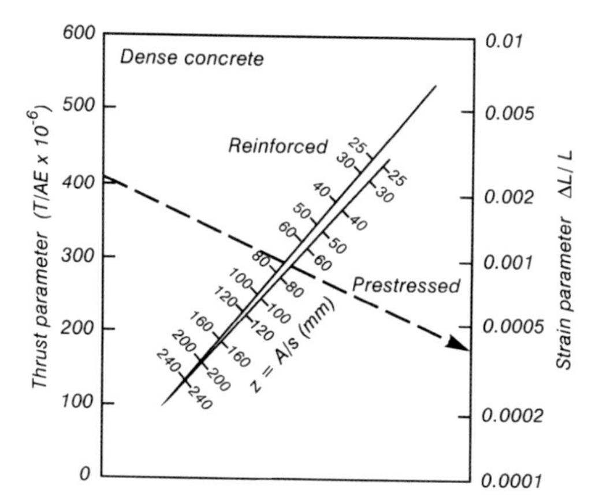

Figure 9.25 Nomogram for thrust in concrete members (Reproduced from (Gustaferro and Martin, 1988) by permission of Prestressed Concrete Institute)

$$a_{\mathrm{f}} = \frac{T + A_{\mathrm{s}}^{+} f_{\mathrm{y,T}}^{+}}{0.85 f'_{\mathrm{c,T}} b_{\mathrm{f}}} \tag{9.21}$$

where $A_{\mathrm{s}}^{+} f_{\mathrm{y,T}}^{+}$ is the tensile strength of the bottom steel at midspan. Repeat steps (6) and (7) if necessary to get convergence.

(8) Calculate the non-dimensional thrust parameter T/AE and the effective thickness $z = A/s$ (mm) where A is the cross-sectional area of the member (mm^2), E is the modulus of elasticity of concrete (usually about 25 GPa), and s is the heated perimeter of the member (mm).

(9) Determine the strain parameter $\Delta L/L$ from Figure 9.25 using T/AE and z. (Enter Figure 9.25 with T/AE, pass through diagonal line at calculated value of z, and read-off $\Delta L/L$. Example shown by dashed line)

(10) Calculate the maximum permitted displacement $\Delta L/L$ (mm) by multiplying the strain parameter $\Delta L/L$ by the heated length L (mm).

(11) Determine independently whether the surrounding structure can withstand the thrust T with a displacement no greater than ΔL. If so, the structure can withstand the fire.

9.6.6 Columns

Columns are more difficult to design than flexural members because of possible instability problems. There are a range of design methods in the literature which will be described briefly. At the most simple generic level, minimum dimensions such as those shown in Table 9.1 can be used. A slight advance on this is Table 9.2 from Eurocode 2 (EC2, 1993) which gives minimum dimensions and cover as a function of

Table 9.2 Generic fire-resistance ratings for concrete columns (EC2, 1993)

	Column exposed on more than one side						Exposed on one side	
Load ratio	0.2		0.5		0.7		0.7	
	Width (mm)	Cover (mm)	Width (mm)	Cover (mm)	Width (mm)	Cover (mm)	Width (mm)	Cover (mm)
30 minutes	150	10	150	10	150	10	100	10
60 minutes	150	10	180	10	200	10	120	10
90 minites	180	10	210	10	240	35	140	10
120 minutes	200	40	250	40	280	40	160	45
180 minutes	240	50	320	50	360	50	200	60
240 minutes	300	50	400	50	450	50	300	60

Width is the minimum dimension of the column and cover is the location of the centre line of the reinforcing relative to the outside of the column. Cover required for durability will control in some cases.

load level, for columns fully or partially exposed to fire. Some of the sizes in Table 9.2 are smaller than those in Table 9.1 and some of these are unsafe (Franssen, 2000).

Some codes such as the Canadian Code (NBCC, 1995) have empirical formulae (from Harmathy, 1993) which are based on the results of fire tests, but give no insight into the fire performance of the column. The recommended conservative design approach is to use the simplified method assuming zero strength for all concrete above 500°C (or 400°C for high-strength concrete) and normal temperature design formulae.

A more sophisticated method proposed by Anderberg (1993) is to carry out a thermal analysis to identify the 400°C and 600°C contours, assuming an approximate rectangular shape, and to design the column using normal temperature design methods, assuming that the concrete over 600°C has no strength, the concrete above 400°C has 70% strength and the inner core has 90% strength. Slenderness can be calculated using the dimensions of the cross section less than 600°C. For columns with one reinforcing bar in each corner (four bars total), Anderberg considers them to have 50% of their normal temperature yield strength. For columns with more than four bars, he ignores the corner bars and considers the remaining bars to have 50% strength. This procedure can be used for columns with concentric or eccentric loading.

Recent fire tests of reinforced concrete columns have been carried out by Lie and Irwin (1993) and Dotreppe *et al.* (1996), who tested columns up to 400×400 mm in cross section and up to 4 m long. Dotreppe *et al.* (1996) found that the tested columns achieved satisfactory fire resistance, but there was rather wide scatter in the results, columns with large diameter bars (25 mm) performed more poorly than expected, and the effect of increasing the concrete cover was less than assumed in the Eurocode 2 tables.

9.6.7 Walls

Concrete walls or partitions which are not part of the main load-bearing structure do not require structural design. Fire-resistance requirements can be met by providing a minimum thickness to meet the insulation criterion. This thickness is often the same as required for slabs, as shown in Table 9.1, but minimum cover requirements may be different. Load-bearing walls should be considered to be axial load-carrying members, and should be designed as slender columns, using the tables or calculation methods given for columns. A difference is that columns are most often designed for fire exposure on all sides, but most walls are exposed to fire on only one side. In rare cases they may be exposed to fire on both sides.

O'Meagher and Bennetts (1991) have shown that the load-bearing capacity of reinforced concrete walls exposed to fire is very sensitive to the top and bottom end conditions. Computer analysis shows that load-bearing walls with pinned connections at the top and the bottom have low fire resistance, but walls built into the structure with some continuity at top and bottom have far greater fire resistance because the deflections of the walls are greatly reduced.

For industrial buildings, Lim (2000) has shown how the SAFIR program can be used to model cantilever walls, either free-standing walls or with the top of the wall connected to a steel roof structure. Free-standing cantilever walls with a single layer of central reinforcing have considerable fire resistance because they tend to deflect outwards, away from the fire, resulting in the compressive face of the wall being on the cool side, with the reinforcing protected by a thick layer of cover concrete.

9.6.8 Frames

There are no hand methods available for the structural design of reinforced concrete frame structures exposed to fires, if frame action, continuity and restraint are to be properly considered. Individual members can be designed by the methods described above, but a special purpose computer program is necessary for detailed analysis and design of significant structures. Available programs include CONFIRE (Forsen, 1982), FIRES-RC-II (Iding *et al.*, 1977(b)), TCD (Anderberg, 1989) and SAFIR (Franssen, 2000). The detailing requirements described above for slabs and beams also apply to reinforced concrete frame structures.

9.7 COMPOSITE STEEL–CONCRETE CONSTRUCTION EXPOSED TO FIRE

'Composite steel–concrete construction' most often refers to concrete slabs cast on permanent steel-deck formwork and steel beams which act compositely with a concrete slab to resist flexure. Calculation of temperatures in composite structures has been described in Chapter 8. This form of construction is illustrated in Figure 8.10. Composite steel–concrete construction also includes hollow steel sections filled with concrete, or other shapes of steel sections fully or partially surrounded by concrete, as shown in Figures 8.11 to 8.14.

9.7.1 Composite slabs

Composite steel–concrete slabs are popular because they eliminate the need for re-useable formwork and are light enough to be installed over large areas without heavy lifting equipment. The steel decking material acts as permanent formwork and as external reinforcing. There are a number of possible profiles available, all of which have deformations to ensure some bond between the steel and the concrete to carry shear forces and to resist delamination. Fire behaviour is discussed under the three categories of integrity, insulation and stability.

Integrity

Composite steel–concrete slabs generally have excellent integrity because even if cracks occur in the concrete slab, the continuous steel deck will prevent any passage of flames or hot gases through the floor system (Figure 8.38(b)).

Insulation

To meet the insulation criterion it is simply necessary to provide sufficient thickness of slab. A solid slab of uniform thickness would require the same thickness as a normal reinforced concrete slab. For trapezoidal or dovetail profiles it is necessary to evaluate an effective thickness. Generic listings are given in some codes including BS 5950 Part 8 (BSI, 1990(a)) and Eurocode 4 (EC4, 1994). All manufacturers of steel decking have proprietary ratings for their products which give this information. It is possible to spray the underside of the steel sheeting with spray-on insulation, but this is rarely economical. Calculation of temperatures has been discussed in Chapter 8.

Stability

The strength of composite steel–concrete slabs is severely influenced by fire because the steel sheeting acting as external reinforcing loses strength rapidly when it is exposed to the fire. However, composite slabs have been shown to have good fire resistance because of three contributing factors: axial restraint, moment re-distribution and fire emergency reinforcing.

Composite slabs often have different fire-resistance ratings for restrained and unrestrained conditions (e.g. UL, 1996). During a fire test, if a composite slab is built into a rigid testing frame which allows almost no axial expansion (Figure 6.2), the slab can achieve a fire-resistance rating with no reinforcing other than the external steel sheeting, because of the thermal thrust developed at the supports (see Chapter 7). Some buildings are sufficiently stiff and strong to provide such restraint to a fire-exposed floor system, but this is difficult to assess accurately, so it is usual to rely on some reinforcing within the slab.

Composite steel–concrete slabs usually have nominal reinforcing consisting of welded wire mesh or tied bars to control any cracking caused by shrinkage or overloading. If this is placed near the top of the concrete and if the slab is continuous over several supports, the slab can develop significant negative moment capacity over the supports (hogging moment) through moment redistribution, and hence retain sufficient load capacity during the fire.

If a slab is simply supported, or if moment redistribution is insufficient to resist the applied loads, it is common practice to place 'fire emergency reinforcing' in the slab, consisting of steel reinforcing bars in the troughs of the sheeting, with sufficient cover (25 to 50 mm) from the bottom surface to control temperature rise in the bars. If the temperature of the bars is known, the flexural strength of the composite slab can be calculated as for a conventional reinforced concrete slab. Design recommendations are given by ECCS (1983), Lawson (1985) and EC4 (1994). As described in Chapter 8, composite steel–concrete slabs have been observed to behave well in fires when they are part of a large composite structure. If large deflections occur, such slabs can develop tensile membrane action as described by Wang (1996).

9.7.2 Composite beams

Hot-rolled steel beams

Composite steel–concrete beams generally consist of reinforced concrete slabs supported by hot-rolled structural steel beams, connected together to obtain composite structural action. The most common system is for composite steel-deck slabs to run over the top of the steel beams as shown in Figure 8.10, but the beams are sometimes partially embedded in the slab as shown in Figure 8.11 with precast concrete slabs. The required shear connection between the steel beam and the concrete slab is usually provided by shear studs welded to the top of the beam (Figure 8.10). Composite beams act as tee-beams, with the slab in compression under positive (sagging) moments and the slab in tension under negative (hogging) moments.

Several alternative methods of thermal analysis are available, as described in Chapter 8. If the slab consists of a composite steel-deck slab spanning between the beams, there will be intermittent voids between the underside of the slab and the top of the beam, so the beam will be exposed on all four sides in these regions, which must be considered in the thermal calculations (Newman and Lawson, 1991). The structural calculation for fire is essentially the same procedure as in normal temperature conditions. Detailed design methods are given by BS 5950 Part 8 (BSI, 1990(a)) and Eurocode 4 (EC4, 1994), but an approximate design is often sufficient.

For a simply supported composite section consisting of a concrete slab on a steel beam, the flexural strength for positive moment (sagging) can be approximated by assuming that the slab resists all the compression force, and the bottom flange of the beam resists all of the tensile force. The top flange of the steel beam contributes little force because it is close to the neutral axis of the cross section. In this case the bottom flange can be assumed to be an axial tension member, and the strength can be calculated using the methods from Chapter 8. If the strength calculated in this way is insufficient, the tensile contribution of the web can be included to give a

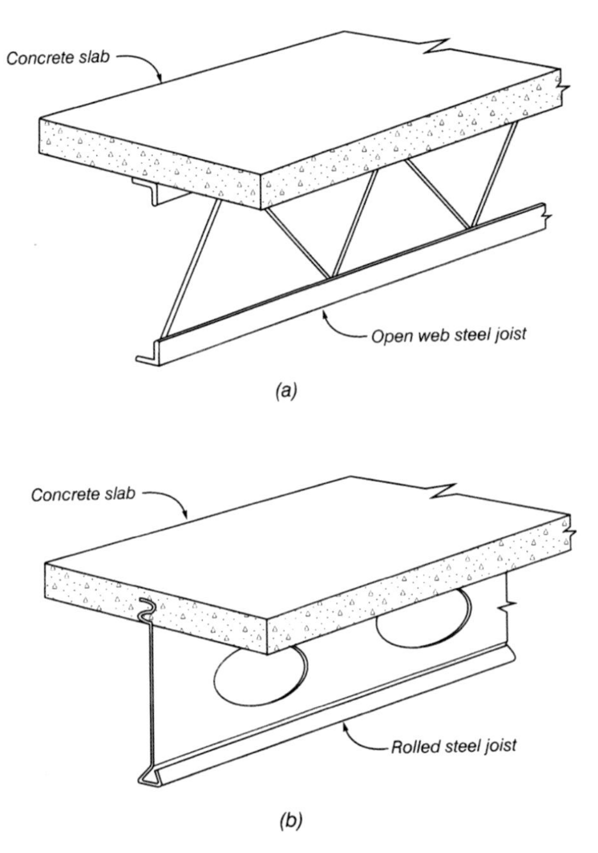

Figure 9.26 Composite construction with light steel joists

more accurate result. It is unlikely that the compressive strength of the slab will become limiting.

For negative moment (hogging) near the supports, this situation is reversed such that the slab is in tension and the bottom flange is in compression. The bottom flange can now be considered to be a compression member. The flange is restrained vertically by the web, but lateral buckling must be considered by using calculation methods for a column or an unrestrained beam, depending on the distance between any lateral restraints. The reinforcing in the slab should be checked for axial tensile capacity, which can be assisted by the top flange of the beam if necessary. If the slab has steel tray decking, that can enhance the flexural capacity of the cross section only if it is running parallel to the beam.

The capacity of the shear studs between the beam and the slab is unlikely to be seriously affected by fire exposure. BS 5950 Part 8 (BSI, 1990(a)) does not require any check on the studs, but Eurocode 4 (EC4, 1994) gives a formula based on normal temperature behaviour with the temperature at the base of the stud being 80% of the top flange temperature, and the concrete temperature being half of that at the base of the stud.

Light steel joists

A common system of composite construction uses open web steel joists or light gauge cold-rolled steel joists combined with concrete, as shown in Figure 9.26. These systems often include a special formwork system that allows the light steel joists to support the boxing (formwork) with no additional propping (shoring). These low cost systems are mainly designed as simply supported spans because the negative flexural capacity is very low. The fire resistance of this type of composite construction is very poor without additional protection to the steel joists, because of the very high section factor (low effective thickness) of the steel, which will heat up and lose strength rapidly when exposed to a post-flashover fire. Additional fire resistance can be provided with a fire resisting ceiling membrane or with fire protection material sprayed on to the steel elements.

9.7.3 Composite columns

Considerable research has been carried out on concrete filling in hollow steel columns exposed to fires (Figure 8.14). The concrete filling has three beneficial effects; it acts as a heat sink to slow the rise in temperature of the steel column, it provides lateral stability to prevent local buckling of the column wall and it can carry some or all of the axial load as the strength of the steel reduces. In some cases of low fire exposure, a thermal analysis may show that the steel column can carry all the axial load, relying on the concrete only to slow the increase in steel temperature. For more serious fire exposure it is more likely that the heated concrete filling can carry the applied loads without any contribution from the hot steel member. Drain holes at least 12.5 mm diameter must be provided at every floor level to prevent bursting due to pressure of steam from evaporating moisture in the heated concrete during a fire.

Many tests have been carried out using the standard fire exposure. Design equations for concrete filled steel columns are available from several sources (e.g. BSI, 1990(a) and ECCS, 1988) and tables of fire-resistance ratings are given in Eurocode 4 (EC4, 1994). Lie and Kodur (1996) have tested many columns, leading to a design formula in the National Building Code of Canada (NBCC, 1995) for one and two hour fire-resistance ratings. Kodur (1999) has extended the applicability of the equation, showing that fibre-reinforced concrete is similar to conventional reinforced concrete, with greater fire resistance than plain concrete filling. The Canadian empirical design equation gives the fire resistance t_r (minutes) of a circular or square steel column completely filled with concrete, as

$$t_r = \frac{f(f'_c + 20)\, d^{2.5}}{(KL - 1000)\sqrt{N}} \tag{9.22}$$

where f is a factor from Table 9.3, f'_c is the strength of the filling concrete (MPa), d is the outside diameter or width of the column (mm), L is the unsupported length of the column (mm), KL is the effective length of the column, considering the end support conditions (mm), and N is the applied load on the column (kN).

Table 9.3 Value of factor f for fire resistance of concrete-filled steel columns (Kodur, 1999)

Filling concrete	Square columns	Circular columns
Plain concrete	0.06	0.07
Bar-reinforced concrete	0.065	0.075
Fibre-reinforced concrete	0.065	0.075

Note that tabulated values of f may be increased cumulatively as follows.
Tabulated values are for siliceous aggregate concrete. For carbonate aggregate concrete, add 0.01.
Bar-reinforced concrete values are for cover <25 mm. For cover $\geqslant 25$ mm, add 0.005.
Bar-reinforced concrete values are for reinforcing $<3\%$. For reinforcing $\geqslant 3\%$, add 0.005.

This formula is valid for fire-resistance times up to 2 hours for plain concrete and 3 hours for reinforced concrete, and for column sizes from about 140 mm to 410 mm, except that bar reinforcing cannot be used in columns smaller than about 200 mm. Square columns with fibre-reinforced concrete can be as small as 100 mm. The width to thickness ratio should not exceed 'class 3' according to the Canadian steel design code.

The design methods described above are for standard fire exposure. Design methods for exposure to real fire time–temperature curves are not well established, but an approximate calculation can be made by carrying out a thermal analysis, then summing the load-resisting contribution of the steel and the filling concrete calculated separately.

WORKED EXAMPLE 9.1
SIMPLY SUPPORTED REINFORCED CONCRETE SLAB

(Refer to Figure 9.18)

For a simply supported reinforced concrete slab with known span, load, geometry and reinforcing, check the flexural capacity after 60 minutes exposure to the standard fire. Use Wickström's formula to calculate the reinforcing temperature.

Given information:

Slab span	$L = 7.0\,\text{m}$	Dead load $G_1 = 0.5\,\text{kN/m}$
Slab thickness	$h = 200\,\text{mm}$	(excluding self weight)
Concrete density	$\rho = 24\,\text{kN/m}^3$	Live load, $Q = 2.5\,\text{kN/m}$
Concrete strength	$f'_c = 30\,\text{MPa}$	
Yield stress	$f_y = 300\,\text{MPa}$	
Bar diameter	$D_b = 16\,\text{mm}$	Bar spacing, $s = 125\,\text{mm}$
Bottom cover	$c_v = 15\,\text{mm}$	

Design a 1 m wide strip $b = 1000\,\text{mm}$

Self weight $G_2 = \rho h b = 4.8\,\text{kN/m}$

Total dead load	$G = G_1 + G_2 = 0.5 + 4.8 = 5.3$ kN/m
Steel area	$A_s = n\pi r^2 b/s = 1608\ \text{mm}^2$
Effective depth	$d = h - c_v - D_b/2 = 177$ mm
Effective cover	$c_e = c_v + D_b/2 = 23\ \text{mm} = 0.023$ m

COLD CALCULATIONS (for a 1 m wide strip)

Strength reduction factor	$\Phi = 0.85$
Stress block depth	$a = A_s f_y / 0.85 f'_c b = 1608 \times 300 / 0.85 \times 30 \times 1000$ $= 18.9$ mm
Internal lever arm	$jd = d - a/2 = 177 - 18.9/2 = 168$ mm
Design load (cold)	$w_c = 1.2\,G + 1.6\,Q = 10.4$ kN/m
Bending moment	$M^*_{cold} = w_c L^2/8 = 10.4 \times 7.0^2/8 = \underline{63\ \text{kN/m}}$
Bending strength	$M_n = A_s f_y\, jd = 1608 \times 300 \times 168/10^6 = 81$ kN/m
	$\Phi M_n = 69$ kN/m
	$\Phi M_n > M^*_{cold}$ so design is OK.

FIRE CALCULATIONS

Revised strength reduction factor	$\Phi = 1.0$
Design load (fire)	$w_f = G + SW + 0.4Q \qquad w_f = 6.3$ kN/m
Bending moment	$M^*_{fire} = w_f L^2/8 = 6.3 \times 7.0^2/8 = \underline{38.6\ \text{kNm}}$
After 60 minutes of standard fire exposure	$t = 60$ min ($t_h = 1.0$ hour)
Fire temperature	$T_f = 20 + 345 \log(8t + 1) = 945°\text{C}$
Surface temperature	$T_w = [1 - 0.0616\ t_h^{-0.88}]\ T_f$ $= [1 - 0.0616 \times 1.0^{-0.88}] \times 945 = 887°\text{C}$
Concrete temperature	$T_c = [0.18 \ln(t_h/c_e^2) - 0.81]\ T_w$ $= [0.18 \ln(1.0/0.023^2) - 0.81] \times 887 = 486°\text{C}$
Steel temperature	$T_s = T_c = 486°\text{C}$
Reduced yield stress	$f_{y,T} = f_y(720 - T_s)/470 = 300(720-486)/470 = 149$ MPa
Stress block depth	$a = A_s f_{y,T}/0.85 f'_c b = 1608 \times 149/0.85 \times 30 \times 1000$ $= 9.4$ mm
Internal level arm	$jd = d - a/2 = 177 - 9.4/2 = 172$ mm
Bending strength	$M_{nf} = A_s f_{y,T}\, jd = 1608 \times 149 \times 172/10^6 = 41.2$ kN.m
	$\Phi M_{nf} = 1.0 \times 41.2 = 41.2$ kN.m
	$\Phi M_f > M^*_{fire}$ so design is OK.

WORKED EXAMPLE 9.2
REINFORCED CONCRETE BEAM
(Refer to Figure 9.20 and Figure 9.27)

For a simply supported reinforced concrete beam with known span, load, geometry and reinforcing, check the positive flexural capacity after 90 minutes exposure to the standard fire.

Given information:

Beam span	$L = 15.0\,\text{m}$	Dead load $G_1 = 6.0\,\text{kN/m}$ (excluding self weight)
Beam width	$b = 400\,\text{mm}$	
Beam depth	$h = 800\,\text{mm}$	Live load $Q = 12.5\,\text{kN/m}$
Bottom cover	$c_v = 25\,\text{mm}$	Concrete density $\rho = 24\,\text{kN/m}^3$
Bar diameter	$D_b = 32\,\text{mm}$	Concrete compressive strength $f'_c = 30\,\text{MPa}$
Number of bars	$n = 8$ (2 rows of four bars)	Steel yield stress $f_y = 300\,\text{MPa}$
Area of one bar	$A_{s1} = \pi r^2 = 804\,\text{mm}^2$	
Total steel area	$A_s = n\pi r^2 = 6434\,\text{mm}^2$	

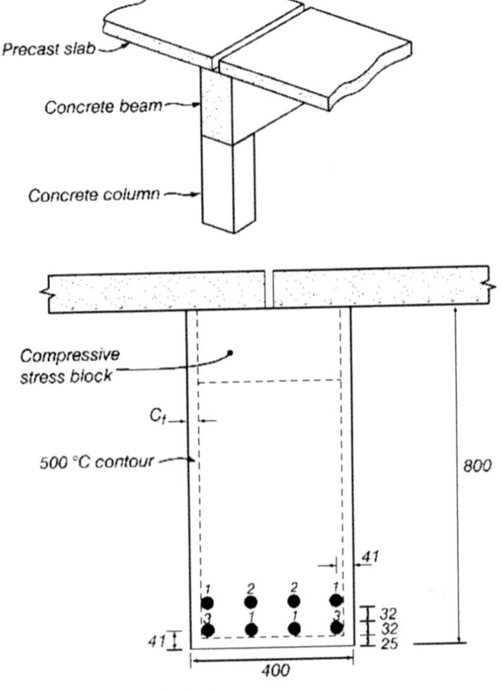

Figure 9.27 Beam for Worked Example 9.2

Effective depth	$d = h - c_v - 1.5\ D_b = 800-25-48 = 727$ mm
Self weight	$G_2 = \rho bh = 24 \times 0.4 \times 0.8 = 7.7$ kN/m
Total dead load	$G = G_1 + G_2 = 6.0 + 7.7 = 13.7$ kN/m

COLD CALCULATIONS

Strength reduction factor	$\Phi = 0.85$
Stress block depth	$a = A_s f_y / 0.85 f'_c b = 6434 \times 300/(0.85 \times 30 \times 400)$ $= 189$ mm
Internal lever arm	$jd = d - a/2 = 727 - 189/2 = 632$ mm
Design load	$w_c = 1.2\,G + 1.6\,Q = 1.2 \times 13.7 + 1.6 \times 12.5 = 36.4$ kN/m
Bending moment	$M^*_{cold} = w_c\ L^2/8 = 36.4 \times 15^2/8 = \underline{1024\ \text{kNm}}$
Bending strength	$M_n = A_s\,f_y\,jd = 6434 \times 300 \times 632/10^6 = 1219$ kNm
	$\Phi M_n = 0.85 \times 1219 = 1036$ kNm
	$M^*_{cold} < \Phi M_n$ so design is OK.

FIRE CALCULATIONS

Design load (fire)	$w_f = G + SW + 0.4Q = 6.0 + 7.7 + 0.4 \times 12.5$ $= 18.7$ kN/m
Bending moment	$M^*_{fire} = w_f L^2/8 = 18.7 \times 15^2/8 = \underline{526\ \text{kNm}}$
Fire duration	$t = 90$ minutes
Depth of 500°C isotherm	$c_f = 33$ mm

(From Figure 9.10 assuming one-dimensional heat transfer at side of beam.)

Reduced width	$b_f = b - 2c_f = 400 - 2 \times 33 = 334$ mm

We assume that the concrete with temperature above 500°C has no compressive strength and concrete below 500°C has full compressive strength.

Steel temperatures from the isotherms in Figure 9.10:

Bar group (1):	450°C
Bar group (2):	<200°C
Bar group (3):	580°C

Reduced yield strength of reinforcing bars at elevated temperatures (from Equation (8.13))

$$f_{y,T1} = 300(720-450)/470 = 172\ \text{MPa}$$
$$f_{y,T2} = 300\ \text{MPa}$$
$$f_{y,T3} = 300(720-580)/470 = 89\ \text{MPa}$$

$$A_s f_{y,T} = (4 \times A_{s1} \times f_{y,T1} + 2 \times A_{s1} \times f_{y,T2} + 2 \times A_{s1} \times f_{y,T3})$$
$$= 804 \times (4 \times 172 + 2 \times 300 + 2 \times 89)/1000$$
$$= 1179\ \text{kN}$$

Stress block depth	$a_f = A_s f_{y,T}/0.85 f'_{c,T}\ b_f$
	$= 1179 \times 1000/(0.85 \times 30 \times 334) = 138$ mm
Internal lever arm	$jd_f = d - a_f/2 = 727 - 138/2 = 658$ mm
Bending strength	$M_{nf} = A_s f_{y,T}\ jd_f = 1179 \times 1000 \times 658/10^6 = 776$ kNm
	$M^*_{fire} < M_{nf}$ so design is OK.

WORKED EXAMPLE 9.3
REINFORCED CONCRETE TEE-BEAM

A reinforced concrete tee-beam is continuous over three supports (two spans). Check the structural adequacy of the beam before and after exposure to 2 hours of the standard fire. Ignore the contribution of compressive reinforcing. The beam is one of a series of beams 400 mm wide by 800 mm deep, at 4.0 m centres, supporting a 150 mm thick concrete slab, as shown in Figure 9.28.

Given information:

Beam span	$L = 13.0$ m	Live load = 3.0 kN/m^2 (storage)
Beam depth	$h = 800$ mm	Concrete density 24 kN/m^3
Web width	$b_w = 400$ mm	Concrete strength $f'_c = 30$ MPa
Tributary width (for load)	$b_t = 4.0$ m	Steel strength $f_y = 300$ MPa
Effective flange width (for positive moment)	$b_e = 2.0$ m	

Reinforcing	**Bottom**	**Top**
Number of bars	$n_b = 5$	$n_t = 18$
Bar diameter	$D_b = 28$ mm	$D_t = 20$ mm
Area of one bar	$A_{sb1} = \pi(D_b/2)^2 = 616$ mm^2	$A_{st1} = \pi(D_b/2)^2 = 314$ mm^2
Total steel area	$A_{sb} = n_b\ A_{sb1} = 3079$ mm^2	$A_{st} = n_t\ A_{st1} = 5655$ mm^2
Cover	$c_v = 25$ mm	$c_v = 25$ mm
Effective depth	$d_b = h - c_v - D_b/2 = 761$ mm	$d_t = h - c_v - D_b/2 = 765$ mm

Loads

Dead load (self weight)	$G = (0.15 \times 4.0 + 0.65 \times 0.4) \times 24$	$= 20.6$ kN/m
Live load	$Q = 3.0\ \text{kN/m}^2 \times 4.0$ m	$= 12.0$ kN/m
Load combination for cold conditions	$w_c = 1.2G + 1.6Q$	$= 44.0$ kN/m
Load combination for fire conditions	$w_f = G + 0.6Q$	$= 27.8$ kN/m

(load combination for fire, storage occupancy, from Chapter 7)

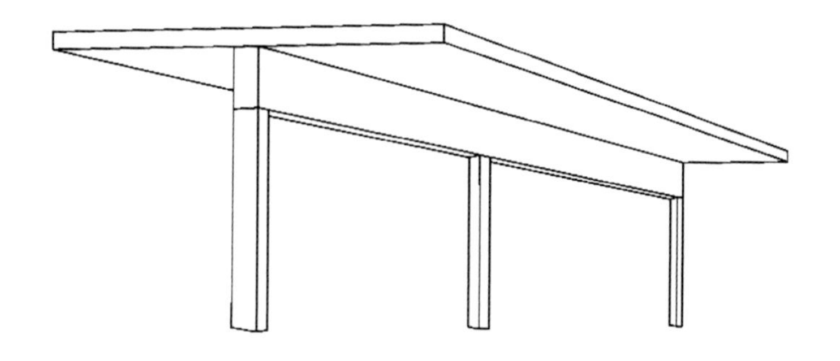

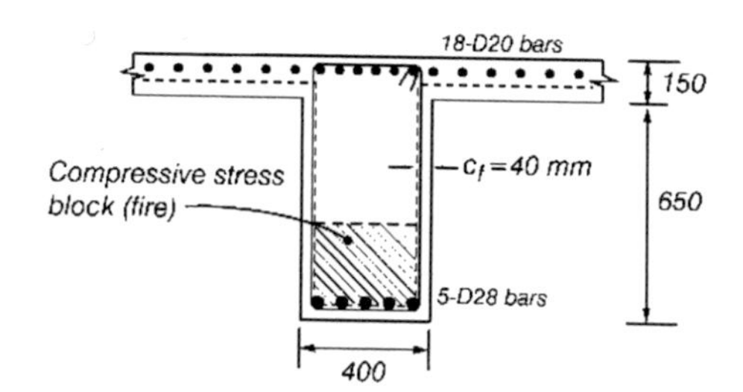

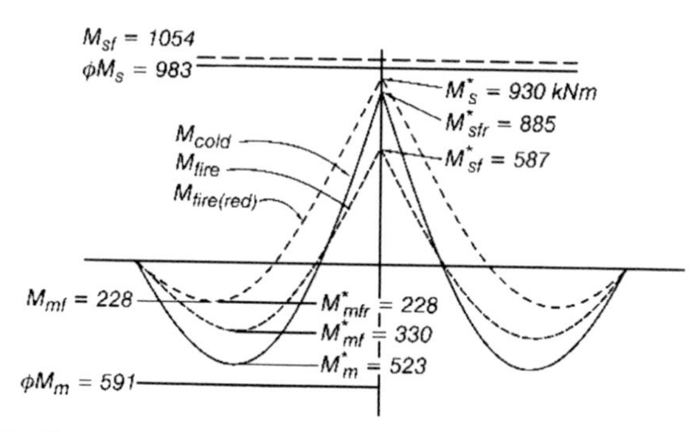

Figure 9.28 Beam for Worked Example 9.3

CHECK COLD CAPACITY — Strength reduction factor $\Phi = 0.85$

	NEAR MID-SPAN (positive moment)	AT SUPPORT (negative moment)
Elastic bending moment	$M^*_{\mathrm{m}} = 9w_{\mathrm{c}}L^2/128 = \underline{523\,\mathrm{kNm}}$	$M^*_{\mathrm{s}} = w_{\mathrm{c}}L^2/8 = \underline{930\,\mathrm{kNm}}$
Stress block depth	$a_{\mathrm{m}} = A_{\mathrm{sb}}\,f_{\mathrm{y}}/0.85\,f'_{\mathrm{c}}\;b_{\mathrm{e}}$	$a_{\mathrm{s}} = A_{\mathrm{st}}\,f_{\mathrm{y}}/0.85\,f'_{\mathrm{c}}\;b_{\mathrm{w}}$
	$= 3079 \times 300/0.85 \times 30 \times 2000$	$= 5655 \times 300/0.85 \times 30 \times 400$
	$= 18.1\,\mathrm{mm}$	$= 166\,\mathrm{mm}$
Internal lever arm	$jd_{\mathrm{b}} = d_{\mathrm{b}} - a_{\mathrm{m}}/2$	$jd_{\mathrm{t}} = d_{\mathrm{t}} - a_{\mathrm{s}}/2$
	$= 761 - 9 = 752\,\mathrm{mm}$	$= 765 - 166/2 = 682\,\mathrm{mm}$

Flexural strength	$M_m = A_{bs} f_y \, jd_b$	$M_s = A_{st} f_y \, jd_t$
	$= 3079 \times 300 \times 752/10^6$	$= 5655 \times 300 \times 682/10^6$
	$= 695$ kNm	$= 1157$ kNm
Design strength	$\Phi M_m = 591$ kNm	$\Phi M_s = 983$ kNm

$$M^*_m < \Phi M_m \text{ and } M^*_s < \Phi M_s \text{ so design is OK.}$$

DESIGN FOR FIRE — Fire duration $t = 120$ minutes

NEAR MID-SPAN

Elastic bending moment $\quad M^*_{mf} = 9 w_f L^2/128 \quad = \underline{330 \text{ kNm}}$

Effective cover to bottom bars $\quad c_e = c_v + D_b/2 = 25 + 28/2 \quad = 39$ mm

Steel bar temperature from the isotherm of Figure 9.10:

Bar group (1) (corner bars) 650°C

Bar group (2) (interior bars) 510°C

Reduced yield strength

$$f_{y,T1} = (1.53 - 650/470) \times 300 = 44.1 \text{ MPa}$$
$$f_{y,T2} = (1.53 - 510/470) \times 300 = 134 \text{ MPa}$$

$$\begin{aligned} A_{sb} f_{y,T} &= (2A_{sb1} f_{y,T1} + 3A_{sb1} f_{y,T2}) \\ &= 616 \times (2 \times 44 + 3 \times 134)/1000 \\ &= 301 \text{ kN} \end{aligned}$$

Stress block depth

$$\begin{aligned} a_{mf} &= A_{sb} f_{y,T}/0.85 f_c \, b_e \\ &= 301 \times 1000/(0.85 \times 30 \times 2000) = 5.9 \text{ mm} \end{aligned}$$

Check concrete temperature:

From Figure 9.10 concrete temperature at depth of 150 mm is less than 200°C, so no reduction in f'_c

Internal lever arm $\quad jd_{bf} = d_b - a_{mf}/2 = 761 - 3 = 758$ mm

Bending strength $\quad M_{mf} = A_{sb} f_{y,T} \, jd_{bf} = 301 \times 1000 \times 758/10^6 = 228$ kNm

$M^*_{mf} > M_{mf}$ so the cross section fails

AT THE SUPPORT

Elastic bending moment $\quad M^*_{sf} = w_f L^2/8 = \underline{587 \text{ kNm}}$

Top bars are less than 250°C so no reduction in strength

Check concrete temperature:

We assume that the concrete with temperature above 500°C has no compressive strength and concrete below 500°C has full compressive strength.

Depth of 500°C isotherm $c_f = 40\,\text{mm}$

(From Figure 9.7 assuming one-dimensional heat transfer near surface.)

Reduced width of stress block $b_{wf} = b_w - 2c_f = 400 - 2 \times 40 = 320\,\text{mm}$

Reduced effective depth $d_{tf} = d_t - c_f = 765 - 40 = 725\,\text{mm}$

Stress block depth

$$a_{sf} = A_{st} f_y / 0.85 f'_{c,T} b_{wf}$$
$$= 5655 \times 300/(0.85 \times 30 \times 320) = 208\,\text{mm}$$

Internal lever arm $jd_{tf} = d_{tf} - a_{sf}/2 = 725 - 208/2 = 621\,\text{mm}$

Bending strength $M_{sf} = A_{st} f_y jd_{tf} = 5655 \times 300 \times 621/10^6 = 1054\,\text{kNm}$

These calculations show that the fire has caused the mid-span flexural capacity to drop below the elastic bending moment which would cause failure if this was a simple supported beam. However, the flexural capacity over the support has increased (due to the change in Φ in fire conditions) so it is necessary to establish whether the beam can survive with the help of moment redistribution.

MOMENT REDISTRIBUTION

For an end span of a continuous beam of length L, with uniformly distributed load w, and known positive moment M^+, it can be shown (Gustaferro and Martin, 1988) that the negative bending moment at the support M^- is given by Equation (7.27) as

$$M^- = wL^2/2 - wL^2\sqrt{2M^+/wL^2}$$

This can be used to calculate the redistributed bending moment at the support when the mid-span moment is just equal to the fire-reduced flexural capacity.

$$M^+ = M_{mf} = 228\,\text{kNm},\ w_f = 27.8\,\text{kN/m and } L = 13\,\text{m},$$
$$M^*_{sfr} = w_f L^2/2 - w_f L^2\sqrt{2M_{mf}/w_f L^2}$$
$$= 27.8 \times 13^2/2 - 27.8 \times 13^2\sqrt{2 \times 228/(27.8 \times 13^2)}$$
$$= 885\,\text{kNm}$$
$$M^*_{sfr} < M_{sf}$$

This shows that the design is OK, because the bending moments can be redistributed to the line shown by $M_{fire(red)}$ in Figure 9.28 where the maximum negative moment is now 885 kNm, less than the flexural capacity of 1054 kNm.

The location of the maximum mid-span moment is given by Equation (7.28) as

$$a = \sqrt{(2M^+/L)}$$
$$= \sqrt{(2 \times 228/13)}$$
$$= 5.9\,\text{m}$$

The termination of the bottom reinforcing bars must be checked to determine if it is possible to develop full flexural strength at this location.

WORKED EXAMPLE 9.4
AXIAL RESTRAINT

Consider a reinforced concrete floor constructed from precast concrete tee-beams as shown in Figure 9.29. The slabs are simply supported over a span of 6.0 m, carrying a live load of 3.0 kPa. The dead load is 4.8 kPa (including the self weight). Calculate the restraint condition necessary to give a fire-resistance rating of 120 minutes.

Given information:

Slab span	$L = 6.0\,\text{m}$	Concrete strength	$f'_c = 25\,\text{MPa}$
Overall depth	$h = 300\,\text{mm}$	Steel strength	$f_y = 350\,\text{MPa}$
Web width	$b_w = 200\,\text{mm}$	Concrete MOE	$E = 25\,\text{GPa}$
Overall width	$b_f = 1200\,\text{mm}$		
Cross-sectional area	$A = 185\,000\,\text{mm}^2$		
Heated perimeter	$s = 1550\,\text{mm}$		

Load combinations:

Dead load per metre	$G = 1.2 \times 4.8$	$= 5.76\,\text{kN/m}$
Live load per metre	$Q = 1.2 \times 3.0$	$= 3.6\,\text{kN/m}$
Load combination for cold conditions:	$w_c = 1.2G + 1.6Q$	$= 12.7\,\text{kN/m}$
Load combination for fire conditions:	$w_f = G + 0.4Q$	$= 7.20\,\text{kN/m}$

Reinforcing:

Number of bars $\quad n = 4$

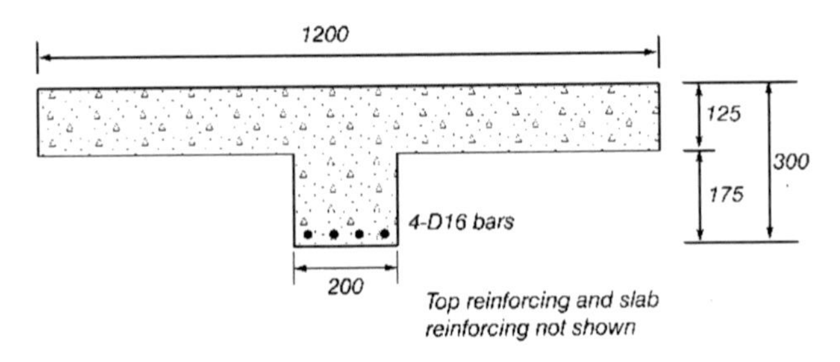

Figure 9.29 Beam for Worked Example 9.4

Bar diameter	$D_b = 16\,\text{mm}$
Single bar area	$A_{s1} = \pi(D_b/2)^2 = 201\,\text{mm}^2$
Total bar area	$A_s = nA_{s1} = 804\,\text{mm}^2$
Cover	$c_v = 20\,\text{mm}$
Effective depth	$d = h - c_v - D_b/2 = 272\,\text{mm}$
Effective cover	$c_e = c_v + D_b/2 = 28\,\text{mm}$

COLD CALCULATIONS — Strength reduction factor $\Phi = 0.85$

Mid-span bending moment: $M^*_{cold} = w_c L^2/8 = 12.7 \times 6.0^2/8 = \underline{57.2\,\text{kNm}}$

Stress block depth:
$$a = A_s f_y / 0.85 f'_c\, b_f = 804 \times 350/0.85 \times 25 \times 1200 = 11.0\,\text{mm}$$

Internal lever arm: $jd = d - a/2 = 272 - 11/2 = 266\,\text{mm}$

Flexural strength:
$$M_c = A_s f_y\, jd = 804 \times 350 \times 266/10^6 = 75.0\,\text{kN·m}$$
$$\Phi M_c = 0.85 \times 75.0 = 63.8\,\text{kN·m}$$

$\Phi M_c > M^*_{cold}$ so cold design is OK.

FIRE CALCULATIONS — Strength reduction factor $\Phi = 1.0$

Mid span bending moment: $M^*_{fire} = w_f L^2/8 = 7.2 \times 6.0^2/8 = \underline{32.4\,\text{kN·m}}$

Steel bar temperature from the isotherms on Figure 9.10:

Bar group (1) (corner bars) $T_{s1} = 830°\text{C}$

Bar group (2) (corner bars) $T_{s2} = 640°\text{C}$

Reduced yield strength

$$f_{y,T1} = 350 \times (720 - 830)/470 = 0\,\text{MPa}$$
$$f_{y,T2} = 350 \times (720 - 640)/470 = 59.5\,\text{MPa}$$

$$A_s f_{y,T} = (2\, A_{s1} f_{y,T1} + 2\, A_{s1} f_{y,T2}) = (2 \times 201 \times 0 + 2 \times 201 \times 59.5)/1000 = 23.9\,\text{kN}$$

Stress block depth:
$$a_f = A_s f_{y,T}/0.85 f'_c\, b_f = 23.9 \times 1000/(0.85 \times 25 \times 1200) = 0.93\,\text{mm}$$

Internal lever arm: $jd_f = d - a_f/2 = 272 - 0.93/2 = 271.5\,\text{mm}$

Flexural strength:
$$\Phi M_f = M_f = A_s f_{y,T}\, jd = 24 \times 271.5/10^6 = 6.4\,\text{kN·m}$$

$\Phi M_f < M^*_{fire}$ so slab will fail unless restraint or continuity is provided.

Provide axial restraint (numbers in brackets are steps from text)

(4) Estimate the mid-span deflection:

Mid-span deflection of the reference specimen $\Delta_o = 65\,\text{mm}$ (from Figure 9.20)

Heated length $L = 6000\,\text{mm}$

Distance from neutral axis to extreme bottom fibre $y_b = 290\,\text{mm}$

(assume that neutral axis is 10 mm from top of slab)

Mid-span deflection $\Delta = L^2\,\Delta_o/89000\,y_b$

$= 6000^2 \times 65/(89000 \times 290) = 90.7\,\text{mm}$

Distance to line of thrust from top of beam $d_T = 0.9h = 280\,\text{mm}$

(assume that the slab is built-in to the surrounding construction, thrust 0.1 h from bottom.)

(6) Calculate the required thrust to prevent collapse

$T = 1000\,(M^*_{\text{fire}} - M_f)/(d_T - a_f/2 - \Delta)$

$= 1000\,(32.4 - 6.4)/(270 - 0.9/2 - 90.7)$

$= 145\,\text{kN}$

(7) Recalculate a_f $\quad a_f = (T + A_s f_{y,T})/0.85\,f'_c\,b_f$

$= (14\,500 + 23\,700)/(0.85 \times 25 \times 1200)$

$= 6.6\,\text{mm}$

Recalculate T $\quad T = 1000\,(32.4 - 16.1)/(270 - 6.6/2 - 90.7)$

$= \underline{148\,\text{kN}}$

(8) Non-dimensional thrust parameter $\quad T/AE = 147\,700/(185\,000 \times 25\,000)$

$= 32 \times 10^{-6}$

Shape parameter $\quad z = A/S = 185\,000/1550 = 119\,\text{mm}$

Strain parameter $\quad \Delta L/L = 0.0065$ (from Figure 9.21)

Maximum permitted displacement $\quad \Delta L = 0.0065 \times 6000$

$= \underline{39\,\text{mm}}$

So, this slab will have a fire resistance of 120 minutes if the surrounding structure at each end is capable of resisting an axial thrust of 148 kN with an axial elongation of less than 39 mm.

WORKED EXAMPLE 9.5

Calculate the fire resistance of a circular steel column filled with concrete (example from Kodur (1999)).

Given information:

Column length	$L = 4800\,\text{mm}$
Effective length factor	$K = 0.67$ (fixed-fixed end conditions)
Axial load	$N = 1344\,\text{kN}$

Concrete strength	$f'_c = 40$ MPa (carbonate aggregate concrete)
Column size	HSS 324×6.4
Column diameter	$d = 324$ mm
Bar reinforcing	3%
Cover	25 mm

From Table 9.3, $f = 0.095$. From Equation (9.22) the fire resistance is given by

$$\begin{aligned} t_r &= f\,(f'_c + 20)\, d^{2.5}/(KL - 1000)\sqrt{N} \\ &= 0.095\,(40 + 20)\; 324^{2.5}/(0.67 \times 4800 - 1000)\sqrt{1344} \\ &= 133 \text{ minutes} \end{aligned}$$

10

Timber Structures

10.1 OVERVIEW

This chapter describes the fire behaviour of timber construction, and gives design methods for heavy timber structural members exposed to fire. Fire behaviour of connections in timber structures is also discussed.

10.2 DESCRIPTION OF TIMBER CONSTRUCTION

Timber structures tend to fall into two distinct categories: 'heavy timber' structures and 'light timber frame' construction ('light wood frame' in North America). Heavy timber structures are those where the principal structural elements are beams, columns, decks, or truss members made from glue-laminated timber or large-dimension sawn timber. Many innovative structures using heavy timber structural members are described by Götz *et al.* (1989).

Light timber frame construction uses smaller sizes of wood framing, as studs in walls, and as joists in floors. Walls and floors are covered with panels of lining materials to provide resistance to impact, sound transmission and fire spread. Fire resistance of light timber frame construction is covered in Chapter 11.

10.2.1 Heavy Timber Construction

'Heavy timber construction' describes all uses of large-dimension timber framing in buildings. Many historic commercial and industrial buildings consist of external load-bearing masonry walls, with internal timber columns and beams supporting thick timber floor decking. The term 'heavy timber construction' or 'mill construction' has a specific meaning in North American fire codes where it applies to beams and columns with a minimum nominal dimension of 150 mm and decks with a minimum nominal thickness of 50 mm. In this book, the term 'heavy timber' generally refers to timber members whose smallest dimension is no less than 80 mm.

10.9.2 Glulam

'Glue laminated timber' (glulam) describes timber members which are manufactured from several laminations glued together. The length of individual laminations is often the full length of the member, joined end-to-end with finger-joints. Glulam members can be manufactured in any size or shape, the major limitation being transportation. The individual laminations must be thin in curved members (10 to 25 mm thickness, depending on the radius of the curve) and are thicker in straight members (usually 35 to 45 mm thickness). The most common adhesives for glulam are based on resorcinol, melamine urea, or casein. Many fire tests have shown that glulam members exposed to fires behave in the same way as solid sawn-timber members of the same cross section, with the possible exception of those manufactured with casein adhesives. Epoxy-based adhesives do not behave well in fires.

In some countries, especially in North America, the laminations at the top and bottom edges of glulam members are made from specially selected high-strength wood, in order to increase the flexural strength and stiffness. This practice must be allowed for in fire design when the outer laminations may be burned away, placing more reliance on the inner laminations which are of lower strength.

10.2.3 Behaviour of Timber Structures in Fire

Heavy timber construction has become recognized as having very good fire-resistance. There are many well-documented examples of structures surviving severe fire exposure without collapse, and many of these have been repaired for re-use (Figures 10.1 and 10.2).

When large timber members are exposed to a severe fire, the surface of the wood initially ignites and burns rapidly. The burned wood becomes a layer of char which insulates the solid wood below. The initial burning rate decreases to a slower steady rate which continues throughout the fire exposure. The charring rate will increase if the residual cross section becomes very small. The layer of char shrinks, making it thinner than the original wood, causing fissures which facilitate the passage of combustible gases to the surface (Drysdale, 1998). The char layer does not usually burn because there is insufficient oxygen in the flames at the surface of the char layer for oxidation of the char to occur. When the wood below the char layer is heated above 100°C, the moisture in the wood evaporates. Some of this moisture travels out to the burning face, but some travels into the wood, resulting in an increase in moisture content in the heated wood a few centimetres below the char front (Fredlund, 1993, White and Schaffer, 1980).

10.2.4 Fire-retardant Treatments

A number of fire-retardant chemicals are available for treating wood to reduce its combustibility (Schaffer, 1992). The main purpose of such chemical treatments is to reduce the rate of flame spread over the surface of the wood, to improve fire safety in rooms lined with wood or wood-based panel products. Pressure impregnation of

(a)

(b)

Figure 10.1 (a) Severe fire damage to an industrial building with curved glulam portal frames. (b) One of the beams being repaired for re-use by sandblasting (Both figures are reproduced from (TRADA, 1976) by permission of New Zealand Timber Industry Federation).

Figure 10.2 Curved glulam roof beams after repair following a severe fire (Reproduced by permission of McIntosh Timber Laminates Ltd)

chemicals is considered more effective than surface painting. The pressure impregnation process is similar to that used for applying chemicals to resist decay, but the retentions of salts required for fire retardancy are much higher. Impregnation by fire-retardant chemicals can have some negative effects including loss of wood strength and corrosion of fasteners, exacerbated by the hydroscopic nature of many of the chemicals (LeVan and Winandy, 1990, Winandy, 1995).

Fire-retardant chemicals do not significantly improve the fire-resistance of timber members, because even though treated wood will not support combustion, it will continue to char if exposed to the temperatures of a fully-developed fire. Some proprietary intumescent paints have been advertised with properties which can increase the fire-resistance of timber members, but insufficient test results have been published to recommend such products for general use. A discussion is given by White (1984). Recently developed fibre-glass reinforced coatings are described by del Senno *et al.* (1998).

10.3 FIRE-RESISTANCE RATINGS

The fire-resistance of timber structures can be assessed using the same general principles as for other materials.

10.3.1 Verification Methods

The design process for fire-resistance requires verification that the provided fire-resistance exceeds the design fire severity. Using the terminology from Chapter 5, verification may be in the *time domain*, the *temperature domain* or the *strength domain*.

The temperature domain is not used for timber structures because there is no critical temperature for fire-exposed timber. In most countries, fire design of heavy timber structures is by *calculation* using the methods outlined in this chapter. The results of these calculations are verified in the time domain by comparing the time of structural collapse with a specified fire-resistance time, or in the strength domain by comparing the residual strength with the applied loads after a certain period of fire exposure.

Some countries have *generic* fire-resistance ratings for heavy timber construction. For example, some US codes allow heavy timber construction to be used in certain classes of buildings, with no calculations required (e.g. UBC, 1997). There are very few *proprietary* ratings for heavy timber, in contrast to light timber construction where there are many proprietary ratings based directly on test results, as described in Chapter 11.

10.4 WOOD TEMPERATURES

When heavy timber members are exposed to severe fires, the outer layer of wood burns and is converted to a layer of char. The temperature of the outer surface of the char layer is close to the fire temperature, with a steep thermal gradient through the char. The boundary between the char layer and the remaining wood is quite distinct, corresponding to a temperature of about 300°C. The commonly accepted charring temperature in North America is 288°C (550 °F), but the precise temperature is not important because of the steepness of the temperature gradient. Below the char layer there is a layer of heated wood about 35 mm thick. The part of this layer above 200°C is known as the pyrolysis zone, because this wood is undergoing thermal decomposition into gaseous pyrolysis products, accompanied by loss of weight and discolouration. Moisture evaporates in the wood above 100°C. The inner core of the member remains at its initial temperature for a considerable time. These layers are shown in Figure 10.3 (from Schaffer, 1967).

Structural design of heavy timber members is based on the rate of charring of the wood surface, so it is not necessary for designers to calculate temperatures within the fire-exposed wood.

10.4.1 Temperatures Below the Char

Temperatures in the wood below the char layer have been measured in many tests. For wood thick enough to be considered as a semi-infinite solid, Eurocode 5 gives the temperature T (°C) below the char layer as

$$T = T_i + (T_p - T_i)(1 - x/a)^2 \tag{10.1}$$

where T_i is the initial temperature of the wood (°C), T_p is the temperature at which charring starts (300°C), x is the distance below the char layer (mm), and a is the thickness of the heat-affected layer (40 mm).

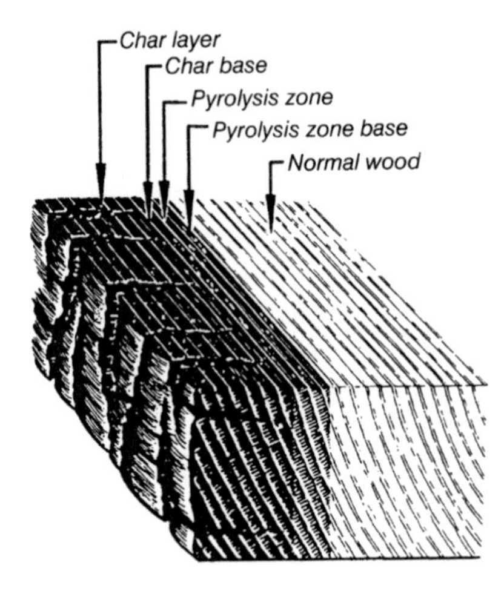

Figure 10.3 Char layer and pyrolysis zone in a timber beam (Schaffer, 1967)

Janssens and White (1994) show that a better fit to experimental data is obtained with $a = 35\,\text{mm}$ The resulting parabolic temperature profile is plotted in Figure 10.19(b).

10.4.2 Thermal Properties of Wood

The temperatures inside fire-exposed timber members can be calculated using finite-element numerical methods. The thermal properties are not well defined, and vary considerably with temperature as moisture is driven off at 100°C and as wood turns to char over 300°C (Janssens, 1994). The values given below are typical average values from recent literature (Thomas, 1997, König and Walleij, 1999).

The density of wood varies significantly between species. It also varies between trees of the same species and within individual trees. The density drops to about 90% of its original value when the temperature exceeds 100°C, and to about 20% of its original value when the wood is converted to char above 300°C.

The thermal conductivity varies greatly between different authors. Figure 10.4 shows the variation of thermal conductivity with temperature as proposed by Knudson and Schneiwind (1975) which is about the average of other published values. König and Walleij (1999) found that they had to increase the thermal conductivity to much higher values at temperatures over 500°C in order to give good predictions of measured behaviour. Figure 10.5 shows the variation of specific heat with temperature as proposed by König and Walleij (1999). The large spike at 100°C represents the heat required to evaporate the moisture in the wood.

10.5 MECHANICAL PROPERTIES OF WOOD

Wood has several significant differences from other common materials such as steel and concrete. For example:

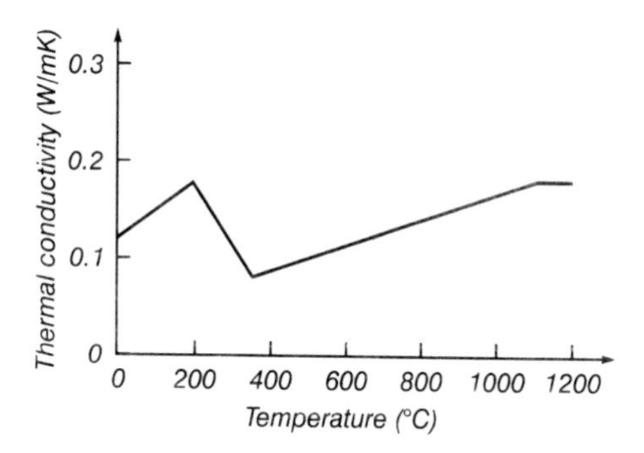

Figure 10.4 Variation of thermal conductivity of wood with temperature

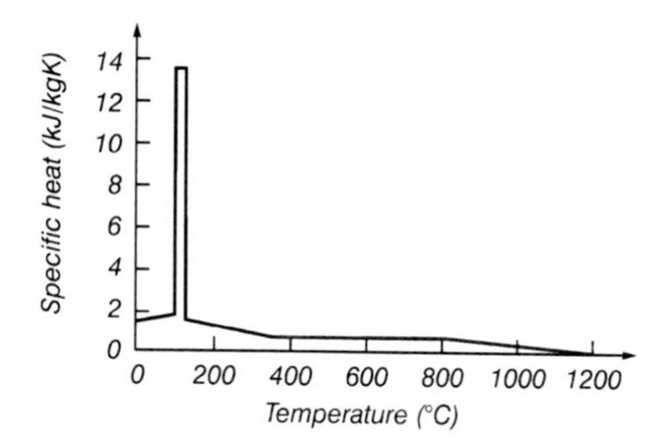

Figure 10.5 Variation of specific heat of wood with temperature

- wood strength is very variable, both within boards and between boards;
- mechanical properties are different in different directions (parallel and perpendicular to the grain);
- strength and ductility are very different in tension and compression;
- failure stresses depend on the size of the test specimens; and
- the strength reduces under long duration loads.

Figure 10.6 shows different ways in which wood can be loaded, each producing a different failure mode. This chapter reviews wood behaviour at normal temperature before describing properties at elevated temperatures.

10.5.1 Mechanical Properties of Wood at Normal Temperatures

Tension and compression behaviour

Figure 10.6 shows typical stress–strain relationships for small clear specimens of wood with no defects. Considering behaviour parallel to the grain, the straight line in tension indicates linear elastic behaviour to brittle failure at a tensile stress f_t. Wood is brittle in tension because there is no load sharing within the wood material, so that a crack can lead to sudden failure as soon as it reaches a certain critical size.

In compression the stress–strain relationship is linear in the elastic region, with the same modulus of elasticity as in tension. The line then curves, indicating yielding (or crushing), it reaches a peak and eventually drops as the wood is crushed further. With larger strains the specimen will continue to deform in a ductile manner. Compression yielding is accompanied by visible wrinkles on the surface of the wood. The compression curves in Figure 10.7 indicate crushing of wood in short columns. Long slender columns have lower load capacity because they will fail by bucking at loads well below the crushing strength, as described in Chapter 7.

In clear wood, the tension strength f_t is usually much greater than the compression strength f_c. In commercial quality timber, the relative strengths are often reversed because growth characteristics such as knots have a severe effect on tensile strength but only a small effect on compressive strength (Bodig and Jayne, 1982).

The dotted line in Figure 10.7 shows the stress–strain relationship for wood loaded perpendicular to the grain direction. The slope indicates a lower modulus of elasticity

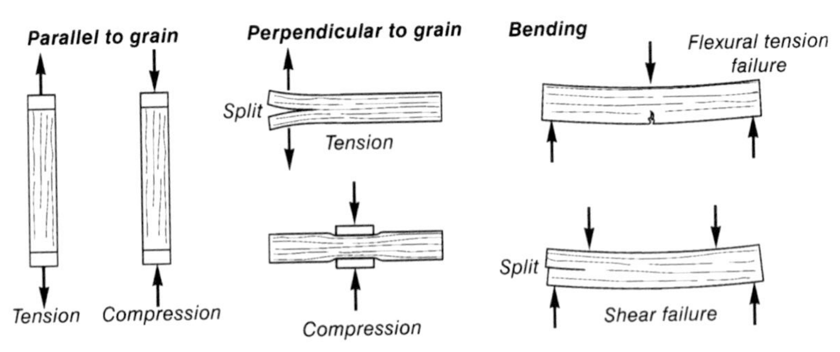

Figure 10.6 Loading of wood in different directions

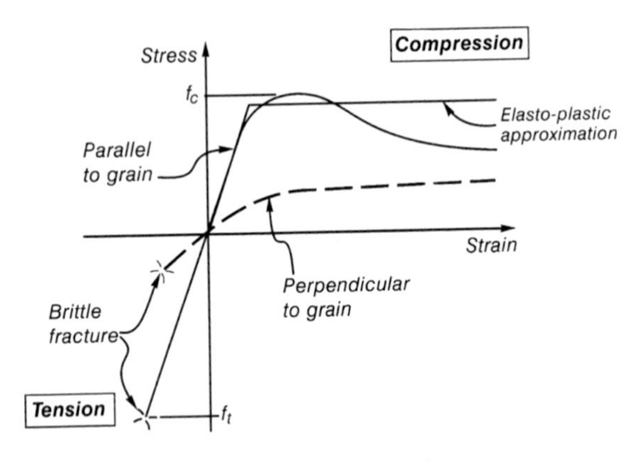

Figure 10.7 Stress–strain relationships for clear wood

than for loading parallel to grain. The wood is ductile in compression, with the load slowly increasing as strains increase. In tension perpendicular to the grain, the strength is very low and unpredictable, with splitting causing brittle fractures. This weakness can lead to structural failure if it is not properly allowed for in design.

Bending behaviour

Bending behaviour is a combination of tension and compression behaviour (Buchanan, 1990). Internal compressive and tensile stresses in a timber beam are shown in Figure 7.2. Commercial quality timber beams tend to fail suddenly due to poor tensile strength at knots in the tension zone. Some ductility is available in timber beams when the material is stronger in tension than in compression, and this ductility can increase during fire exposure because of softening of the wood in the compression zone.

Design values

Structural design calculations require values of the design strength of the wood material. For *limit states design* (LRFD) the design stress, or 'characteristic stress', is the fifth percentile failure stress under short-duration loading, for a typical population of timber boards. Because the strength of timber is much more variable than that of steel or concrete, characteristic stresses are usually obtained from *in-grade* tests of large numbers of representative samples of full size timber members, selected from typical production. This allows the effects of size, grade, defects, and variability to be determined directly. To accommodate a very large number of species and grades, most codes specify characteristic values of strength and stiffness for a number of defined strength classes. The fifth percentile value for design in normal temperature conditions may be modified to the 20th percentile strength value for fire design, as described later in this chapter.

Failure stresses in timber depend on many factors, including the size of the test specimen. Large timber members tend to fail at lower stresses than similar small members, because a large member has a larger number of potential defects, hence a greater probability of a large knot, than a small member (Madsen and Buchanan, 1986). Size effects are recognized in some codes.

The design strength of timber also depends on the duration of the applied load, so that most timber design codes include a duration-of-load factor. In *limit states design* (LRFD) codes, the duration-of-load factor is usually 1.0 for short-duration loads, decreasing to 0.8 or 0.6 for medium- and long-duration loads. In *working stress design* codes, allowable stresses are for long-duration loading, derived from test results of small clear specimens of wood. In this case the duration-of-load factor is usually 1.0 for long-duration loads, increasing to 1.25 or 1.6 for medium- and short-duration loads. The duration-of-load factor for fire design should be the appropriate value for short-duration loads, because the duration of the load during the fire is likely to be less than one hour.

10.5.2 Mechanical Properties of Wood at Elevated Temperatures

Sources

The most comprehensive review on the effect of moisture content and temperature on the mechanical properties of wood is by Gerhards (1982) who reported the results of many previous studies. Some of this information is summarized by Buchanan (1998) and the Wood Handbook (1987). Wood properties are affected by steam at 100°C, wood begins to pyrolyse at about 200°C and turns into char by 300°C. The range of interest for fire engineering is therefore from room temperature to 300°C.

Effect of moisture content

The strength of wood at elevated temperatures is not well understood. In addition to temperature, the interaction with moisture content is very important, making the range of testing options even more difficult than shown in Figure 7.5. When testing timber at elevated temperatures, the moisture content is sensitive to the test method and the size of the test specimen. Some test specimens are maintained at constant moisture content throughout the test with a climate-controlled testing facility or an oil bath (Östman, 1985). In other tests the specimen is at a certain moisture content before the test and allowed to dry out when heated, either before or during the test, in which case some moisture may migrate into the interior of the specimen and the moisture gradients will depend on the size of the specimen. If wood is heated to a temperature above 100°C in dry air, all moisture will evaporate after some time, depending on the permeability of the particular species.

Plasticity

The strength of a structural timber member is reduced in fire, because the wood converted to char has no strength and the temperature and moisture gradients below the char layer reduce the strength and increase the plasticity of that wood. The increase in wood plasticity is very important, especially for tension and bending members which would have brittle failures at room temperatures. When a timber beam is tested in bending at normal temperatures, it usually fails suddenly, with fracture at a weak point on the tension edge when a small crack reaches a critical size. If heated wood were to lose strength with no increase in plasticity, cracks would occur in the heated tension zone of the beam, leading to premature failure in fire. The excellent performance of large timber beams in fire results from plastic behaviour in the heated wood, allowing redistribution of stresses into the cooler wood further from the char layer.

Steam softening

It is well known from the furniture and boat building industries that hot moist wood can be bent into curved shapes using steam bending. Steam bending occurs because wood becomes plastic in compression under certain combinations of temperature and moisture content. There is very little literature on this subject, other than Stevens and Turner (1970) and the Wood Handbook (1987) which states the following:

> 'Wood at 20 to 25 percent moisture content needs to be heated without losing moisture; at lower moisture content, heat and moisture must be added. As a consequence, the recommended plasticizing processes are steaming or boiling for about 1/2 hour per inch of thickness for wood at 20 to 25 percent moisture content and steaming or boiling for about 1 hour per inch of thickness for wood at lower moisture content values.'

Because most of the deformation must take in compression, it is often necessary to apply a net compressive force on the member, using a steel strap to reinforce the tension side of the cross section during the bending process.

When wood is heated in a fire, the conditions which produce softening of the wood may occur for only a short period of time. If the moisture content subsequently decreases, the wood will harden, even if temperatures continue to increase. The effect of wood softening will be very different for large and small members. In a large member, conditions to produce softening may occur in a thin layer which progresses into the wood at about the same velocity as the rate of charring, having little effect on the overall strength or stiffness of the member. Small members may experience these conditions over a large proportion of the cross section, in which case the member may deform plastically in compression or bending, leading to premature failure (Young and Clancy, 1998). These conditions may only occur for a short time period, so if the assembly can resist the applied loads during this short period, with the help of other load paths or cladding materials, the wood may regain strength after it dries and be able to survive a much longer time of fire exposure.

Parallel to the grain properties

Modulus of elasticity Figure 10.8 shows the modulus of elasticity of wood at elevated temperatures from the Wood Handbook (1987). This is similar to Figure 10.9. which combines results from Nyman (1980) and Östman (1985) with those from Schaffer (1973) and Preusser (1968) (quoted by Gerhards, 1982). Other results shown by Gerhards (1982) fit into the same envelope. The effect of temperature on modulus of elasticity parallel to the grain is roughly linear up to 200°C. There is increased scatter over 200°C where Preusser shows a much more sudden drop than that measured by Schaffer or Östman.

Figure 10.9 also shows recent results derived by König and Walleij (2000) from tests of 145 × 45 mm timber studs in insulated wall assemblies, exposed to the ISO 834 standard furnace fire while loaded in bending. Similar results were obtained by Young (2000) who used even lower values of modulus of elasticity to model the results of full-scale fire-resistance tests of timber stud walls.

Tensile strength Östman (1985) tested samples of spruce (1 mm by 10 mm cross section) at a range of temperatures and moisture contents, obtaining the stress–strain

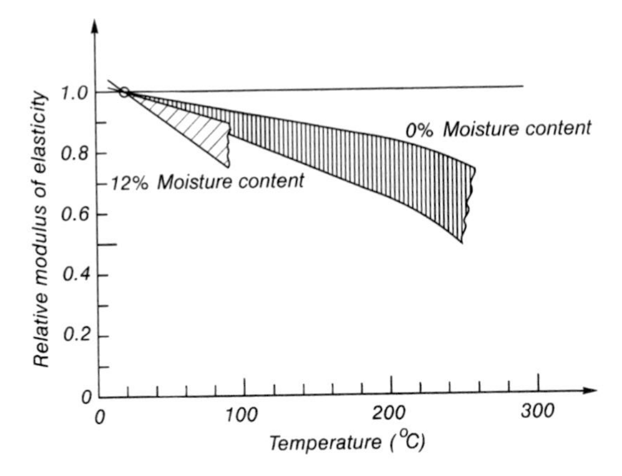

Figure 10.8 Modulus of elasticity of wood parallel to the grain *versus* temperature (Wood Handbook, 1987)

relationship redrawn in Figure 10.10 for temperatures of 25°C and 90°C at low and high moisture content. It can be seen that the failure stress at 90°C and 29.5% moisture content is about 60% of that of dry cool wood. All the lines are curved, indicating a small amount of plastic behaviour before failure.

The only reported 'in-grade testing' at elevated temperatures is by Lau and Barrett (1997) who tension tested a large number of 90 × 35mm boards, 25 minutes after heating the surfaces to temperatures up to 250°C. They show that the tension behaviour remains brittle, and failure is governed by the weakest link in the test specimen, generally a knot. Lau used a damage accumulation model to predict the tensile strength under constant temperature, finding that the strength reduction over 150°C depends on the duration of loading. A comparison with other test data is shown in Figure 10.11 where the results of Östman (1985) lie between those of Schaffer (1973) and Knudson and Schneiwind (1975). Nyman (1980) has similar results to Östman. The upper curves are for dry specimens, Knudson and

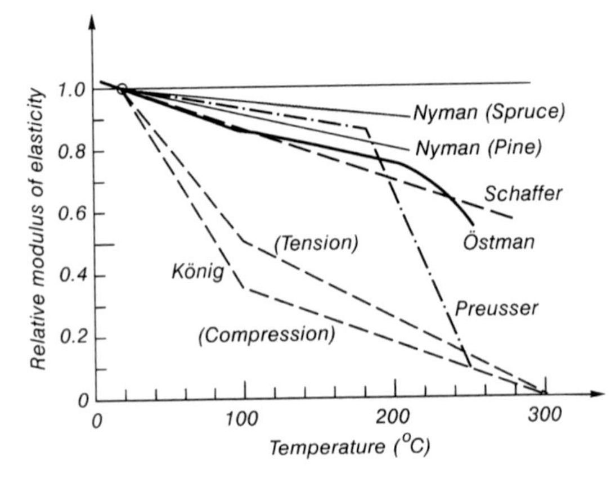

Figure 10.9 Modulus of elasticity of wood parallel to the grain *versus* temperature (Adapted from (Gerhards, 1982) with permission of Society of Wood Science and Technology)

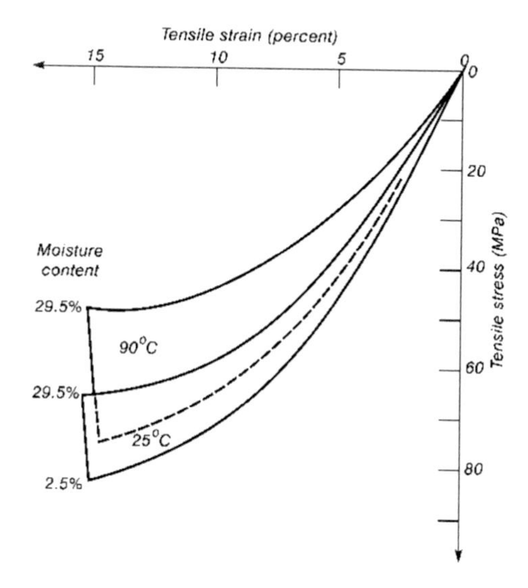

Figure 10.10 Stress–strain relationships for wood in tension parallel to the grain (adapted from Östman, 1985)

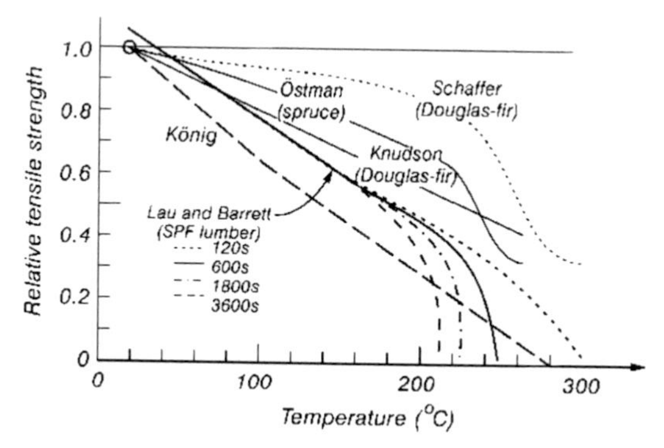

Figure 10.11 Tensile strength parallel to the grain *versus* temperature (Adapted from (Gerhards, 1982) with permission of Society of Wood Science and Technology)

Schneiwind's test specimens were at 12% moisture content before rapid (30 s) heating at the time of testing, and Lau and Barrett's specimens were at 7–11% moisture content before heating. Figure 10.11 also shows the relationship derived by König and Walleij (2000).

Compressive strength The effect of temperature on compressive strength parallel to the grain is shown in Figure 10.12, which is the envelope of all the results quoted by Gerhards (1982). All of these results are for dry wood except the marked shaded region which shows results for tests with moisture content between 12% and the fibre saturation point. There are very few reported compression test results for moist wood over 60°C where plastic behaviour is expected (Young and Clancy, 1998). As discussed earlier in this chapter, plastic behaviour of wood in compression becomes very important in some timber structures exposed to fire. Figure 10.12 also shows the

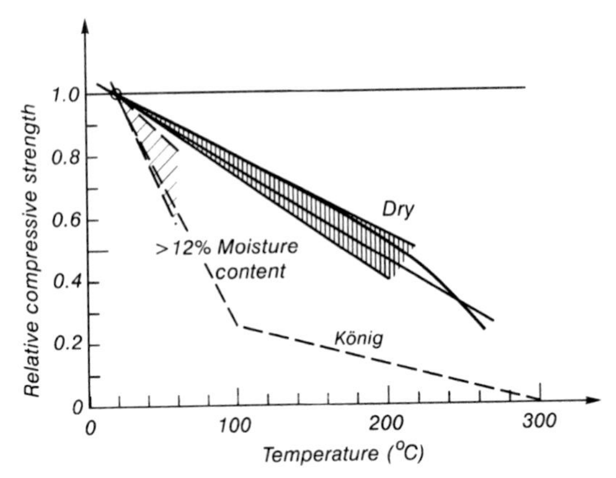

Figure 10.12 Compression strength parallel to the grain *versus* temperature (Adapted from (Lau and Barrett, 1977) by permission of IAFSS)

relationship derived by König and Walleij (2000) which follow the earlier results for moist wood to 100°C, followed by a straight line to zero-strength at 300°C. The question as to whether the strength of initially moist wood increases again after the moisture is driven off at temperatures over 100°C requires further research, because this information is required for finite-element modelling of timber structures exposed to fires.

Bending strength Bending behaviour in wood can be best described from an understanding of the tension and compression behaviour. The effect of elevated temperatures on bending behaviour is, in theory, predictable from the information presented in Figures 10.8 to 10.12. Dry wood will lose strength and stiffness at the rate given in those figures. The situation is much more complicated for moist wood, because if conditions become suitable for plastic behaviour, large strains will occur in the compression zone resulting in a relocation of the neutral axis as shown in Figure 7.2, leading to large deformations.

There are limited test results available. The shaded areas on Figure 10.13 show the results collected by Gerhards (1982) which include only one test series for temperatures over 80°C (from Okuyama, 1974). The upper dashed line shows the results obtained by Glos and Henrici (1991) for bending strength of 70 × 150 mm beams at temperatures of 100°C and 150°C. The moisture content at the time of the tests was in the range 7–10% for the 100°C beams and 3–6% for the 150°C beams. The lower dashed line shows the previous German design values (Kordina and Meyer-Ottens, 1983) obtained from tests by Kollman and Schulz (1944). The wood used in the earlier tests may have been more moist which would explain the different slope. The German design values given by Kordina and Meyer-Ottens (1995) have been changed to reflect the results of Glos and Henrici.

König (1995) performed fire-resistance tests on single joists with the narrow edge unprotected and the wide sides protected by rock wool insulation. The joists were tested in bending with the fire-exposed side in tension or compression. Extreme plastic behaviour was observed, with a very large shift in the neutral axis location, especially for those members with the fire-exposed edge in compression. Fire tests of

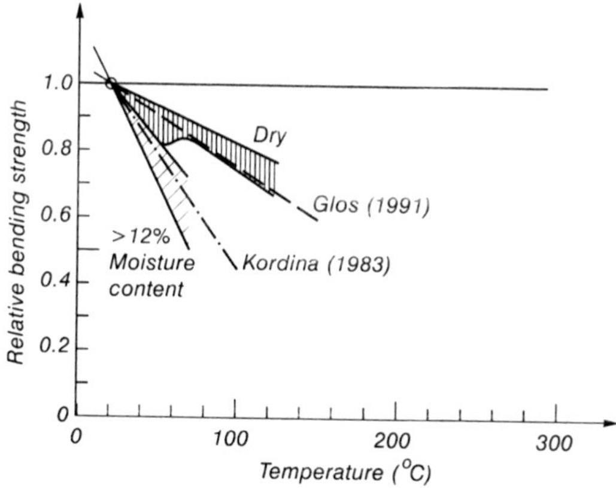

Figure 10.13 Bending strength of wood *versus* temperature (Adapted from (Gerhards, 1982) with permission of Society of Wood Science and Technology)

unprotected timber joist floors by Woeste and Schaffer (1979) and tests of studs by Norén (1988) showed a reduction of variability and an increase in load sharing during fire, resulting from increased ductility. Tests of glulam beams by Bolonius Olesen and Hansen (1992) showed continued softening of the wood during the cooling period, even after charring had stopped.

Perpendicular to the grain properties

Modulus of elasticity For modulus of elasticity perpendicular to the grain, Gerhards (1982) reports eight studies which all lie in the wide range shown in Figure 10.14, for temperatures up to 100°C. The dependence on temperature tends to be greater for moisture content above 20%, but there is a lot of overlap between the studies. Much of the data show negligible stiffness for moist wood as the temperature approaches 100°C, indicating plastic behaviour as reported for parallel to the grain behaviour.

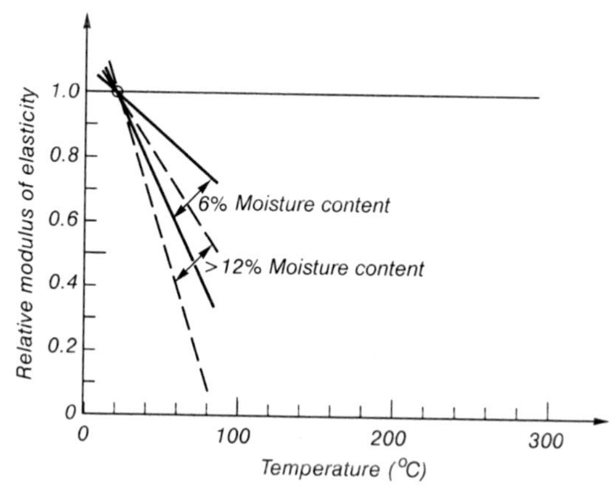

Figure 10.14 Modulus of elasticity perpendicular to grain *versus* temperature (Adapted from (Gerhards, 1982) with permission of Society of Wood Science and Technology)

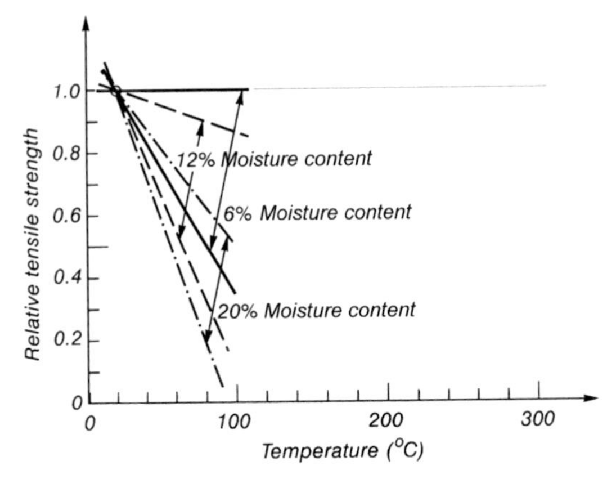

Figure 10.15 Effect of temperature on tensile strength perpendicular to the grain (Adapted from (Gerhards, 1982) with permission of Society of Wood Science and Technology)

Tensile strength The effect of temperature on tensile strength perpendicular to the grain is shown in Figure 10.15, where all the data are from Gerhards (1982). It can be seen that there is a wide range of results with much overlap for the different moisture contents, but a trend of a greater strength reduction as the moisture content increases. There are no results of tests over 100°C. Tensile strength perpendicular to the grain is an indication of resistance to splitting, which is a very unpredictable wood property, even in the best of conditions.

Compressive strength Figure 10.16 shows the effect of temperature on strength in compression perpendicular to the grain. This shows data from five studies reported by Gerhards (1982) which all overlap with much scatter and even less dependence on moisture content than for tension. The measured strength in these tests was the

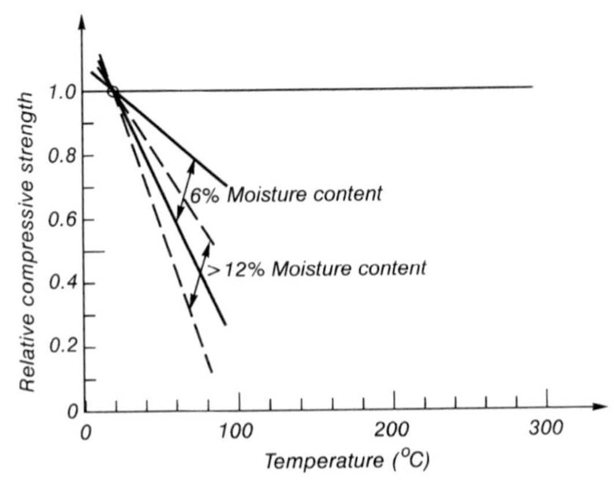

Figure 10.16 Effect of temperature on compression strength of wood perpendicular to the grain (Adapted from (Gerhards, 1982) with permission of Society of Wood Science and Technology)

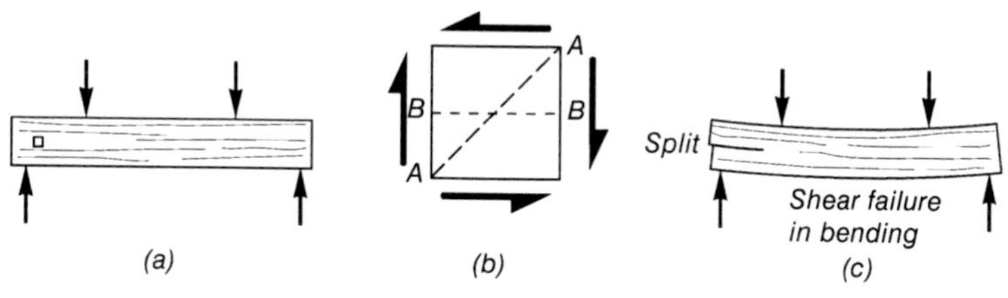

Figure 10.17 Shear stresses and shear failure in timber beam

proportional limit which is very difficult to define when the stress–strain relationship may be curved from very low loads as shown in Figure 10.7. The proportional limit is not an important strength property, because some local compression yielding is expected in situations such as highly loaded studs bearing on bottom plates in light frame construction. The ultimate crushing strength is more useful, but also difficult to measure because it requires very large strains, and there may be no maximum value, also shown in Figure 10.7.

Shear

Shear in beams Figure 10.17(a) shows a beam with applied loads. The elemental volume near the left-hand support is enlarged in Figure 10.17(b). In a homogeneous isotropic material, the stresses shown in Figure 10.17(b) would usually produce a diagonal tension failure along the dotted diagonal line A–A. However, in wood, which has a well-defined longitudinal grain structure, a shear failure will usually result in a horizontal split along the grain as shown by the line B–B and the split in Figure 10.17(c). Standard tests are available to measure the shear strength and shear modulus (shear stiffness) of wood. Shear failures rarely occur in timber beams, and shear only becomes critical if there are pre-existing splits at the ends of the beam or if very high shear stresses are developed near connections.

Gerhards (1982) only reports two studies on shear strength of wood at elevated temperatures. Figure 10.18, reproduced from Gerhards, shows shear strength

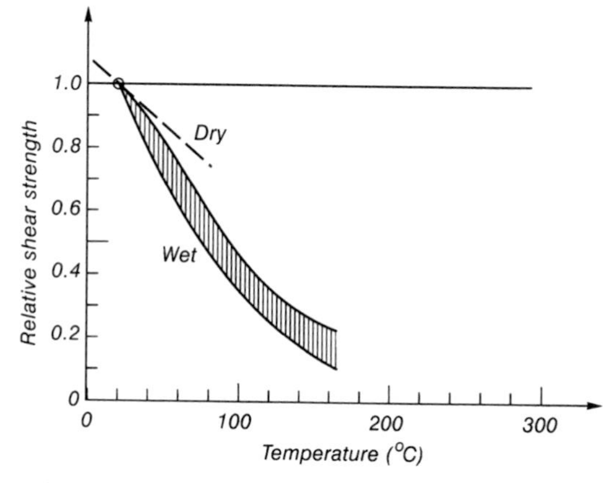

Figure 10.18 Effect of temperature on shear strength of wood

dropping rapidly to 150°C in wet wood. The data points are from Ohsawa and Yoneda (1978) and Sano (1961). For shear modulus (or modulus of rigidity) Gerhards (1982) only reports one study, by Okuyama *et al.* (1977) who observed the shear modulus dropping to 20–50% of the 20°C values at 80°C.

Derived results

Reduction factors The temperature and moisture gradients in heated wood (below the char layer in a glulam beam) affect the strength and stiffness of the member. An approximate estimate of the reduced strength and stiffness can be made by combining the predicted temperatures of the wood beneath the char with the effects of temperature on strength. Figure 10.19(b) shows the temperature profile from Janssens and White (1994). Figure 10.19(a) shows the assumed effect of temperature on mechanical properties, estimated from the test results summarized earlier. The modulus of elasticity is assumed to drop linearly to 50% of its normal temperature value at 300°C. The tension strength follows the same relationship to 200°C, then drops to zero at 300°C. These properties are assumed to be the same for wet and dry wood. For compression strength, dry wood drops linearly to zero at 300°C. Wet wood is assumed to drop to 50% of its normal temperature value at 100°C and remain at that value until it reaches a temperature of 160°C, after which it follows the relationship for dry wood.

Figure 10.19(c) shows the resulting drop in strength of the wood below the char layer. All three properties are significantly reduced in the 25 mm of wood below the char, the greatest reduction being in the compression strength, with a plateau at 15 mm depth which corresponds with wood temperatures at or above 100°C.

Stress–strain relationship Figure 10.20 shows stress–strain relationships derived by König and Walleij (2000) from computer modelling of bending tests carried out by König (1995). These relationships are consistent with the lines shown on Figures 10.9, 10.11 and 10.12. They are similar to relationships derived by Thomas *et al.* (1995) who investigated the structural performance of light timber frame walls and floors exposed to fire using finite-element models for thermal and structural analysis. These derived properties were used in a finite-element model to predict the results of fire-resistance tests of timber stud walls. The moisture content was not monitored during König's tests, but the wood studs had a typical initial moisture content of about 12%.

The relationships in Figure 10.20 are idealized in a simple way to allow prediction of overall behaviour. It can be seen that in the tension region, linear elastic behaviour has been assumed until failure. There may be some plasticity in tension as shown in Östman's results (Figure 10.10) but that will be of little significance in bending and compression members where compression yielding dominates the behaviour. In the compression region, idealized elasto-plastic behaviour has been assumed, which has been shown by Buchanan (1990) to give good results when modelling flexural behaviour.

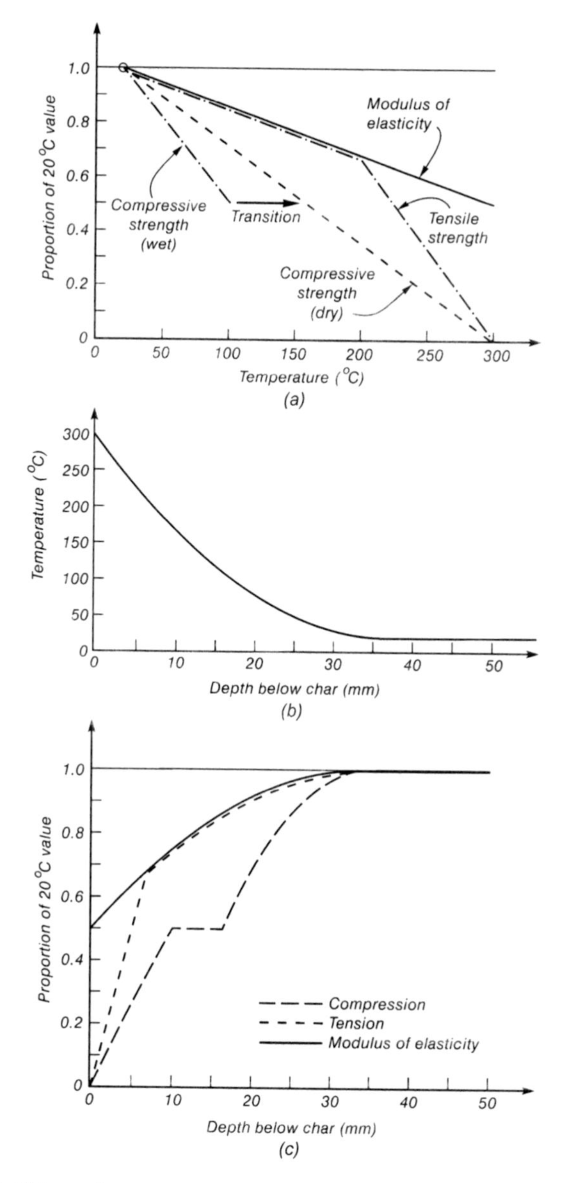

Figure 10.19 (a) Effect of temperature on mechanical properties of wood; (b) Temperature profile below the char layer; (c) Reduction in strength of wood below the char layer

The derived curves shown in Figure 10.20 include the creep that takes place in the duration of typical fire-resistance testing. The different moduli of elasticity in tension and compression at elevated temperatures are a result of greater creep in compression than in tension (Thomas, 1997). Similar results are reported by Young (2000) who included an increase in compressive strength as the heated wood dried out over 100°C.

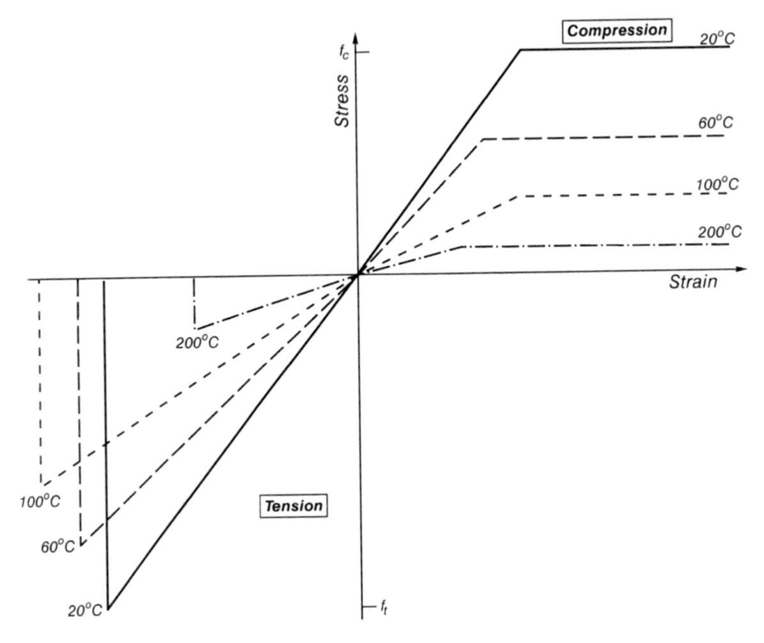

Figure 10.20 Derived stress–strain relationships for wood at elevated temperatures

10.6 DESIGN CONCEPTS FOR HEAVY TIMBER EXPOSED TO FIRE

Large timber members exposed to fire have excellent fire-resistance (Figures 10.21 and 10.22). The fire-resistance is easily calculated because of the predictable rate of charring on surfaces exposed to the standard fire. Figure 10.23 shows the common cases of three- and four-sided fire exposure of a rectangular member. The original cross section $b \times d$ is reduced to the *residual cross section* $b_f \times d_f$ as a result of charring. The depth to the char front is shown as the dimension c (mm) which is equal on all exposed surfaces, given by

$$c = \beta t \tag{10.2}$$

where β is the rate of charring (mm/min), and t is the time of fire exposure (minutes).

The dimensions of the residual cross section are given by

$$\begin{aligned} b_f &= b - 2c \\ d_f &= d - c \quad \text{(three-sided exposure)} \\ d_f &= d - 2c \quad \text{(four-sided exposure)} \end{aligned} \tag{10.3}$$

The boundary between the char layer and the remaining wood is quite distinct, corresponding to a temperature of about 300°C. There is a layer of heated wood about 35 mm thick below the char layer, and the inner core remains at room temperatures. The residual cross section is capable of supporting loads, providing a

Figure 10.21 Fire-resistance test of glulam beams; the beams span a 4 m long furnace with loads applied using concrete blocks

Figure 10.22 Residual cross section of a large glulam beam after a fire test (Reproduced by permission of American Institute of Timber Construction)

level of fire-resistance which depends on the load ratio (see Chapter 7). Failure occurs when the residual cross section is stressed beyond its ultimate strength.

Structural design of timber members is based on the strength and stiffness of the residual member considering the depth of char, the temperature and moisture profile of the wood below the char line, and the mechanical properties of wood at elevated temperatures. This is a simple concept, but national codes give many different methods of making the calculations. The main methods are described below.

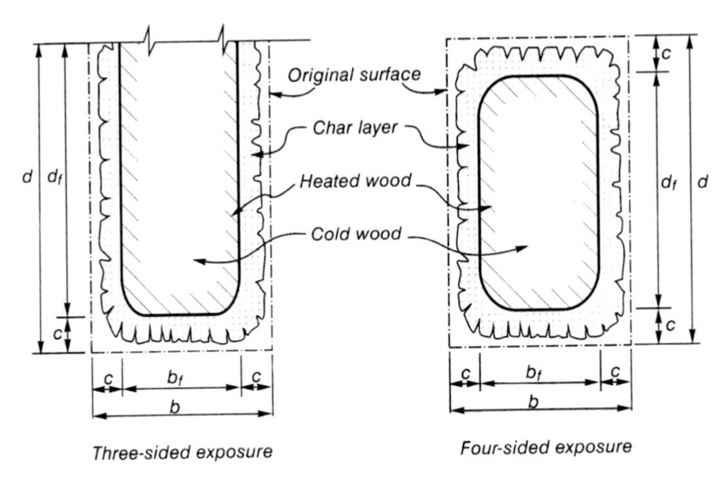

Figure 10.23 Design concepts for large timber members

10.6.1 Verification

As with other materials, verification of the strength during fire exposure requires that

$$U^*_{\text{fire}} \leqslant R_{\text{fire}} \tag{10.4}$$

where U^*_{fire} is the design force resulting from the applied load at the time of the fire, and R_{fire} is the load capacity during the fire situation.

Design forces are obtained from the applied loads by conventional structural analysis. Calculations of the load-bearing capacity are described below, based on the mechanical properties of wood at elevated temperatures. The design force U^*_{fire} may be axial force N^*_{fire}, bending moment M^*_{fire}, or shear force V^*_{fire} occurring singly or in combination, with the load capacity calculated accordingly as axial force N_{f}, bending moment M_{f} or shear force V_{f} in the same combination. Calculation of the load capacity is described below.

Simply supported beams

Design concepts are discussed here with reference to beams because they are the most commonly used heavy timber members. For a member submitted to a bending moment, the design is verified by satisfying the design equation

$$M^*_{\text{fire}} \leqslant M_{\text{f}} \tag{10.5}$$

where M^*_{fire} is the bending moment at the time of the fire, and M_{f} is the design flexural capacity under fire conditions, given by

$$M_{\text{f}} = Z_{\text{f}} f_{\text{f}} \tag{10.6}$$

where f_{f} is the design strength of wood in fire conditions (MPa), and Z_{f} is the elastic section modulus (mm^3) reduced for fire exposure.

The value of the design strength of wood in fire conditions f_f is handled differently in different codes, but it should always be the strength under short-duration loads because the duration of the fire exposure is short. For a rectangular section with no corner rounding, the elastic section modulus is

$$Z_f = b_f d_f^2/6 \qquad (10.7)$$

Note that that Equation (10.5) does not include a partial safety factor for mechanical properties γ_M (or a strength reduction factor Φ) because both have a value of 1.0 in fire conditions, as described in Chapter 7.

10.6.2 Charring rate

Investigations in many fire-resistance tests have shown that the rate of charring of timber is predictable in the standard test fire, depending on the density and moisture content of the wood. Many national codes specify a constant charring rate in the range 0.60–0.75 mm per minute for softwoods and about 0.5mm per minute for hardwoods (BSI, 1978, SAA, 1990(b)). Glulam and solid wood are usually considered to char at the same rate. The charring rate may reduce after prolonged fire exposure due to the increasing thickness of the insulating layer of char, but this is not usually recognized in design codes.

The effect of density and moisture content on the charring rate is shown in Figure 10.24, (from Lie, 1992, after Schaffer, 1967). The Australian code (SAA, 1990(b)) gives the following equation for charring rate β (mm/min) as a function of wood density, which gives similar values to those shown in Figure 10.24 for moisture content between 10 and 15%:

$$\beta = 0.4 + (280/\rho)^2 \qquad (10.8)$$

where ρ is the wood density (kg/m^3).

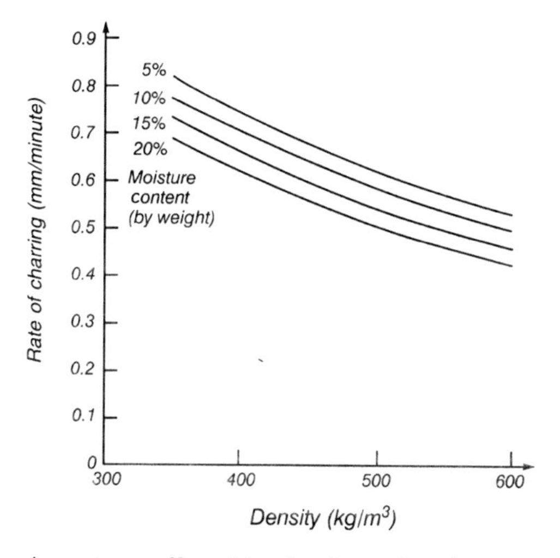

Figure 10.24 Charring rate as affected by density and moisture content (Lie, 1972)

Table 10.1 Charring rates for design

Material	Minimum density (kg/m³)	Char rate β (mm/minute)	Char rate β_l (mm/minute)
Glue-laminated softwood timber	290	0.64	0.70
Solid or glue-laminated hardwood timber	450	0.50	0.55
Softwood panel products (plywood, particle board) minimum thickness 20 mm	450	0.9	

More recent charring studies have been carried by a number of people including Schaffer (1977), Mikkola (1990), Hadvig (1981) and White and Nordheim (1992), as summarized by White (1995). König and Walleij (1999) experimentally studied the charring rates of glulam in the standard fire and in parametric fires, and investigated the effect of protective materials before and after they fell off the surface of the wood.

The New Zealand code (SNZ, 1993) specifies a charring rate of $\beta = 0.65$ mm per minute, derived from Collier (1992) who carried out charring tests on radiata pine glulam beams, and used the equations of White and Nordheim (1992) to relate charring rate to density for typical radiata pine which has a density of 550 kg/m^3 at 12% moisture content, as described by Buchanan (1994(b)). Correlations with density must be made with care, because wood density varies considerably both within and between trees and sites, and there are several different definitions of wood density depending on how the moisture is included in the calculation (Collins, 1983). Charring of small size wood members has been studied by Lau *et al.* (1999).

Table 10.1 shows recommended charring rates based on Eurocode 5 (EC5, 1994). The measured charring rate β is intended to be used with the actual cross section with rounded corners, and a 10% larger notional charring rate β_l is to be used if there is no allowance made for corner rounding in the calculations. It is more accurate to use the values of β and modify the section properties for corner rounding, but the effect of rounding becomes very small and can be ignored for large members. Many other national codes have β values similar to those in Table 10.1.

In North America, recommendations for the charring rate are given by AFPA (1999), based on the non-linear model of White (1988). The proposed charring rate β is the average charring rate (mm/minute) over the period to time t (minutes), given by

$$\beta = 2.58\ \beta_n / t^{0.187} \tag{10.9}$$

where β_n is the nominal charring rate obtained from the char depth measured after 1 hour of fire exposure ($\beta_n = 0.635$ mm per minute), and t is the time (minutes). The resulting char layer thickness c (mm) at time t (minutes) is given by

$$c = \beta t = 2.58\ \beta_n\ t^{0.813} \tag{10.10}$$

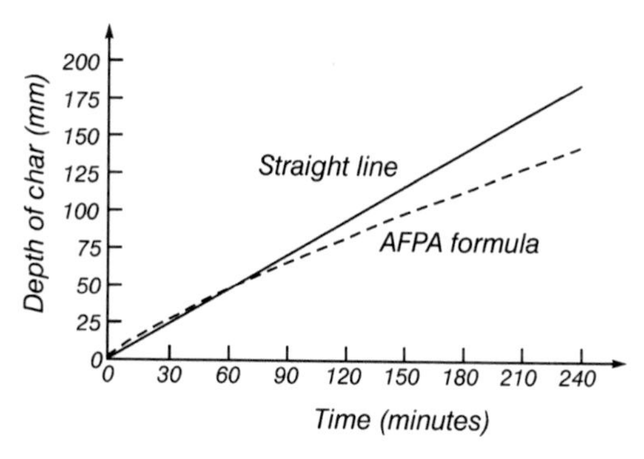

Figure 10.25 Depth of char from North American recommendations

Equation (10.9) includes a 20% increase in charring rate over measured rates to allow for rounding at the corners and the reduction of strength of the heated layer below the char front. It has been converted from imperial units in the original publication, based on $\beta_n = 1.5$ inches per hour. The curve in Figure 10.25 shows the resulting depth of char during 4 hours of standard fire exposure, compared with the straight line which is the depth of char for a uniform charring rate of 0.762 mm per minute (1.2×0.635). It can be seen that there is little difference up to 1 hour where the curve crosses the line, but the AFPA non-linear equation gives less depth of char than the uniform charring rate for exposure times over 2 hours.

All the charring rates given above are for timber exposed to the standard fire-resistance test. Charring rates in more realistic fires are given below in the description of the Eurocode design method for parametric fires (Section 10.6.6).

10.6.3 Corner Rounding

All fire tests of large rectangular timber sections show some rounding of the corners, because the corners are subjected to heat transfer from two surfaces. Figure 10.26 shows the shape of a typical charred cross section, from BS 5268 (BSI, 1978). Most design codes use the simple relationship whereby the radius of the rounding is equal to the depth of the charred layer. This is supported by several studies, including Hadvig (1981). Majamaa (1991) proposed a radius equal to 80% of the char depth. The Eurocode (EC5, 1994) gives a radius approximately equal to the char depth in the early stages of the fire, but less after 30 minutes as shown in Figure 10.27.

If corner rounding is taken into account, the section properties will be affected slightly, depending on the size of the member. For a beam exposed to fire on three sides, the section modulus $Z_{f,r}$ of the reduced cross section is given approximately by

$$Z_f = b_f \, d_f^{\,2}/6 - 0.215\, r^2 d_f \tag{10.11}$$

where b_f is the residual width of the beam, d_f is the residual depth of the beam, and r is the radius of the charred corner.

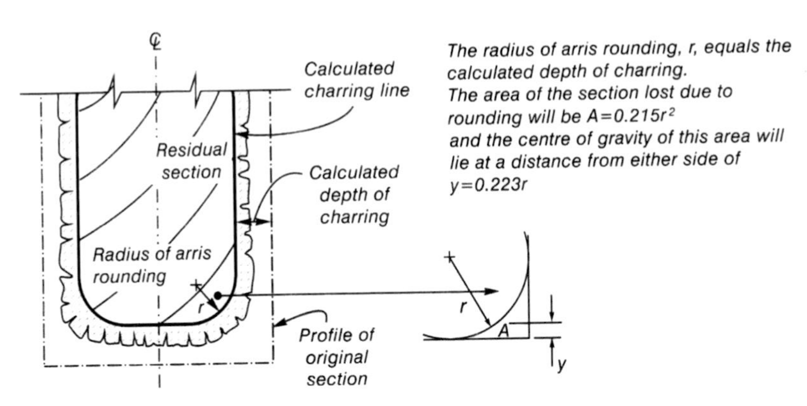

Figure 10.26 Residual cross section of timber beam exposed to fire (Reproduced from (BSI, 1978) by permission of BSI)

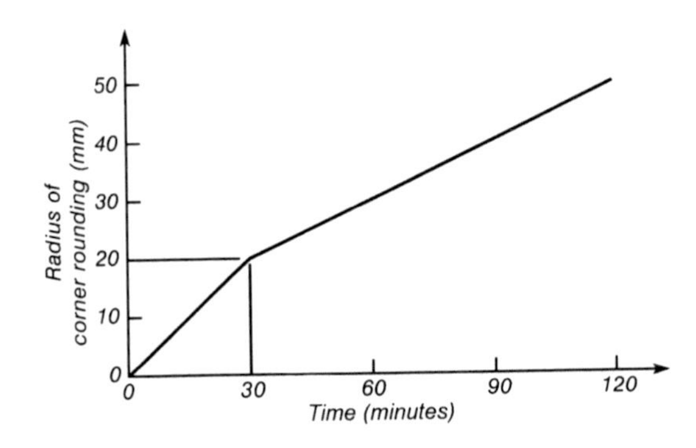

Figure 10.27 Time-dependent radius of the char line at corners

10.6.4 Effect of Heated Wood Below the Char Line

There are several alternative design methods to allow for heated wood below the char line. Some codes (BSI, 1978, SNZ, 1993) ignore any reduction of wood strength below the char, which can lead to unsafe results for small cross sections (less than about 100 mm thick). The *effective cross section method* uses a cross section smaller than the residual cross section, effectively assuming a layer of zero-strength wood below the char line, but assuming that the material properties in the inner part of the member are unaffected by temperature. The *reduced properties method* uses the residual cross section dimensions with an average reduction in the material properties over the whole residual cross section.

Effective cross section method

The effective cross section method allows for heated wood below the char line by subtracting a nominal layer of zero-strength wood from the fire-reduced cross section. The wood in the effective cross section is assumed to have normal temperature properties with no reduction for temperature.

The design flexural capacity M_f should be calculated using Equation (10.6), where (assuming no corner rounding), the section modulus Z_f for three-sided fire exposure is given by

$$Z_{f,z} = (b_f - 2z)\,(d_f - z)^2/6 \tag{10.12}$$

where z is the thickness of the zero-strength layer (mm).

For this method, the design strength of the wood in Equation (10.5) is the strength at normal temperature, so that

$$f_f = f_b \tag{10.13}$$

where f_b is the design strength of the wood at normal temperature (MPa).

In Eurocode 5, the thickness of the zero-strength wood layer is $z = 7$ mm, for fire exposure greater than 20 minutes. For exposure less than 20 minutes, the 7 mm thickness is reduced proportionately to zero. The Australian code (SAA, 1990(b)) specifies a thickness $z = 7.5$ mm for the zero-strength layer. The justification for the 7 mm thickness is given by Schaffer *et al.* (1986) who considered the reduction in modulus of elasticity of the heated wood below the char layer.

The proposed AFPA North American design method increases the nominal charring rate by 20% to allow for the heated wood below the char line. This implies a zero-strength layer of wood with increasing thickness during the fire exposure, the thickness being about 8 mm for each hour of exposure which becomes increasingly significant for long periods of fire-resistance.

Using the effective cross section method in accordance with Eurocode 5, the charring rate β_l should be taken from Table 10.1 and the section modulus Z_f should be calculated from Equation (10.12), with no corner rounding considered in the calculation.

Reduced properties method

The reduced properties method, from Eurocode 5, is based on a strength reduction factor k_f applied to all of the wood below the char zone, rather than assuming a hot layer with zero strength. As before, the design flexural capacity M_f must be calculated using Equation (10.6). The section modulus Z_f should preferably be calculated from Equation (10.11) (including corner rounding) with the charring rate β from Table 10.1. A simpler but less accurate method permitted by Eurocode 5 is to calculate Z_f from Equation (10.7) and use β_l from Table 10.1.

For the reduced properties method, the design strength of the wood under fire conditions is given by

$$f_f = k_f\, f_b \tag{10.14}$$

where k_f is the strength reduction factor for the residual cross section.

The value of the strength reduction factor k_f cannot be determined precisely, because it depends on the temperature and moisture content of the heated layer, and the size of the heated layer as a proportion of the whole member. Lie (1977) used a

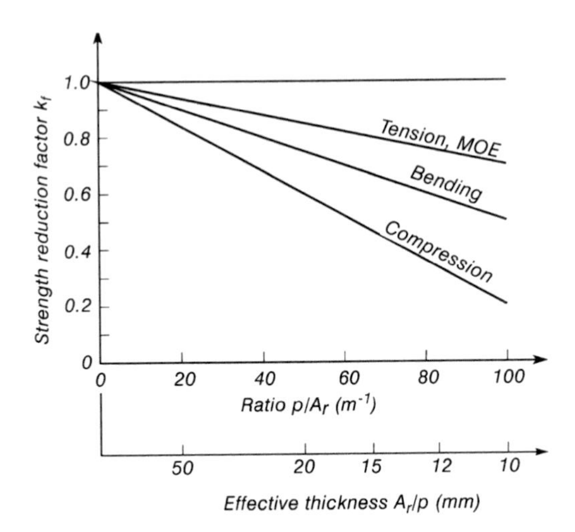

Figure 10.28 Reduction in mechanical properties for standard fire exposure

constant value of $k_f = 0.8$ in his derivation of the North American design equations, taking no account of the section size. In Eurocode 5, k_f is given by

$$k_f = 1.0 - 1/g(A_r/p) \qquad (10.15)$$

where p is the perimeter of the fire-exposed residual cross section (m), A_r is the area of the residual cross section (m^2), g is a factor (m^{-1}), with the value of 200 for bending, 125 for compression and 330 for tensile strength and modulus of elasticity.

The strength reduction factor k_f is plotted in Figure 10.28. The reduction is most severe for compression strength, intermediate for bending, and least severe for modulus of elasticity and for tension strength. Equation (10.15) needs to be verified by more tests (König 1998(a)). The ratio A_r/p (lower horizontal scale in Figure 10.28) is the effective thickness of wood in the residual cross section. For a slender deep beam, the effective thickness A_r/p is approximately half of the breadth of the residual beam. Most heavy timber members will have an effective thickness greater than about 30 mm whereas light timber framing will have effective thickness less than 20 mm, so the relative difference in strength loss between these two categories of construction can be deduced from the graph.

An alternative formula for reduced properties of the residual section is given below in the design method for parametric fires.

Characteristic strength of wood

A significant difference between wood and other materials such as steel and concrete is the variability of wood strength between pieces of wood. This can have an effect on structural design in fire conditions, depending on how the design strength is derived from test results. For normal temperature design, the characteristic design strength of a population of boards is taken as the fifth percentile value. In most limit states design (LRFD) formats, the fifth percentile value of strength obtained from in-grade testing $f_{0.05}$ is listed in the code and used directly in the design calculations. For fire design, some codes (e.g. SAA, 1990(b), SNZ, 1993) use the fifth percentile value of strength $f_{0.05}$ so that the strength f_b to be used in the fire calculation is given by

$$f_b = f_{0.05} \tag{10.16}$$

where $f_{0.05}$ is the characteristic design strength (fifth percentile value) in the code for normal temperature design.

Some other codes (e.g. EC5, 1994) modify the fifth percentile strength $f_{0.05}$ used for normal temperature design to the 20th percentile strength for fire design, justified by the low probability of occurrence of serious fires. The design strength f_b for fire conditions is given by

$$f_b = k_{20}\, f_{0.05} \tag{10.17}$$

where k_{20} is a correction factor to convert fifth percentile to 20th percentile values (1.25 for solid timber, 1.15 for glulam in Eurocode 5).

The proposed American method (AFPA, 1999) uses the mean value of wood strength for fire design. This method is based on working stress design, so the allowable stress in the code f_a is modified to give an allowable stress in fire conditions $f_{a,f}$ using

$$f_{a,f} = k_{mean}\, f_a \tag{10.18}$$

where f_a is the allowable stress in the code (MPa), k_{mean} is a correction factor to convert allowable stresses to mean values (2.85 for tension and bending, 2.58 for compression, 2.03 for buckling failures (AFPA, 1999)).

Using the mean strength rather than a lower percentile implies that there is little or no safety factor in this method. However, the 'working stress' design methods in North American codes require fire design for full dead and live load without the reduction in live load which would be found in limit states design methods (see Chapter 7) so the end results may be more conservative than other codes. The Australian code (SAA, 1990) allows the design load to be reduced by 25% for fire design, if the working stress method is used.

10.6.5 Summary

This section has reviewed several different design methods for heavy timber members exposed to the standard fire. The methods are all based on the same concept, but the assumptions and details are different. The methods will give different results depending on the beam geometry and different ratios of dead load to live load. It is recommended that the design be carried out in limit states design format using the Eurocode 5 *reduced properties method* using Equations 10.5, 10.6, 10.11, 10.14, 10.15 and 10.17 with the charring rate β from Table 10.1. This is considered to be the most accurate method, but the others will give similar results. A design method for parametric fires is given below.

10.6.6 Design for Real Fires

All of the above design methods are based on the charring rate of timber exposed to the standard test fire. For realistic fires the only proposed design method is that in Annex A of Eurocode 5 (EC5, 1994) which gives charring rates and strength reduction factors for a particular class of real fires known as 'parametric fires'. These fires have been described in Chapter 4.

This method is based on the work of Hadvig (1981) who published the predicted depth of char for timber members exposed to the fires shown in Figure 4.8. This approach has been confirmed by Bolonius Olesen and Hansen (1992) who carried out full-scale tests on glulam beams exposed to such fires. The charring rates in this method have been further confirmed by König and Walleij (1999) who tested 95 mm thick blocks of glulam exposed to fire on one side.

For parametric fires, the initial charring rate β_{par} is related to the notional charring rate β_l according to the amount of ventilation available during the fire

$$\beta_{par} = k_p \, \beta_l \tag{10.19}$$

where k_p is the parametric char factor, given by

$$k_p = 1.5(5F_v - 0.04)/(4F_v + 0.08) \tag{10.20}$$

where F_v is the opening factor ($m^{1/2}$) given by

$$F_v = A_v\sqrt{H_v}/A_t \tag{10.21}$$

where A_v is the area of vertical ventilation openings (m^2), A_t is the total area of the internal bounding surfaces of the compartment (m^2), and H_v is the weighted average height of the vertical openings (m).

The parametric char factor k_p by which the parametric charring rate exceeds the notional charring rate, is plotted in Figure 10.29 for a range of opening factors. It

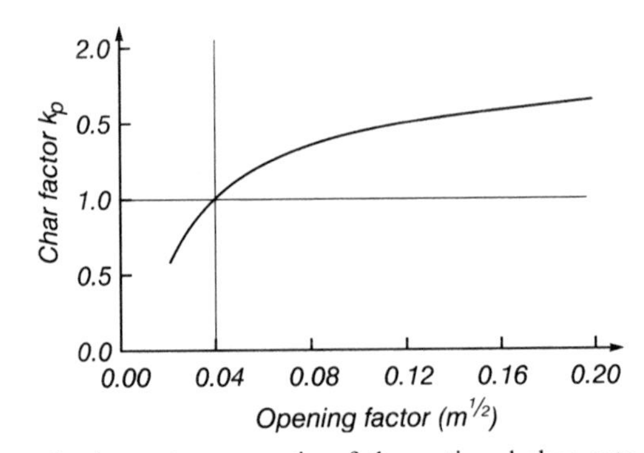

Figure 10.29 Parametric char rate as a ratio of the notional char rate

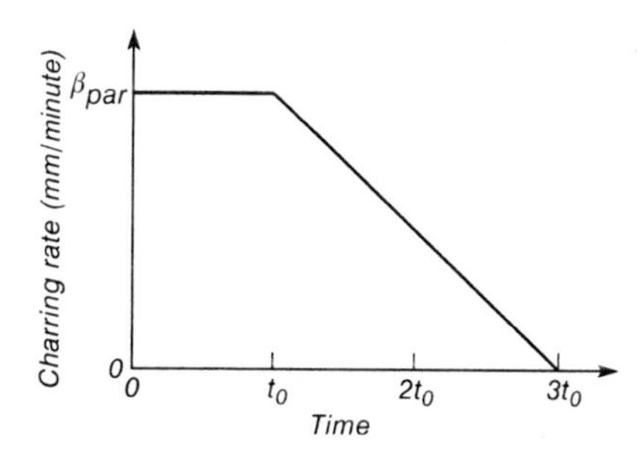

Figure 10.30 Charring rate with time for parametric fire exposure

can be seen that the charring rate is greater than the notional charring rate for opening factors greater than 0.04. The charring is assumed to proceed at the initial rate β_{par} for a time t_o then drop linearly to zero over subsequent time of $2t_o$ as shown in Figure 10.30 so the total depth of char at the end of the exposure (time $3t_o$) is

$$c = 2\beta_{par}\, t_o \tag{10.22}$$

The initial char time t_o is obtained from

$$t_o = 0.006 e_t / F_v \tag{10.23}$$

where e_t is the design fire load density in MJ/m^2 of internal surface area of the fire compartment. This can be compared with the ventilation controlled burning rate of wood using Equation (4.2) to show that burning of wood with calorific value 15 MJ/kg at a rate of $5.5A_v\sqrt{H_v}$ kg/minute would have a duration of time $2t_o$.

Table 10.2 shows results from the above equations. Column 2 gives the charring rate from Equation (10.13) for the valid range of opening factors. Opening factors greater than 0.12 are exceptional. The initial char time is shown for a range of fuel loads from Equation (10.17) and the total depth of char c from Equation (10.16) for the same fuel loads assuming a nominal char rate of $\beta_1 = 0.7$ mm/minute. This char

Table 10.2 Char rate, char time and char depth for parametric fire exposure

Opening factor ($m^{1/2}$)	Char rate β_{bar} (mm/minute)	Initial char time t_o (min) — Fuel load (MJ/m^2 total area) 80	160	240	320	400	Total char depth c (mm) — Fuel load (MJ/m^2 total area) 80	160	240	320	400
0.02	0.39	24					19				
0.04	0.70	12	24	36			17	34	50		
0.06	0.85	8.0	16	24	32	40	14	27	41	55	68
0.08	0.95	6.0	12	18	24	30	11	23	34	45	57
0.12	1.05	4.0	8.0	12	16	20	8.4	17	25	34	42
0.20	1.15	2.4	4.8	7.2	10	12	5.5	11	16	22	27
0.30	1.20	1.6	3.2	4.8	6.4	8.0	3.8	7.7	11	15	19

depth is to be used with the residual rectangular cross section ignoring corner rounding.

To allow for the reduced strength of the residual cross section exposed to a parametric fire, Eurocode 5 specifies that the load capacity of the remaining wood should be calculated using a reduction factor k_f from Bolonius Olesen and König (1992) given by

$$k_f = 1 - 3.2\ c/b \tag{10.24}$$

where c is the depth of char, and b is the original minimum dimension of the cross section. This relationship is shown in Figure 10.31.

This design method for parametric fires suggests that the timber members could survive a complete burnout of the fire compartment, but it has been observed after fire tests that charring often continues and the beams continue to lose strength and stiffness, even though the fire is out. Hence, even though the members are designed to resist a complete burnout, it is necessary for fire fighters to cool the beams with water at the end of the fire to terminate the charring and the creep deflections.

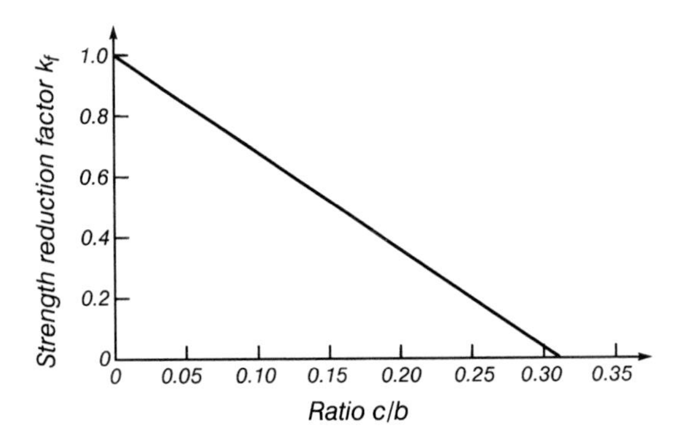

Figure 10.31 Reduction in mechanical properties for parametric fire exposure

10.6.7 Empirical Equations

Most North American building codes include a set of simple equations for calculating fire-resistance of large timber beams and columns, based on the work of Lie (1977). In the US the design method applies to glulam and sawn timber (e.g. UBC, 1997) but in Canada it only applies to glulam (NBCC, 1995). Lie assumed a uniform charring rate of 0.6 mm per minute, and allowed for the reduced strength of the hot wood layer under the char by assuming that the section remains rectangular and the entire residual core has 80% of its initial strength. The derivation assumes that the ultimate strength of the wood is three times the allowable design stress. The resulting equations are non-linear and must be solved in an iterative manner to determine the fire-resistance time, so Lie approximated them by a set of simple equations that allow a straightforward calculation of fire-resistance time as a function of member size and load ratio.

The derivation is based on a beam loaded with the maximum allowable load, so that failure in fire occurs when the ultimate flexural strength of the residual section is equal to the allowable design strength of the original section. Comparing the section moduli of the two sections gives the failure condition as

$$\alpha b_f d_f^2 = k_a b d^2 \tag{10.25}$$

where α is the ratio of hot wood strength to cold wood strength ($=0.8$), and k_a is the ratio of allowable strength to ultimate strength ($=0.33$).

The fire-resistance time can be calculated by eliminating $b_f d_f$ and c from Equations (10.2), (10.3) and (10.25) and solving for t. Lie solved these equations for a realistic range of member sizes, also introducing a load factor z to allow for the ratio of actual to allowable load on the member. For dimensions in millimetres, the approximate solution to these equations for beams gives the time to failure t_f (minutes) as

$$\begin{aligned} t_f &= 0.1zb(4 - b/d) \quad \text{(3-sided exposure)} \\ t_f &= 0.1zb(4 - 2b/d) \quad \text{(4-sided exposure)} \end{aligned} \tag{10.26}$$

with

$$z = 0.7 + 0.3/R_a \tag{10.27}$$

where R_a is the ratio of actual to allowable load at normal temperature.

For columns, the failure mode depends on the slenderness. Short columns fail when the ultimate compressive stress is exceeded, so comparing the cross-sectional area of the two sections gives the failure condition as

$$\alpha b_f d_f = k_a b d \tag{10.28}$$

Long columns fail by buckling. Assuming that d is the smaller dimension and buckling occurs in that direction, comparing the moment of inertia of the two sections gives the failure condition as

$$\alpha b_f d_f^3 = k_a b d^3 \tag{10.29}$$

The fire-resistance time can be calculated by eliminating $b_f\, d_f$ and c and solving for t. For columns, Lie used Equation (10.25) as an average between Equations (10.28) and (10.29). The resulting equations are

$$\begin{aligned} t_f &= 0.1zb(3 - d/2b) \quad \text{(3-sided exposure)} \\ t_f &= 0.1zb(3 - d/b) \quad \text{(4-sided exposure)} \end{aligned} \tag{10.30}$$

For long columns, z is calculated from Equation (10.27). For short columns the value of z is increased to give better agreement with experimental results for columns of low slenderness ratio, using

$$z = 0.9 + 0.3/R_a \tag{10.31}$$

In the current version of the North American codes, Equations (10.27) and (10.31) can only be used for values of R_a greater than or equal to 0.5, which ignores the increase in fire resistance for loading less than 50% of the allowable load. These equations have been derived for use in working stress design format. They can be used in limit states (LRFD) format if R_a is taken as the ratio of the applied moment to the design flexural resistance at normal temperatures. These equations may be used to satisfy North American code requirements, but they are not recommended because they give the user no insight into the processes which occur when timber members are exposed to fires.

10.7 DESIGN OF HEAVY TIMBER MEMBERS EXPOSED TO FIRE

This section describes the factors to be included in the structural design of various large timber components exposed to fire. The approach taken here is that design from first principles, using the information presented in this book, will lead to a better understanding of fundamental behaviour and less room for error than the use of formulae or design charts. A summary of fire performance of timber structures is given by Jönsson and Pettersson (1985).

10.7.1 Beams

Large timber beams exposed to fire have demonstrated excellent predictable behaviour. Beams can be designed using the same design equations as for normal temperature conditions, with modifications for strength and cross section as described above.

It is important to determine which surfaces of the beam are exposed to fire. Most beams provide support to fire-resisting floor systems, so that the top edge of the beam is not exposed to the fire, as shown in Figure 10.32(b). In some cases more of the beam may be protected from fire as shown in Figure 10.32(c), in which case this can be included in the calculations.

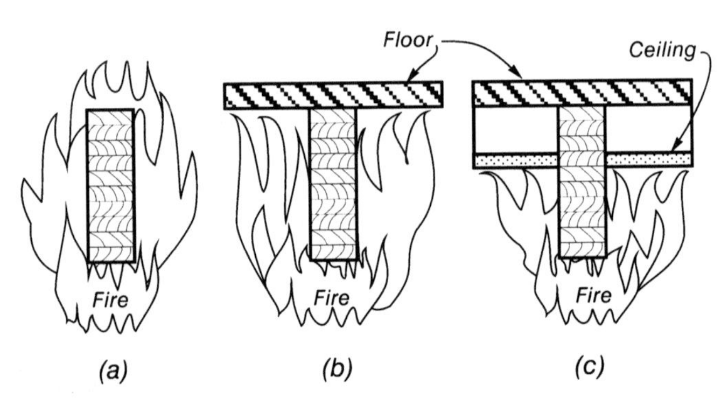

Figure 10.32 Three- and four-sided beam exposure

In addition to the calculations of flexural strength given above, beams must be checked to ensure that they are not likely to fail by lateral torsional buckling. No buckling check is necessary if the compression edge of the beam is provided with continuous lateral restraint such as, for example, a fire-rated floor system restraining the top edge of a simply supported beam. If lateral restraint is missing or intermittent, normal calculations from national timber design codes should be used, based on the residual charred cross section of the beam. Buckling resistance depends on the torsional rigidity of the cross section, which can drop to low levels as charring proceeds, especially for slender rectangular beams. Fredlund (1979) has studied the lateral stability of unrestrained timber beams exposed to fire.

Shear stresses are not normally a critical design consideration in rectangular beams, but may become important for I-beams or beams with large holes for services to pass through. Design can be made using the fire-reduced cross section with allowance for reduced strength of the residual cross section. Eurocode 5 (EC5, 1994) states that shear may be disregarded in solid cross sections, and for notched beams it should be verified that the residual cross section in the vicinity of the notch is at least 60% of the cross section required for normal temperature design. Kordina and Meyer-Ottens (1995) describe fire tests of glulam beams with rectangular openings reinforced with extra glulam material glued around the opening.

Beam deflections can be calculated using the reduced loads and reduced section modulus as shown above, and a reduced modulus of elasticity. Deflections are not usually of concern because the strength limit state is more important than serviceability limit states during fire exposure.

10.7.2 Tension Members

Tension members are not affected by the possibility of buckling. The tensile load capacity of a fire-reduced cross section can be calculated in the same way as for a flexural member, using one of the Eurocode design methods. There are no reported fire test results of large timber tension members, but there is no reason why a tension member will not behave in the same way as the tension edge of a deep beam, many of which have been tested.

10.7.3 Columns

The strength of a short column depends on the crushing strength of the material. Under fire exposure this can be calculated from the reduction of cross section and the reduced strength of the wood in the residual cross section, in the same way as for tension members. Long columns are susceptible to buckling failures, so the failure load depends on the moment of inertia and modulus of elasticity of the residual cross section. Lateral stability is much more important for columns than for beams. The likelihood of buckling will increase as the fire progresses because the reduced cross-sectional dimensions increase the slenderness of the column. Free-standing columns have no lateral support over the full height. For columns with mid-height bracing, the bracing must have sufficient fire resistance to provide restraint for the duration of the fire. Columns built into walls may have better fire-resistance than free-standing columns as a result of partial protection against charring, and lateral restraint in the direction of the wall.

A few full-scale fire tests of large timber columns are reported in the literature. A well documented series of 16 column tests is described by Malhotra and Rogowski (1970). The columns achieved fire-resistance ratings between 30 minutes and 90 minutes, depending on load and slenderness ratio. In all cases the fire resistance was greater than predicted by simple analysis of the charring rate using the code formulae. Precise analysis is difficult because of the partial fixity at the column ends, and unknown ultimate strength of the wood. A report by AFPA (1999) analyses these and several tests to verify the North American design equations.

Schaffer (1984) describes several studies of large timber columns exposed to fire, all of which use the same principles of considering buckling of a fire-reduced cross section. One of these is the German design equation which gives results 'in reasonably good accordance with the results of fire-resistance tests' (Meyer-Ottens, 1983). Calculations can be made using the column design formulae from national codes, with allowances for the fire-reduced cross section and the degree of fixity at the ends. The modulus of elasticity used in column calculations should be the lower fifth percentile value, not the mean value used for deflection calculations.

10.7.4 Beam-columns

A 'beam-column' is a member subjected to combined bending and axial loading. This may be a beam with some axial load, or more often a column with some bending moment.

Some design equations for beam-columns are based on a limiting maximum stress in the extreme fibre of the cross section. The so-called secant formula uses this approach to give an exact expression for maximum stress in an elastic column. Formulae such as these are difficult to use in fire situations because neither the location nor the strength of the extreme fibre are readily identified. It is more appropriate to follow the normal temperature design approach of codes (e.g. SAA, 1997) which give a general interaction formula including both flexural strength and axial load capacity, such as

$$(N/N_u)^2 + M/M_u \leqslant 1 \qquad (10.32)$$

where N is the applied axial load (kN), N_u is the axial load capacity, including the effects of buckling (kN), M is the applied bending moment, including a moment magnifier for slender members (kNm), and M_u is the flexural capacity including the effects of lateral buckling (kNm)

10.7.5 Decking

Solid wood decking must be considered differently from beams and columns, because it is required to perform a containing function as well as a load-bearing function. Assessment of fire resistance of decking must consider all three possible

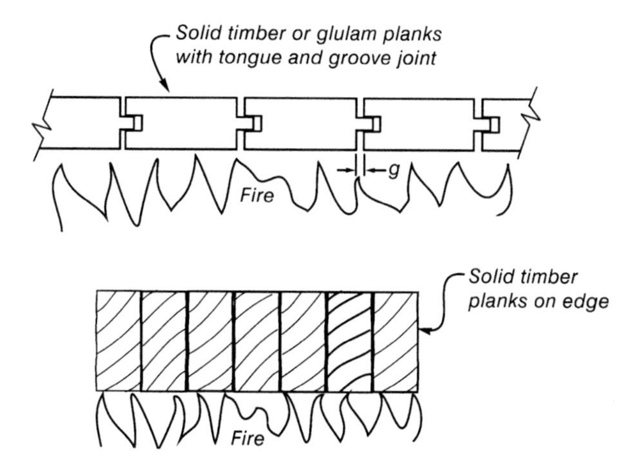

Figure 10.33 Tongue and groove decking and solid plank decking exposed to fire

failure criteria of stability, integrity and insulation. Solid wood decking as considered in this chapter includes solid timber or glulam timber planks laid flat and butted together with tongue and groove edges, and timber planks set on edge and nailed together, as shown in Figure 10.33.

This discussion is for fire exposure to the underside of timber decking. Fire exposure to the top surface of floors is not generally considered a serious problem because room temperatures are lower near the floor, convective flows are always up, not down, and the top surface of floors are often protected with furniture, floor coverings, and debris from the fire.

Stability

The strength, or stability, criterion can be assessed in the same way as for beams and columns. If the planks are fitted tightly together, the fire exposure will cause charring only on the lower surface. This results in gradually decreasing thickness as the fire proceeds. Calculations for a timber deck must ensure that the residual decking material can carry the applied loads during the fire exposure. The strength can be calculated, considering the reduced thickness and the increased temperature of the remaining wood, as described for beams. There are no problems of lateral stability for decking members. If the deck consists of two or more layers, the strength calculation must consider the structural properties of each layer separately, as the charring proceeds. For decking continuous over several spans, moment redistribution can be included in the calculations, as described in Chapter 7.

To avoid such calculations, Janssens (1997) has proposed an empirical design formula for structural performance of solid wood decks. This is based on a temperature and charring model considering the solid deck as a transformed section made up of a number of thin layers. The time to structural failure t_{sf} (minutes) is given by

$$t_{sf} = 1.25d(1 - \sqrt{0.4R_a}) - 11.3 \tag{10.33}$$

where d is the thickness of the deck (mm), and R_a is the ratio of the applied load to the allowable design load.

Integrity

The integrity criterion may be the most difficult to satisfy for wood deck systems. In order to meet this criterion, it is essential that no flames or hot gases pass through the floor, because these could lead to ignition of items on the upper surface.

The difficulties arise at the junctions between the planks, which may increase in width due to shrinkage of wood which often occurs during the life of a building. Tongue and groove joints between the planks are the best solution. If the planks have gaps large enough for the tongue to be exposed to fire temperatures, it can be assumed that the tongue will char at the same rate as the other exposed surfaces. This rate can be used to calculate whether the tongue will burn through during the design fire exposure. Fire exposure of the tongue can be reduced by painting the edges of the planks with an intumescent paint before assembling the deck, or using fire-resistant caulking.

The width of shrinkage gaps between the decking planks depends on the width of the individual planks and the moisture content of the wood at the time of installation of the planks. Wide planks installed with high moisture content will have wide shrinkage gaps. If the gap between the planks (dimension g in Figure 10.33) is small enough, the tongue will not be exposed to fire temperatures. Carling (1989) reports two studies (Aarnio, 1979, Aarnio and Kallioniemi, 1983) which studied the effect of a gap between two glulam beams or between a glulam beam and a concrete slab, as shown in Figure 10.34. These studies showed that temperatures within the gap remained low enough to prevent charring if the gap was less than 5 mm wide. A maximum gap of 3 mm is recommended for design purposes (Kordina and Meyer-Ottens, 1995).

The problem is similar but different with narrow planks of solid timber on edge. The gaps between the planks are likely to be less than with tongue and groove flooring because the distance between each gap is smaller, but there is no tongue to prevent hot gases from rising through the deck. Once hot gases are driven through a gap by convective forces, temperatures within the gap will rise rapidly and charring will increase the width of the gap even further. Possible methods of preventing convective flows through the assembly are to use an intumescent paint or fire-resistant putty within the gaps, or to provide a sheet of plywood or other sheet flooring material on the top surface.

If the gap is less than 1 mm, Eurocode 5 permits the total thickness of the deck or thin wall panel to be used in a charring calculation, with a reduction coefficient which depends on the type of tongue and groove between adjacent planks. The calculation gives the failure time t_{pr} (minutes) as

$$t_{pr} = \xi d/\beta \tag{10.34}$$

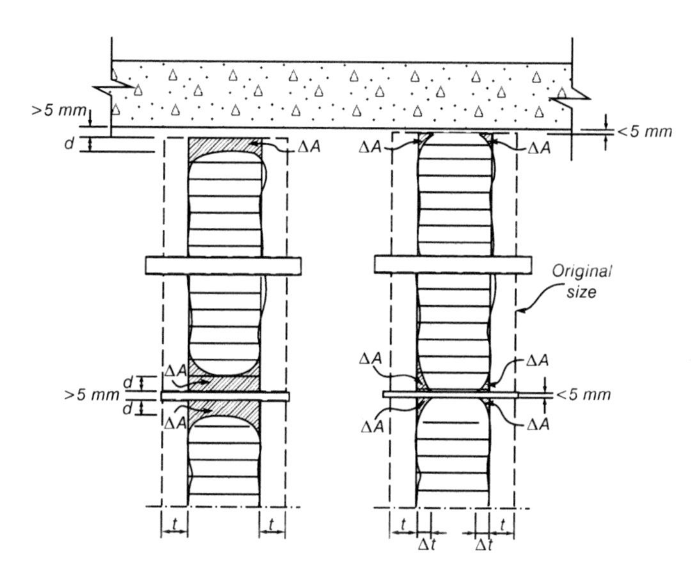

Figure 10.34 Effect of gap width on charring (Reproduced from Carling (1989) by permission of Building Research Association of New Zealand)

where ξ is the reduction coefficient shown in Figure 10.35, d is the thickness of the planks (mm), and β is the rate of charring from Table 10.1 (mm/minute).

Insulation

If the integrity and stability criteria are satisfied, there will be no problem meeting the insulation criterion, because the thickness of remaining wood required to carry applied loads will be greater than that required to prevent excessive temperature rise on the top surface.

10.7.6 Timber–concrete Composite Structures

Composite structures using both timber and concrete are used for floor construction. Most systems consist of a structural concrete topping over timber planks or timber beams, with some type of shear connection between the concrete and the wood. The shear connections are usually steel bolts or screws projecting from the top surface of the timber. The timber beam may also be notched to enhance the shear resistance. Timber–concrete composite structures can be used to strengthen historical buildings by casting a new concrete topping on existing timber floors after providing shear connections to the existing timber beams. Another form of composite construction consists of concrete topping over a laminated deck made up of narrow timber boards nailed side by side. Fire resistance of timber–concrete composite structures has been investigated by Fontana and Frangi (1999) who showed that the fire performance can be calculated based on the charring rate of the timber part of the structure. The shear connection will not perform well in fire if it relies on epoxy, because of the low strength of epoxy adhesives when heated.

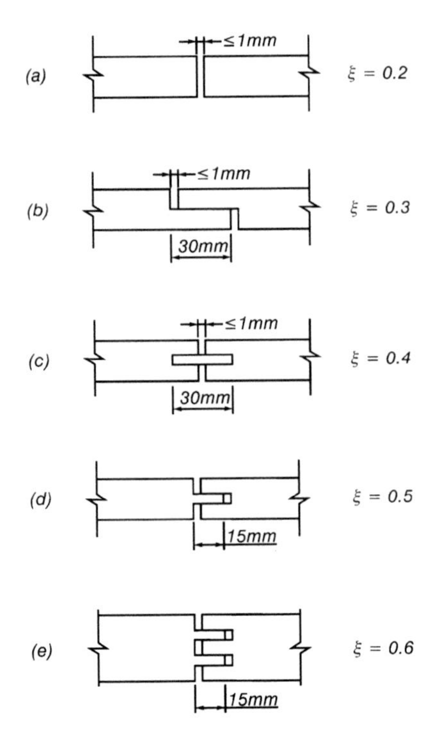

Figure 10.35 Reduction coefficient for joints in floors. (Reproduced from (EC5, 1994) by permission of CEN)

10.8 BEHAVIOUR OF TIMBER CONNECTIONS IN FIRE

The ability of any structure to carry loads and perform its function depends on the strength and stiffness of the structural members and the connections between those members. In a fire-resistant structure, both the members and their connections must perform throughout the fire exposure.

Some connections are not vulnerable to fire exposure. For example, a simple bearing support of a beam on a wall or a column on a foundation is unlikely to fail before the beam or column itself fails. Many other connections are much more vulnerable than the members themselves. It is much less easy to predict connection strength in fire, so the design philosophy should be to ensure that the connections have better fire resistance than the main members. A selection of typical connections in timber structures are shown in Figure 10.36.

Most connections are either metal fasteners or adhesives, which have very different fire performance. Very little research has been done on the performance of connections in timber structures exposed to fire. Most publications on this subject describe what works and what does not work, with little understanding of the underlying science. The most comprehensive review of the fire performance of connections in timber structures exposed to fire is by Carling (1989).

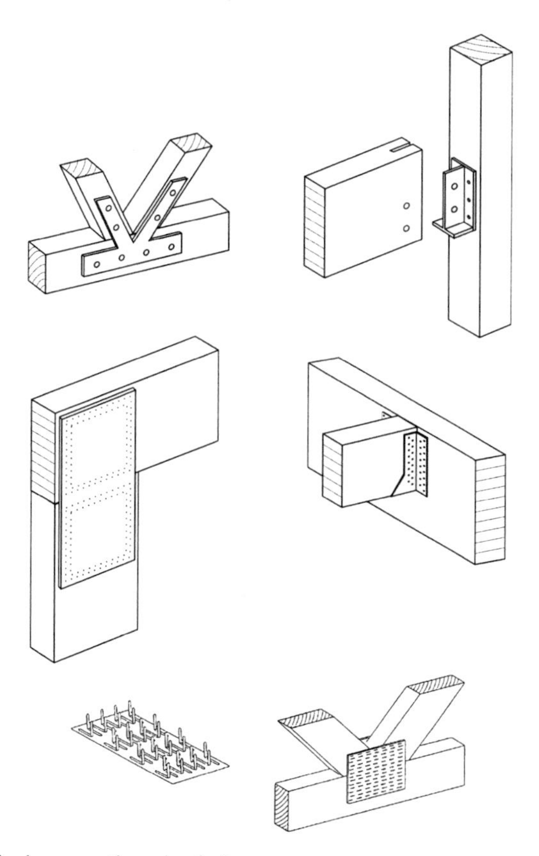

Figure 10.36 Typical connections in timber structures

10.8.1 Metal Fasteners

The behaviour of metal fasteners depends on the temperature of the metal, because that affects the strength of the fastener itself, and high temperatures lead to charring or loss of strength of wood in contact with the metal. Metals are much better conductors of heat than wood, so metal fasteners can conduct heat from the surface into the interior of a connection.

Geometry

As with steel structures, a geometrical arrangement of metal fastener with a high ratio of heated perimeter to cross-sectional area will heat up much more rapidly than one with a low ratio. The high temperatures will result in heat being conducted into the wood, causing softening or charring which will reduce the load carrying capacity (Figure 10.44).

This principle is illustrated in a series of tests by Leicester *et al.* (1979). Figure 10.37(a) shows four tension splice connections which were subjected to two fire exposures. Figure 10.37(b) shows the load–deflection results for the standard fire exposure, where it can be seen that the truss plate connector failed after only 5

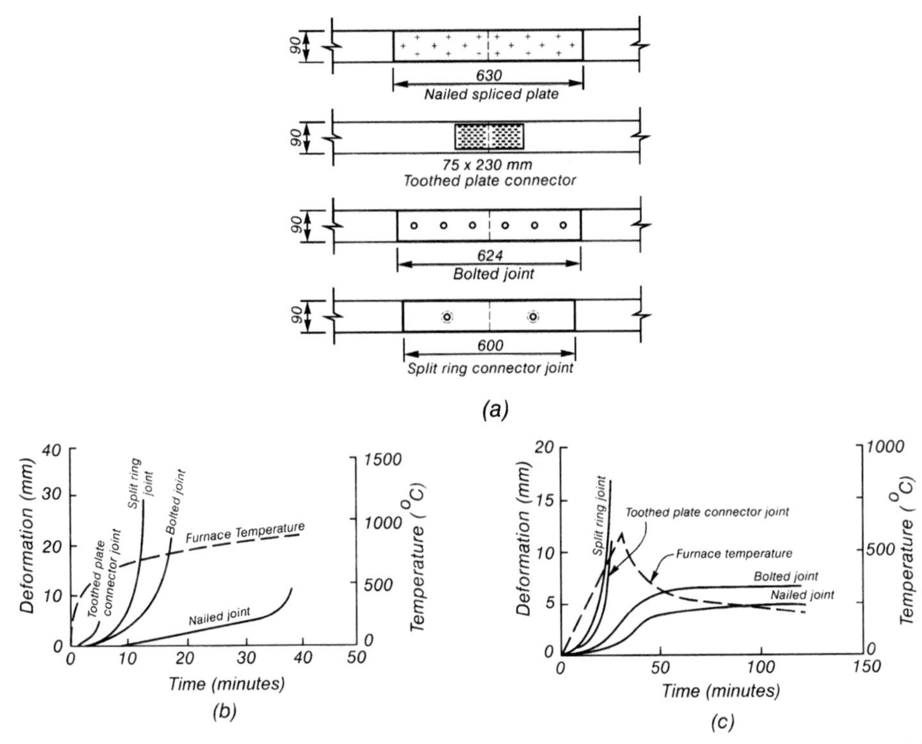

Figure 10.37 Tension tests of nailed and bolted joints (Reproduced from Leicester *et al.*, 1979 by permission of CSIRO Australia)

minutes, as a result of rapid heating of the steel and conduction of heat to the wood around the teeth of the plate. The bolted joint resisted the load for almost 20 minutes, and the nailed joint almost 40 minutes. When subjected to a less severe fire as shown in Figure 10.37(c) both the bolted and nailed joints were able to carry the load for 2 hours, but the truss plate failed after 20 minutes. The area of steel exposed to the fire is much less for bolts than for truss plates, and even less again for nails. For this reason, connections using steel plates will have very much better fire resistance if the plates are slotted into the timber members exposing only one edge, rather than being on the outside where the whole face of the steel is exposed to the fire.

Protection

Metal fasteners can provide excellent fire resistance if they are sufficiently well protected from the fire. Protection can either be achieved by burying the metal fastener within the wood section, or by applying a layer of sheet material such as solid wood or gypsum plasterboard.

Figure 10.38 shows test results and the requirements of the German Standard DIN 4102 part 4 for protecting nailed and bolted connections (Kordina and Meyer-Ottens, 1995). The British Standard and some American codes simply require that

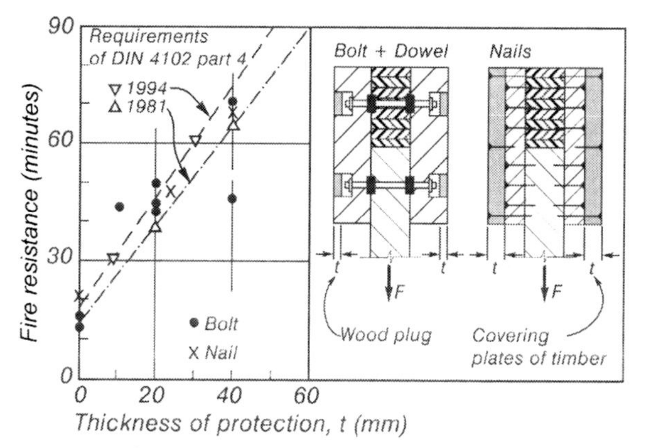

Figure 10.38 Relationship between fire resistance and thickness of wood protection (Kordina and Meyer-Ottens, 1995)

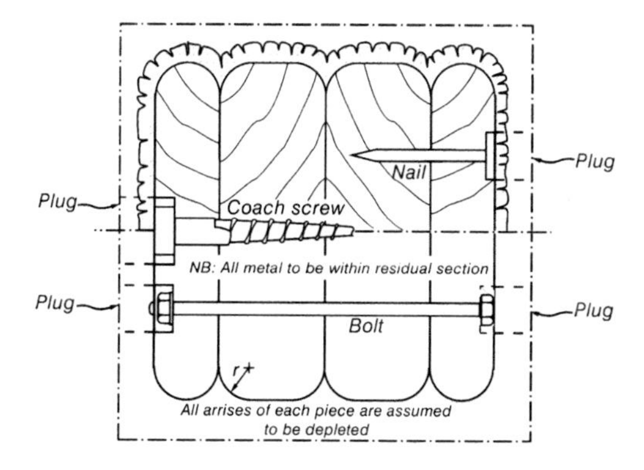

Figure 10.39 Protection of metal fasteners in BS 5268 (Reproduced from (BSI, 1978) by permission of BSI)

any metal connection should be protected with sufficient wood for it to be within the residual uncharred section as shown in Figure 10.39 (BSI, 1978). This requirement is very conservative, as is the char penetration at the gluelines which is usually much less than shown in Figure 10.39.

Eurocode 5 (EC5, 1994) does not provide specific design methods for connections, but relies on a more empirical approach. All unprotected wood-to-wood joints, and steel-to-wood joints where the steel plate is within or between the timber members, are deemed to have a 15 minute fire-resistance rating with no further treatment. The fire-resistance rating can be increased progressively to 60 minutes provided that the joint is protected so that member thickness and end and edge distances are increased by a thickness of wood a_{fi} (mm) given by

$$a_{fi} = \beta_1 (t_r - 15) \tag{10.35}$$

where β_1 is the charring rate given by Table 10.1 (mm/minutes), and t_r is the time of the fire-resistance rating (minutes).

As a compromise between protected and unprotected joints, joints may have 30 minutes fire-resistance rating if certain minimum dimensions are complied with and the joints are loaded at less than their full-load capacity.

Nailed gusset connections are often used to make moment resisting connections between large glulam members, using steel or plywood plates nailed to the sides of the glulam members. Buchanan and King (1991) showed that these gusset connections have poor fire resistance with no applied protection, but a layer of fire-resistant gypsum plaster board can be used to increase the fire resistance of the connection beyond that of the connected members. Intumescent paint applied to the steel gusset plate provided only a small increase in fire resistance, but this topic has not been well researched.

Carling (1989) reports similar tests (Aarnio, 1979, Aarnio and Kallioniemi, 1983) who achieved good fire resistance of gusset joints by protecting them with boxes made with glulam timber or particle board and mineral wool.

10.8.2 Nails and Screws

Nails are one of the best type of connection in timber structures because they penetrate the wood much better than surface adhesives, they do not weaken the wood with drilled holes and they can distribute forces over a larger part of the surface than bolts. Large nailed connections often have many nails passing through perforated steel plates, which provide excellent structural behaviour but poorer fire behaviour because of the large surface area of steel exposed to the fire.

Screws have many of the advantages of nails. In addition they have much better gripping capacity than nails because of the threaded shaft. A disadvantage in some situations is the poorer ductility of screws compared with nails. The fire performance of screwed connections in wood has not been studied extensively, but many of the conclusions below will apply.

Temperatures in nailed connections exposed to fire can be calculated using a finite-element model, as described by Fuller *et al.* (1992). If the nail temperatures and applied loads are known, Norén (1996) has shown that the structural performance of nailed wood-to-wood connections can be calculated using the yield theory which is the basis for design of fasteners at normal temperatures in many countries. Norén tested nailed splice joints in tension, exposed to the ISO 834 standard test fire. Typical results are shown in Figure 10.40 where the nail slip deformations are plotted against time for applied loads varying from 10% to 60% of the load capacity at normal temperature. Time to failure was inversely proportional to applied load, varying from 6 minutes to 21 minutes.

Figure 10.41 shows the test results compared with calculations for mode 1 and mode 4 failures. The longer duration tests (lower loads) failed in mode 1, with the nail shank crushing the wood in the cover plate and no nail bending. The shorter duration tests (higher loads) failed in mode 4 with flexural yielding of the nail shank. Figure 10.42 shows the effect of temperature on compressive strength of the wood as used by Norén. The two straight lines are the relationships for wet and dry wood

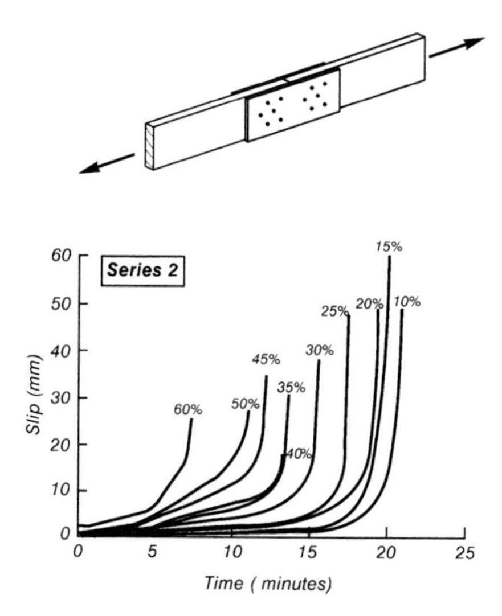

Figure 10.40 Measured nail slip in fire tests of nailed connections (Reproduced from Norén (1996) by permission of John Wiley & Sons Ltd)

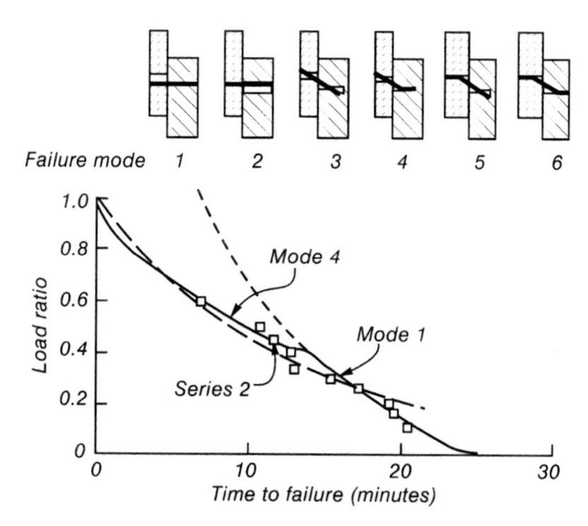

Figure 10.41 Comparison of measured and calculated results (Reproduced from Norén (1996) by permission of John Wiley & Sons Ltd)

from the literature. To allow for the transition from wet to dry, as wood starts to dry out at 100°C, Norén found that the dotted curve (marked III) produced a much better fit of calculated to measured strength than either of the straight-line relationships. Because temperatures of the nail remained below 200°C at failure, the yield strength of the nail was not modified for elevated temperature. These calculations are too cumbersome for normal design office practice, but will be very useful for producing design charts for practical use.

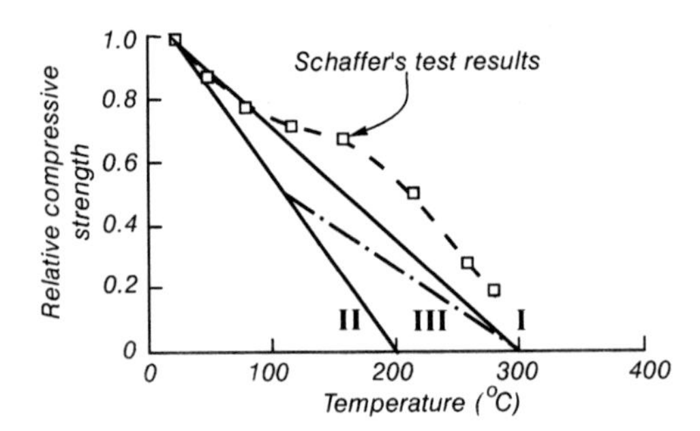

Figure 10.42 Effect of temperature on compressive strength (Reproduced from Norén (1996) by permission of John Wiley & Sons Ltd)

10.8.3 Bolted Connections

Bolted connections are widely used in timber structures with excellent results (Figure 10.43). Connections with many small bolts are stronger than connections with a few large bolts. Dowel connections are similar to bolted connections except that the dowels have no axial capacity. Fire behaviour of bolt and dowel connections depends on the amount of heat able to enter the wood via the bolt. The theory described above for nails could be applied to dowel or bolt connections, but no comprehensive studies have been published. Carling (1989) reports tests by Hviid (1979) who investigated the rate of embedment deformation of heated dowels into wood at various temperatures up to 540°C, both parallel and perpendicular to the grain. This information is useful for assessing the strength of large timber connections where the bolts reach high temperatures.

10.8.4 Truss Plates

Truss plates have a poor reputation for fire-resistance because they have been associated with some premature failures of fire-exposed timber truss roof structures (Dunn, 1988). Figure 10.37 confirms this poor behaviour of unprotected truss plate connections. White and Cramer (1994) have made a comprehensive experimental investigation of the fire performance of truss plates. They made tension tests of 38 × 89 mm timber members connected by a range of different truss plates, using the ASTM E-119 standard fire exposure, constant elevated temperatures of up to 300°C and also simulated plenum temperatures. In the E-119 tests, unprotected plates failed in less than 6 minutes, compared with almost 13 minutes for solid timber with no connection. Various combinations of protection increased the fire resistance by different amounts, the best being slightly over 30 minutes fire-resistance rating when all four sides of the member were protected with 13 mm Type X gypsum plaster with taped edges.

In the tests simulating temperatures within the plenum space of a truss assembly protected by a fire-resisting ceiling (reaching 327 C after 60 minutes) unprotected truss plates had a fire-resistance rating just under 60 minutes, and various forms of

Figure 10.43 Bolted connection between timber members, after fire exposure

Figure 10.44 Nailplate connection between timber members, after fire exposure

protection increased this to over 100 minutes. Shrestha *et al.* (1995) have developed a model for predicting the stiffness of truss plates at elevated temperatures. The model, supported by test results, shows a gradual reduction in stiffness up to 200 or 250°C, followed by a rapid reduction in stiffness to 300°C. Further discussion of trusses is in Chapter 11.

10.8.5 Timber Connectors

The term 'timber connector' refers to steel components such as shear plates and split ring connectors, usually slotted into butting wood surfaces to improve shear

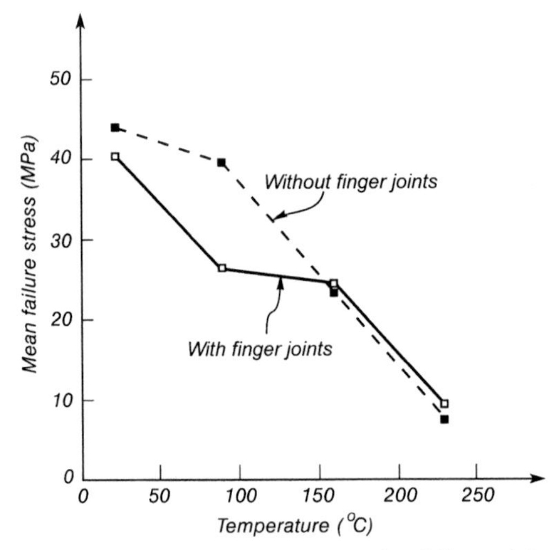

Figure 10.45 Effect of elevated temperature on strength of finger joints (Reproduced from (Nielsen and Olesen, 1982) by permission of Department of Building Technology and Structural Engineering, Aalborg University)

resistance. Figure 10.37 shows poor performance of the split ring connectors in fire tests. Schaffer (1984) reports tests in which split ring connectors failed after less than 5 minutes of fire exposure. This type of connector relies heavily on the shear strength of wood, so the poor performance may be attributable to rapid loss of shear strength of heated wood (Figure 10.18) or loss of clamping force due to charring under the bolt head.

10.8.6 Glued Connections

Many timber structures and timber members are connected with adhesives. When exposed to fire, glued wood members generally behave in the same way as solid wood provided that thermosetting adhesives are used, such as the resorcinol or melamine adhesives most often used for glulam members. Some adhesives such as elastomerics and epoxies are sensitive to elevated temperatures and should not be relied on in fire conditions.

Finger joints

Nielsen and Olesen (1982) tested sawn timber in axial tension at four different temperatures, with and without finger joints. Their results, shown in Figure 10.45, indicate very similar strength for the two groups of material, except at 90°C where the finger jointed material is significantly weaker than the unjointed material. Finger joints are used in most glulam members, so the strength of glulam beams and tension members depends on the strength of the finger joints, especially in the outer laminations.

Epoxied connections

Epoxied connections have poor fire-resistance because of the rapid loss of strength of epoxy at elevated temperatures. Buchanan and Barber (1994) carried out tension tests on deformed steel bars epoxied into the end grain of glulam timber members. The fire-resistance was not good because the epoxy adhesives started losing strength at 50°C, and had very little strength at temperatures over 70°C. Epoxied connections can only achieve fire resistance if the epoxy itself is protected from elevated temperatures. This can be assisted by providing a large thickness of additional wood cover and ensuring that steel connecting brackets are not exposed directly to fire.

Reinforced glulam

There are several methods of reinforcing glulam beams to increase the strength and stiffness. One method is to epoxy one or more steel reinforcing bars into grooves in the outer laminations during the manufacturing process. Any system such as this which relies on epoxy adhesives will not have good fire-resistance. Another method of reinforcing glulam beams is to glue a high-strength fibre reinforced polymer (FRP) laminate to the outside edge of a beam or between the outer laminate and the rest of the beam. Martin and Tingley (2000) show that such reinforcing can double the bending strength of glulam beams under normal temperatures, but fire tests show that the FRP has poor fire performance, such that the fire-resistance of the reinforced beam is similar to that of an unreinforced glulam beam.

WORKED EXAMPLE 10.1

Consider a softwood glulam beam, 130 mm wide by 720 mm deep, spanning 7.5 m with a dead load $G=4.0$ kN/m (including self weight) and live load $Q=7.0$ kN/m. The beam is laterally restrained with timber decking nailed to the top edge. Check the design for normal conditions and for 60 minutes fire-resistance rating, exposed to fire on three sides. Use the Eurocode method with the charring rates from Table 10.1 and the factor $k_{20}=1.15$.

The characteristic flexural strength is $f_b=17.7$ MPa. The strength reduction factor is $\Phi=0.8$ for normal design and $\Phi_f=1.0$ for fire design. The duration-of-load factor is $k_d=0.8$ for cold design and $k_d=1.0$ for fire design.

Check design for normal conditions

Design load	$w_c = 1.2G + 1.6Q$	$= 1.2 \times 4.0 + 1.6 \times 7.0 = 16.0$ kN/m
Bending moment	$M^* = w_c L^2/8$	$= 16.0 \times 7.5^2/8 = 112$ kNm
Section modulus	$Z = bd^2/6 = 130 \times 720^2/6 = 11.2 \times 10^6$ mm^3	
Nominal strength	$M_n = k_d f_{0.05} Z = 0.8 \times 17.7 \times 11.2 = 159$ kNm	
Design strength	$\Phi M_n = 0.8 \times 159 = 127$ kNm	
	$M^* \leqslant \Phi M_n$ so design is OK.	

Loads for fire conditions

Design load	w_c	$= 1.0G + 0.4Q = 1.0 \times 4.0 + 0.4 \times 7.0 = 6.8$ kN/m
Bending moment	M^*_{fire}	$= w_c L^2/8 = 6.8 \times 7.5^2/8 = 47.8$ kNm

Method I (effective cross section, no corner rounding)

Rate of charring	β_1	$= 0.7$ mm/minute
Depth of char	c	$= 60 \times 0.7 = 42$ mm
Reduced breadth	b_f	$= 130 - 2 \times 42 = 46$ mm
Reduced depth	d_f	$= 720 - 42 = 678$ mm
Thickness of zero-strength layer	z	$= 7$ mm
Effective breadth	b_e	$= 46 - 2 \times 7 = 32$ mm
Effective depth	d_e	$= 678 - 7 = 671$ mm
Section modulus	Z_f	$= b_e\, d_e^2/6 = 32 \times 671^2/6 = 2.40 \times 10^6$ mm^3
Flexural strength	M_f	$= k_d\, f_f Z_f = k_d\, k_{20} f_{0.05}\, Z_f = 1.0 \times 1.15 \times 17.7 \times 2.4$ $= 48.9$ kNm

$M^*_{fire} \leqslant M_f$ so design is OK.

Method II (reduced properties, no corner rounding)

Rate of charring	β_1	$= 0.7$ mm/minute
Depth of char	c	$= 60 \times 0.7 = 42$ mm
Reduced breadth	b_f	$= 130 - 2 \times 42 = 46$ mm
Reduced depth	d_f	$= 720 - 42 = 678$ mm
Section modulus	Z	$= b_f\, d_f^2/6 = 46 \times 678^2/6 = 3.52 \times 10^6$ mm^3
Beam area	A	$= b_f\, d_f = 46 \times 678/10^6 = 0.0312$ m^2
Beam perimeter	p	$= b_f + 2d_f = (46 + 2 \times 678)/10^3 = 1.40$ m
Reduction factor	k_f	$= 1 - p/200A = 1 - 1.40/(200 \times 0.0312) = 0.775$
Flexural strength	M_f	$= k_f k_d k_{20} f_{0.05}\, Z_f = 0.775 \times 1.0 \times 1.15 \times 17.7 \times 2.4$ $= 55.6$ kNm

$M^*_{fire} \leqslant M_f$ so design is OK.

Method III (reduced properties, with corner rounding)

Rate of charring	β_1	$= 0.64$ mm/minute
Depth of char	c	$= 60 \times 0.64 = 38.4$ mm
Reduced breadth	b_f	$= 130 - 2 \times 38.4 = 53.2$ mm
Reduced depth	d_f	$= 720 - 38.4 = 682$ mm
Section modulus	Z	$= b_f\, d_f^2/6 - 0.215\, c^2\, d_f$ $= 53.2 \times 682^2/6 - 0.215 \times 38.4 \times 682 = 3.90 \times 10^6$ mm^3
Beam area	A	$= b_f\, d_f = 53.2 \times 682/10^6 = 0.0363$ m^2
Beam perimeter	p	$= b_f + 2d_f = (53.2 + 2 \times 682)/10^3 = 1.42$ m
Reduction factor	k_f	$= 1 - p/200A = 1 - 1.42/(200 \times 0.0363) = 0.805$
Flexural strength	M_{nf}	$= k_f\, k_d\, k_{20}\, f_{0.05}\, Z_f = 0.805 \times 1.0 \times 1.15 \times 17.7 \times 2.4$ $= 63.9$ kNm

$M^*_{fire} \leqslant M_f$ so design is OK.

WORKED EXAMPLE 10.2

The beam worked in Example 10.1 is in a room 7.5 m by 5 m, 3 m high, with one window 4 m wide and 2 m high. The fuel load is 640 MJ/m² floor area. Check whether the beam can resist a burnout of the room.

Length of room	$l_1 = 7.5$ m
Width of room	$l_2 = 5.0$ m
Floor area	$A_f = l_1 l_2 = 7.5 \times 5.0 = 37.5$ m²
Height of room	$H_r = 3.0$ m
Area of internal surfaces	$A_t = 2\ (l_1 l_2 + l_1 H_r + l_2 H_r)$
	$= 2\ (7.5 \times 5 + 7.5 \times 3 + 5 \times 3) = 150$ m²
Height of window	$H_v = 2.0$ m
Width of window	$B = 4.2$ m
Area of window	$A_v = B H_v = 4.2 \times 2.0 = 8.4$ m²
Ventilation factor	$F_v = A_v \sqrt{H_v} / A_t = 8.4 \times \sqrt{2.0}/150 = 0.08$ m$^{-1/2}$
Fuel load (floor area)	$e_f = 640$ MJ/m²
Fuel load (total area)	$e_t = e_f\ A_f / A_t = 640 \times 37.5/150 = 160$ MJ/m²

Initial char time	$t_o = 0.006\ e_t / F_v = 0.006 \times 160/0.08 = 12$ minutes
Notional charring rate	$\beta_1 = 0.7$ mm/minute
Parametric char factor	$k_p = 1.5\ (5F_v - 0.04)/(4F_v + 0.08) = 1.35$
Initial char rate	$\beta_{par} = 1.35\ \beta_1 = 1.35 \times 0.7 = 0.945$ mm/minute
Total char depth	$c = 2\ \beta_{par}\ t_o = 2 \times 0.945 \times 12 = 23$ mm

(Check these figures in Table 10.2.)

Reduced breadth	$b_f = 130 - 2 \times 23 = 84$ mm
Reduced depth	$d_f = 720 - 23 = 697$ mm
Section modulus	$Z = b_f\ d_f^2/6 = 84 \times 697^2/6 = 6.80 \times 10^6$ mm³
Beam area	$A = b_f\ d_f = 84 \times 697/10^6 = 0.0585$ m²
Beam perimeter	$p = b_f + 2d_f = (84 + 2 \times 697)/10^3 = 1.48$ m
Reduction factor	$k_f = 1 - 3.2\ c/b = 1 - 3.2 \times 23/130 = 0.434$
Nominal strength	$M_f = k_f\ k_d\ k_{20}\ f_{0.05}\ Z_f = 0.434 \times 1.0 \times 1.15 \times 17.7 \times 6.8$
	$= 60.1$ kNm
Bending moment from above	$M^*_{fire} = 47.8$ kNm

$M^*_{fire} \leqslant M_f$ so design is OK.

WORKED EXAMPLE 10.3

Repeat Example 10.1 using the North American charring rate in the working stress design format. The allowable stress under long duration loading in flexure is $f_a = 8.0$ MPa. The factor to convert allowable stress to mean failure stress is $k_{mean} = 2.85$.

Check design for normal conditions

Design load	$w = G + Q = 4.0 + 7.0 = 11.0$ kNm
Bending moment	$M^*_w = wL^2/8 = 11.0 \times 7.5^2/8 = 77.3$ kNm
Section modulus	$Z = bd^2/6 = 130 \times 720^2/6 = 11.2 \times 10^6$ mm^3
Flexural stress	$f^*_b = M^*_w/Z = 77.3 \times 10^6/11.2 \times 10^6 = 6.91$ MPa
	$f^*_b \leqslant f_a$ so design is OK.

Fire design (North American char rate, no corner rounding)

Time of calculation	$t = 60$ minutes
Depth of char	$c = 2.58\ \beta_n\ t^{0.813} = 2.58 \times 0.635 \times 60^{0.813} = 45.7$mm
Reduced breadth	$b_f = 130 - 2 \times 45.7 = 38.6$ mm
Reduced depth	$d_f = 720 - 45.7 = 674$ mm
Section modulus	$Z = b_f\ d_f^2/6 = 38.6 \times 674^2/6 = 2.92 \times 10^6$ mm^3
Flexural stress	$f^*_{b,f} = M^*_w/Z = 77.3 \times 10^6/2.92 \times 10^6 = 26.4$ MPa
Allowable stress	$f_{a,f} = k_{mean}\ f_a = 2.85 \times 8.0 = 22.8$ MPa
	$f^*_b > f_a$ so the beam fails in fire.

WORKED EXAMPLE 10.4

Calculate the time to failure for the beam in Worked Example 10.1 using the North American empirical design equation.

Design bending moment	$M^* = 112$ kNm
Design strength	$\Phi M_n = \Phi k_1\ f_b\ Z = 127$ kNm
Load ratio	$R_a = M^*/\Phi M_n = 112/127 = 0.882$
z factor	$z = 0.7 + 0.3/R_a = 0.7 + 0.3/0.882 = 1.04$
Time to failure	$t_f = 0.1\ zb\ (4 - b/d)$
	$= 0.1 \times 1.04 \times 130\ (4 - 130/720) = 50.1$ minutes

Time to failure is less than 60 minutes, so the beam fails in the fire.

WORKED EXAMPLE 10.5

A solid timber deck consists of 150 mm thick planks joined with central splines as shown in Figure 10.35(c). The deck spans 5 m with a superimposed dead load of 1.25 kN/m^2 and live load $Q = 5.0$ kN/m^2. Calculate the failure time using Janssen's formula. Use the Eurocode reduced properties method to calculate if the deck has a 90 minute fire-resistance rating. Check for integrity failure.

The characteristic flexural strength of the decking timber is $f_b = 25.0$ MPa. The density of the wood is 5.0 kN/m^3. The strength reduction factor is $\Phi = 0.8$ for normal design and $\Phi_f = 1.0$ for fire design. The duration of load factor is $k_d = 0.8$ for cold design and $k_d = 1.0$ for fire design. The factor k_f is 1.15 for fire design.

Check for normal conditions

Thickness of deck $d = 150$ mm
Self weight of deck $w_s = \rho d/1000 = 5 \times 0.15 = 0.75$ kN/m^2
Total dead load $G = 0.75 + 1.25 = 2.0$ kN/m^2
Design load $w_c = 1.2G + 1.6Q = 1.2 \times 2.0 + 1.6 \times 5.0 = 10.4$ kN/m^2
Design a strip 1 m wide. Uniformly distributed load $= 1.0 \times 10.4 = 10.4$ kN/m
Bending moment $M^* = w_c L^2/8 = 10.4 \times 5^2/8 = 32.5$ kNm
Section modulus $Z = bd^2/6 = 1000 \times 150^2/6 = 3.75 \times 10^6$ mm^3
Design strength $\Phi M_n = \Phi k_1 f_b Z = 0.8 \times 0.8 \times 25 \times 3.75 = 60.0$ kNm
$M^* \leqslant \Phi M_n$ so design is OK.

Janssen's formula

Load ratio $R_a = M^*/\Phi M_n = 32.5/60 = 0.54$
Time to failure $t_{sf} = 1.25\ d(1 - \sqrt{(0.4 R_a)}) - 11.3$
$= 1.25 \times 150\ (1 - \sqrt{(0.4 \times 0.54)}) - 11.3 = 89$ minutes

Eurocode reduced properties method

Design load $w_c = 1.0G + 0.4Q = 1.0 \times 2.0 + 0.4 \times 5.0 = 4$ kN/m^2
Design a strip 1 m wide. Uniformly distributed load $= 1.0 \times 4.0 = 4.0$ kN/m
Bending moment $M^*_{fire} = w_c L^2/8 = 4 \times 5^2/8 = 12.5$ kNm
Rate of charring $\beta = 0.64$ mm/minute
Depth of char $c = 90 \times 0.64 = 57.6$ mm
Reduced depth $d_f = 150 - 57.6 = 92.4$ mm
Section modulus $Z = bd_f^2/6 = 1000 \times 92.4^2/6 = 1.42 \times 10^6$ mm^3
Section area $A = bd_f = 1000 \times 92.4/10^6 = 0.00924$ m^2
Exposed perimeter $p = b = 1.0$ m
Reduction factor $k_f = 1 - p/200A = 1 - 1.0/(200 \times 0.00924) = 0.46$
Design strength $M_f = k_f\ k_d\ k_{20}\ f_{0.05}\ Z_f$
$= 0.46 \times 1.0 \times 1.15 \times 25 \times 1.26 = 16.6$ kNm
$M^*_{fire} \leqslant M_f$ so design is OK.

Check for integrity failure

Notional charring rate $\beta_0 = 0.7$ mm/minute
Time to char through solid deck $t_s = d/\beta_0 = 150/0.7 = 214$ minutes
Reduction coefficient $\xi = 0.4$ (Figure 10.27)
Integrity failure time $t_{pr} = \xi d/\beta_0 = 0.4 \times 214 = 86$ minutes

11
Light Frame Construction

11.1 OVERVIEW

The objective of this chapter is to describe the fire behaviour of light frame construction using timber and steel components, and to review available design methods. This chapter also describes fire performance of lightweight sandwich panel construction.

11.2 DESCRIPTION

Light timber frame construction ('wood frame construction' in North America) is widely used in one- to four-storey buildings, most often for residential occupancies. Walls and partitions are usually constructed with sawn timber studs. Floors consist of plywood or particle board sheeting nailed or screwed to joists which may be sawn timber or engineered products such as LVL joists, wood I-joists or parallel chord trusses. Figure 11.1 shows a perspective view of a single-storey timber framed house. Multi-storey construction uses similar components as shown in Figure 11.2. Typical timber floor construction is shown in Figure 11.3.

Light frame construction in this chapter also includes light steel framing, such as cold-rolled steel channel-sections and composite construction incorporating light steel joists.

Because of the small size of timber and steel members used in this style of construction, fire-resistance must be based on protective materials, by far the most common being gypsum board (often called 'drywall' or 'sheetrock' in North America). The gypsum board is used as wall and ceiling linings, where it provides a wearing surface as well as contributing to the acoustic, thermal and fire separation of the barrier. Other lining materials, used less often, include a variety of wood-based panel products, fibre cement panels and calcium silicate board. Gypsum board has fire-resisting properties superior to most other similar materials, because of the moisture in the gypsum crystals, described below. This chapter assumes the use of gypsum board as the lining material, but the same principles apply to other materials.

In most wall construction, the gypsum board is fixed directly to the studs using nails or screws. Elastomeric adhesives are used in some situations, but these should

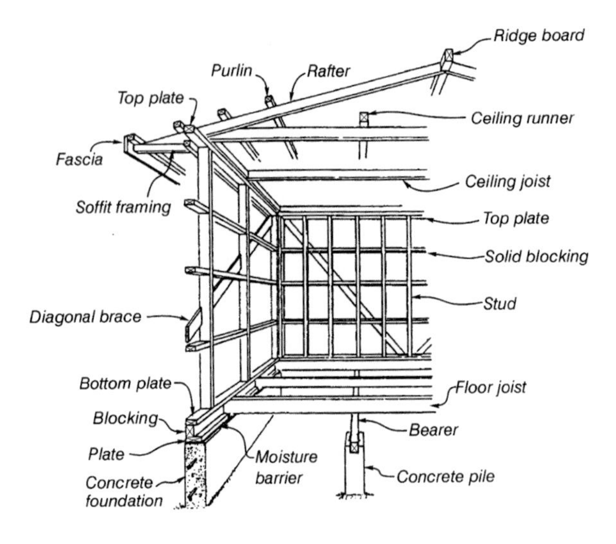

Figure 11.1 Typical light timber house framing

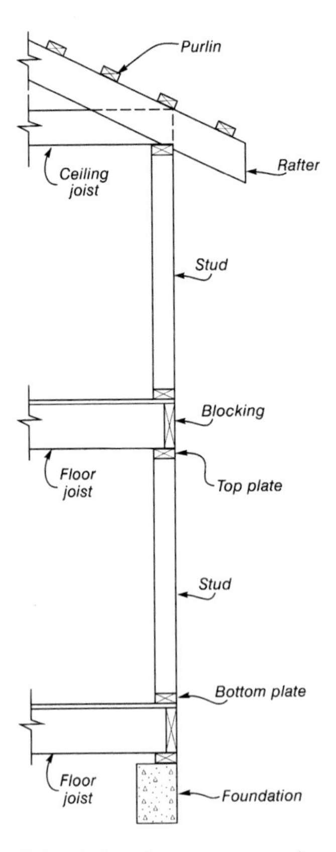

Figure 11.2 Multi-storey light timber frame construction

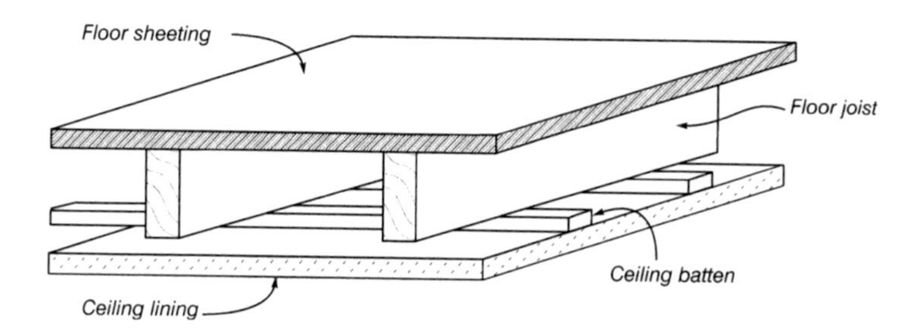

Figure 11.3 Light timber frame floor construction

be ignored under fire conditions. Some wall and floor systems have the gypsum board lining spaced off the studs or joists with a thin steel resilient channel to improve acoustic performance. Insulating batts are often placed in the cavities to improve thermal and acoustic insulation. Other methods of improving acoustic performance are to use double stud walls or staggered stud walls where the wall lining material is only fixed to one side of each stud, to eliminate a direct path for transmission of sound.

Ceiling linings may be connected directly to the underside of the joists, but they are more often attached to timber ceiling battens (Figure 11.3) or steel channels fixed to the joists in order to give more tolerance for erection and better acoustic performance. Ceilings are sometimes suspended on a steel framing system, forming a large ceiling cavity.

Light steel frame construction uses thin steel wall studs (Figure 11.4) or floor joists. These are manufactured by cold rolling from thin steel strip, 0.5 to 1.5 mm thick. Light steel frame construction is similar to light timber frame construction in that the lining is an essential part of the fire-resistive construction, and the quality of the gypsum board and its fixings are most important. In addition to the typical stud wall construction shown in Figure 11.4, there are many other light steel frame systems for walls and floors. Wall designs include office partitions, elevator shafts, and exterior walls, some with multiple layers, insulated cavities and solid gypsum walls with no cavities (e.g. Gypsum Association, 1994). A common type of floor

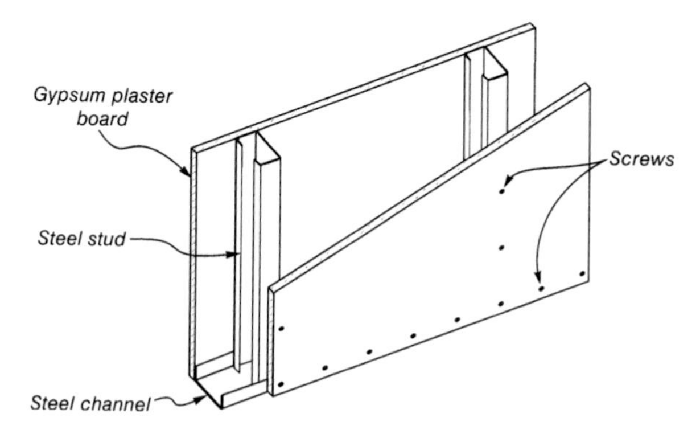

Figure 11.4 Light steel framing

construction in North America consists of a concrete slab on open-web steel joists, fire protected by a ceiling membrane of gypsum board or similar material. Open-web steel joists are light trusses made from steel angles and rods as shown in Figure 9.26.

11.3 FIRE BEHAVIOUR

Light frame construction can have excellent fire behaviour, provided that it is well constructed from the correct materials. Experience in many severe fires had shown that gypsum board linings can prevent fire spread and protect the load-bearing light steel frame or light timber frame for the duration of a severe fire, for example see Figure 2.6(c).

Fire resistance is assigned to complete assemblies of light frame construction, not to the individual components. The performance of the lining material exposed directly to the fire is most important because fire penetration into the cavity can result in premature fire spread or structural collapse of the barrier. Assemblies with no fire protection fail in a few minutes (White *et al.*, 1984).

Figure 11.5 Full-scale fire-resistance test of a light timber frame wall

When gypsum board lining is heated during a fire, temperatures on the exposed face will increase steadily until about 100°C is reached, at which time there will be a delay while the water of crystallization is driven off. As the heating continues, the 100°C temperature plateau will progress slowly through the board, until the entire board has been dehydrated. Temperatures within the board will rise steadily after dehydration is completed, leading to increased temperatures in the cavity and in the framing members.

Gypsum plaster has very low strength after dehydration because it is converted to a powdery form. Any residual strength depends on glass fibre reinforcing and other additives which hold the board together and prevent it from falling off the wall or ceiling. There is a wide range in performance between different boards, with fire-resistant boards having much better fire resistance than unreinforced boards. Resistance to fire spread also depends on how much heat is transferred across the cavity and through the lining on the unexposed side.

As a fire progresses, timber framing will begin to char and steel framing will lose strength due to increased temperatures, but long periods of fire resistance can be achieved if the lining on the fire side remains in place. Critical factors are the thickness of the gypsum board, the quality of the board material and the details of the construction and the fixings.

11.3.1 Walls

In timber stud walls, the load capacity depends on the residual size, temperature and moisture content of the studs as charring occurs. Charring of studs will be greatest on the edge which is in contact with the fire-exposed lining, with lesser charring on the wide faces of the studs. There is usually no charring on the edge fixed to the unexposed gypsum board. The strength of steel stud walls depends on the temperature of the studs and the level of lateral stability provided to the studs by the lining materials.

Fire resistance is improved if the wall cavity is filled with well-fitting mineral wool insulation, because this protects the studs, and the lining on the unexposed face, after the fire-exposed lining falls off. Glass fibre insulation is less beneficial because the glass melts when exposed to post-flashover fire temperatures.

11.3.2 Floors

Timber joist floors exposed to fire behave similarly to walls. The critical factors are the thickness and integrity of the ceiling material exposed to the fire. Fixings are very important because horizontal sheets of ceiling lining are more prone to falling off during a fire than vertical sheets of wall lining material. To achieve good fire performance it is necessary to limit the rate of charring of the floor joists with good protection from the ceiling lining. Manufactured timber I-joists are more efficient than solid sawn timber joists under normal temperature conditions, but may not perform as well in fire if the webs are thin, because of strength loss due to charring of the web. Steel joist floors behave similarly, except that strength loss is from elevated temperatures in the steel rather than loss of cross section by charring.

Floors with parallel chord timber trusses behave similarly to floors with timber or steel joists, provided that the fire-resisting ceiling remains in place. The strength and stiffness of the assembly also depends on the behaviour of the truss plate connections, which has been described in Chapter 10. Unprotected light wood trusses have very little fire resistance because of the vulnerability of truss plates exposed to fire conditions. Even if the truss plates are protected with sheets of gypsum board or similar, the small cross section timber members will fail after a short time of fire exposure.

11.3.3 Buildings

Full scale fire tests in real buildings have shown that light timber frame construction can be designed with excellent fire resistance. The most important factors in preventing fire spread is the use of high quality fire-resistant gypsum board for all linings, and ensuring the integrity of all junctions, doors and penetrations. Full scale tests of a three-storey apartment building exposed to an external fire are described in detail by Hayashi *et al.* (1999). A full scale six-storey building at the Cardington test facility in England was subjected to a post-flashover wood-crib fire in one apartment (Lennon *et al.* 2000). The fire was extinguished after one hour of intense burning. The building behaved as expected, with no fire spread or loss of load carrying capacity for the duration of the fire. Important findings were that the standard of workmanship is of crucial importance in providing the necessary fire resistance, especially nailing of the gypsum board, and the correct location of cavity barriers and fire-stopping is important for maintaining the integrity of the building.

11.4 FIRE-RESISTANCE RATINGS

Fire resistance of light frame structures can be assessed using the same general principles as for other materials.

11.4.1 Verification Methods

The design process for fire resistance requires verification that the provided fire resistance exceeds the design fire severity. Using the terminology from Chapter 5, verification may be in the *time domain*, the *temperature domain* or the *strength domain*.

The temperature domain is not used for timber structures because there is no critical temperature for fire-exposed timber, but it could be used for light steel framing members. Most often, verification of fire resistance of light frame structures is in the time domain, where proprietary ratings are compared with the code-specified fire resistance, or with the calculated equivalent time of a complete burnout.

11.4.2 Fire Severity

Because the fire resistance of most light frame structures is based on generic or proprietary ratings, the fire severity must be assessed in terms of standard fire exposure. This will either be the required fire-resistance rating from a code or the equivalent fire severity for burnout of the compartment, calculated as described in Chapter 5. Time equivalent formulae were not developed explicitly for light frame construction, but they can be used for this purpose. Thomas (1997) showed that the Eurocode time equivalent formula may be a little unconservative, but no more so than for steel or concrete construction.

11.4.3 Listings

Fire design of light frame assemblies is usually by direct reference to results of standard fire-resistance tests or listed approvals based on such tests. Many full-scale fire-resistance tests have been carried out on wall and floor assemblies. Listings of approved fire-resistance ratings are produced and maintained by approval organizations (NBCC, 1995, SNZ, 1991), trade organizations (Gypsum Association, 1994), manufacturers (Winstone Wallboards, 1997) or testing and approval agencies (UL, 1996). Other countries have similar listings. The listed fire-resistance ratings are derived either directly from tests or from expert opinions based on successful tests. Manufacturers of gypsum board or other proprietary products may make their test results available on request.

11.4.4 Failure Criteria

Fire-resistance ratings are assigned to completed assemblies of light frame construction, and not to the individual components. For an assembly to be given a fire-resistance rating, all the relevant criteria must be met. All walls and floors are barriers which must meet the integrity and insulation criteria, as described in Chapter 6. Floors and load-bearing walls must also meet the stability criterion.

Assessment of integrity must be done in full-scale testing because small-scale tests cannot assess factors such as shrinkage in large sheets of gypsum board or cracking due to structural deformations. Large-scale testing is also necessary to assess the resistance of the gypsum board to falling off walls or ceilings during fire. A timber stud wall during and after a full scale fire resistance test is shown in Figures 11.5 and 11.12 respectively.

Generic ratings

Generic ratings, or 'tabulated ratings' are those which assign a time of fire resistance to materials with no reference to individual manufacturers or to detailed specifications. Many listed ratings in North America are generic ratings for non-proprietary products, such as regular gypsum board or Type X gypsum board manufactured by many companies.

Proprietary ratings

Most manufacturers of gypsum board have proprietary fire-resistance ratings for timber and steel framed assemblies containing their products. These fire-resistance ratings usually include a specification of framing members, lining material and fixing methods, all of which must be followed if the assembly is to meet the intended rating. A typical specification for a proprietary rating is shown in Figure 11.6 (Winstone Wallboards, 1997).

Typical fire-resistance ratings

Gypsum board on its own does not have a fire-resistance rating. A fire-resistance rating is assigned to a building system such as a wall or a floor which is an assembly of products including the protective board. A summary of typical ratings for light frame walls and floors is shown in Table 11.1, giving the minimum thickness of gypsum board required to achieve various fire-resistance ratings for assemblies with uninsulated cavities. This applies to symmetrical walls with gypsum board fixed to each face of the studs, and wood-panel floors with gypsum board ceiling fixed to the underside of the joists.

As expected, the table shows increasing fire-resistance with increasing thickness of gypsum board lining, and thicker lining required for load-bearing ratings. Thicknesses of 19 mm or less are single layers of board, and larger thicknesses are the sum of two or more layers, with staggered joints between the layers.

This indicative table is not intended for design. Each listing has specific requirements for stud spacing, blocking between studs, type of board and fixings, none of which are shown in the table. The complete listing and specification must be consulted for design purposes.

Some anomalies in Table 11.1 require discussion. The New Zealand figures are similar to those in Australia. The North American ratings are similar in parts of Europe. It can be seen that in New Zealand thinner boards can generally be used than in North America, which results from several factors. All of the New Zealand listings are proprietary ratings using special purpose Gib® 'Fyreline' board which is of higher density than many other typical boards. All the North American listings are generic listings for 'Type X' board, except for the 180 and 240 minute ratings which are for special purpose 'Type C' boards. For timber stud walls, fire tests in New Zealand were conducted with solid blocking (sometimes called nogs or dwangs) between the studs to provide lateral stability and to protect the joins between the sheets, but such blocking is not generally used in North America. There are also some important differences between the standard fire-resistance tests in the two regions, as described in Chapter 6.

Specification number	Load-bearing capability	Fire-resistance rating	Lining requirements	Sound transmission class	System weight approx
GBT 90	**NLB**	**-/90/90**	**1 x 16mm Gib® Fyreline each side**	**STC 37**	**36kg/m²**
GBTL 90	**LB**	**90/90/90**			

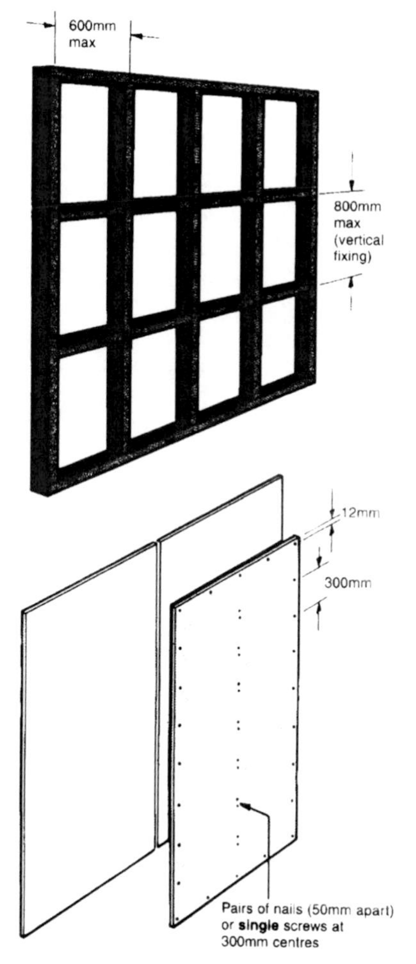

Framing

GBT90 Non Loadbearing - No. 1 framing grade H1 treated Radiata Pine nominal dimensions 75mm x 50mm minimum.

GBTL90 Loadbearing - No. 1 framing grade H1 treated Radiata Pine nominal dimensions 100mm x 50mm minimum.

Studs at 600mm centres maximum.

Nogs at 800mm centres maximum for Vertical fixing.

Nogs at 1200mm centres maximum for Horizontal fixing.

Wall Height

GBT90 Non Loadbearing - Framing dimensions and height as determined by NZS3604 stud tables for non-loadbearing partitions.

GBTL90 Loadbearing - Framing dimensions and height as determined by NZS3604 stud and top plate tables for load-bearing walls.

Lining

1 layer of 16mm Gib® Fyreline each side of the frame.

Vertical or Horizontal fixing permitted.

Sheets shall be touch fitted.

When fixing vertically, full height sheets shall be used where possible.

All sheet joints must be formed over solid timber framing.

Fastening the Lining

Fasteners

50mm x 2.5mm Gib® Clouts or 50mm x 7g bugle head gypsum drywall screws.

Fastener Centres

300mm centres around the sheet perimeter, 12mm from the sheet edge.

Pairs of nails, (50mm apart), or single screws at 300mm centres to intermediate studs.

Jointing

All fastener heads stopped and all sheet joints tape reinforced and stopped in accordance with the publication entitled 'Gib® Stopping and Finishing Systems 1992' or a later approved revision.

Figure 11.6 Specification for light timber frame proprietary rating (Reproduced from (Winstone Wallboards, 1997) by permission of Winstone Wallboards Ltd)

11.4.5 Additive Methods

Although fire-resistance ratings are assigned to completed assemblies of light frame construction, and not to the individual components, some approval organizations permit fire-resistance ratings to be estimated by adding up a contribution from each of the main components. These are crude methods which must be used with caution. The Canadian code (NBCC, 1995) gives an additive method which has been adopted by some US codes. Typical values are shown in Table 11.2 where a

Table 11.1. Minimum gypsum board thickness (mm) to give fire-resistance ratings for cavity walls and floors

Fire-resistance ratings (minutes)		New Zealand				North America			
		Wood		Steel		Wood		Steel	
		Non-load bearing	Load bearing	Non-load bearing	Load bearing	Non-load bearing	Load bearing	Non-load bearing	Load bearing
Walls	30	9.5	9.5	12.5	16.0				
	45					(12.7)	(12.7)		
	60	12.5	12.5	12.5	19.0	15.9	15.9	15.9	
	90	16.0	16.0	16.0	28.5	[25.4]	[25.4]	[25.4]	
	120	19.0	32.0	19.0		31.8	31.8	25.4	31.8
	180	32.0						38.1	
	240							50.8	
Floors	30		12.5						
	60		16.0				12.7		25.4
	90		32.0				31.8		
	120		38.0		38.0				

North American listings are from Gypsum Association (1994) except (12.7) from UL (1996) and [25.4] from NBCC (1995). The New Zealand listings are for 'Gib® Fyreline' board, from Winstone Wallboards (1997).

Table 11.2 Component additive method in the Canadian code

Description	Assigned time (minutes)
Gypsum board	
12.7 mm Type X gypsum board	25
15.9 mm Type X gypsum board	40
Wood frame	
Wood studs at 400 mm centres	20
Wood studs at 600 mm centres	15
Wood joists at 400 mm centres	10
Wood trusses at 600 mm centres	5
Insulation	
Rock fibre batts	15
Glass fibre batts	5

time of fire resistance has been assigned to the lining material on the fire side, the studs and the insulation in the cavity. The lining on the unexposed face is not included in the calculation. Other additive methods include the Swedish method described by Östman *et al.* (1994) and the UK method described in BS 5268, Section 4.2 (BSI, 1990(b)).

As an example of the use of Table 11.2, an assembly with 15.9 mm Type X gypsum board on each face of wood studs at 400 mm centres with rock fibre (mineral wool) batts would be assigned a fire-resistance rating of 40 + 20 + 15 = 75 minutes.

11.4.6 Onset of Char Method

Some approval organizations permit protected timber assemblies to be assigned a fire-resistance rating if it can be shown that the protected wood will not begin to char during the time of fire exposure. This is explicitly permitted in New Zealand (SNZ, 1991). The listings published by UL (1996) include a 'finish rating' for sheet materials fixed to timber studs, defined as the time at which the wood surface closest to the fire reaches an average temperature rise of 121°C or an individual temperature rise of 163°C. These temperatures are lower than the usually accepted charring temperature of 250°C to 300°C, so the 'finish ratings' would be very conservative estimates of initiation of damage to protected wood members. Most Type X gypsum boards have finish ratings of 15 minutes for 12.7 mm board or 20 minutes for 15.9 mm board (UL, 1996).

Design to prevent the onset of char in protected timber members is a conservative approach which may have several applications if used with a realistic assessment of the expected fire severity. Design using this approach will ensure that the structure will remain standing for a complete burnout of the fire compartment. It will also ensure repairability of the structure after a severe fire without having to replace charred wood. This may be appropriate for buildings containing very valuable items or essential services.

11.5 PROPERTIES OF GYPSUM PLASTER BOARD

Gypsum plaster has been used as a construction material for many centuries. *Gypsum plasterboard* or *gypsum board* refers to rigid sheets of building material made from gypsum plaster and other materials. Gypsum board is widely used for interior linings in domestic housing and commercial office buildings, and is the most common lining material used to provide light frame structures with fire resistance. In North America this is called *drywall* construction. Gypsum plaster can also be applied in a wet condition, trowelled over light wood or metal laths to make a smooth interior finish. Gypsum plaster often contains additives such as sand, perlite or vermiculite. One of the most critical factors affecting fire performance of an assembly lined with gypsum board is the quality of the board itself. This aspect is often overlooked by designers who assume that all gypsum boards are the same.

11.5.1 Manufacture

Gypsum is a crystalline mineral found in sedimentary rock formations in many parts of the world. De-hydrated gypsum is well known as 'Plaster of Paris', a white powder which sets hard after being mixed into a paste with water. The manufacturing process involves mining of the raw mineral, crushing and grinding it into a fine powder and heating to about 175°C, driving off three-quarters of the chemically bound water in a process called calcining. To produce gypsum plasterboard, the calcined gypsum powder is mixed with water, some additives and an air-entraining agent to reduce the density, and the plaster sets hard in about 10 minutes. Typical gypsum board has a density between 550 and 850 kg/m^3.

Most gypsum plaster boards consist of a sandwich of a gypsum core between two layers of paper, chemically and mechanically bonded to the core. Most gypsum boards are made with a thickness between 10 mm and 20 mm. The external paper provides tensile reinforcing to the board. Some boards, known as *fibrous plaster*, have no paper facing, relying on glass fibre or sisal reinforcing within the plaster to provide strength.

Many countries have national standards for manufacture of gypsum board. In the US, gypsum board is manufactured in accordance with ASTM C-36 (ASTM, 1995) which specifies dimensional tolerances, minimum flexural strength, hardness and nail-pull resistance under normal temperature conditions, as well as minimum fire-resistance for Type X board.

11.5.2 Types of Gypsum Board

Production and marketing of gypsum boards is different in various countries, but generally follows a similar pattern. There are three broad types of gypsum board, usually known as *Regular* board, *Type X* board, and *Special purpose* boards. All three categories are available in North America. Some parts of Europe and Asia have only the first two categories, and smaller market areas such as New Zealand and Australia only have regular board and special purpose boards.

Regular gypsum board is a generic product sold very competitively for residential construction. Regular board is not required to have any fire-resistance rating, so it usually has a low density gypsum core with no reinforcing (except the external paper). The low-density results from air entrainment during manufacture. Regular gypsum board has poor performance in fire-resistance tests compared with Type X or special boards, because the board tends to crack and fall off the wall or ceiling when the face paper has burned away and the gypsum becomes dehydrated. Regular board has approved fire-resistance ratings for certain assemblies, but the required thicknesses are greater than for the equivalent assemblies with improved boards.

The term *Type X* is used in North America for generic fire-resistant gypsum board (Type GF in Europe). Type X board is defined by performance rather than by a manufacturing specification. The definition of Type X board is that it will give a 60 minute load-bearing fire-resistance rating when one layer of 15.9 mm board is fixed to each side of a wood or steel stud wall assembly (or a 45 minute rating for 12.7 mm board). All Type X boards contain some glass fibre reinforcing and may have other additives to improve fire performance.

Special purpose boards are proprietary products made by many manufacturers, to obtain enhanced fire or structural performance over regular or Type X boards. Some are marketed as *Type C* boards. Special purpose boards are often manufactured in non-standard thicknesses and formulations to meet special market needs for fire-resistance or other performance. Special purpose boards usually have more glass fibres and more core additives than Type X boards. Industry listings (Gypsum Association, 1994) indicate that special purpose boards in the US do not give significantly better fire ratings than generic Type X board, but some special purpose boards manufactured in other countries have better fire performance.

11.5.3 Chemistry

The chemistry of gypsum at its simplest level is described below. Solid gypsum plaster and gypsum rock is calcium sulphate dihydrate $CaSO_4{\cdot}2H_2O$ with two water molecules for each calcium sulphate molecule. The manufacturing process first involves driving the moisture out of the gypsum rock to create the powdery white material of calcium sulphate hemihydrate $CaSO_4{\cdot}\frac{1}{2}\,H_2O$. The dehydration reaction (calcining) is an endothermic decomposition reaction which occurs between 100°C and 120°C:

$$CaSO_4{\cdot}2H_2O \rightarrow CaSO_4{\cdot}\tfrac{1}{2}H_2O + 1\tfrac{1}{2}H_2O \quad (11.1)$$

When the powder is mixed with water and formed into flat sheets of gypsum plaster, the reaction is reversed to become a hydration reaction:

$$CaSO_4{\cdot}\tfrac{1}{2}H_2O + 1\tfrac{1}{2}H_2O \rightarrow CaSO_4{\cdot}2H_2O \quad (11.2)$$

The resulting gypsum is 21% water by weight. Moisture in gypsum plaster is very important because it contributes to the excellent fire-resisting behaviour. Gypsum plaster also contains about 3% free water, depending on the ambient temperature and relative humidity. When gypsum plaster is heated in a fire, the dehydration follows the reaction in Equation (11.1) as solid gypsum is converted back to a powdery form. Significant energy is required to evaporate the free water and make the chemical change which releases the water in the crystal structure. Complete dehydration does not occur until the temperature reaches about 700°C, requiring additional energy input. Gypsum plaster can be recycled relatively easily compared with other building materials, because reactions (11.1) and (11.2) can be repeated indefinitely.

11.5.4 Thermal Properties

Thermal properties of gypsum plaster are required if finite-element thermal calculations are to be made. There is some scatter in published values, but most are similar to those shown below, based on tests of Canadian Type X gypsum board by Sultan (1996). Specific heat of gypsum plaster as a function of temperature is shown in Figure 11.7. The two peaks indicate chemical changes as moisture is driven off during heating, the first being the main reaction at about 100°C described in Equation (11.1), which results in a delay in the temperature rise of protected wood or steel framing members.

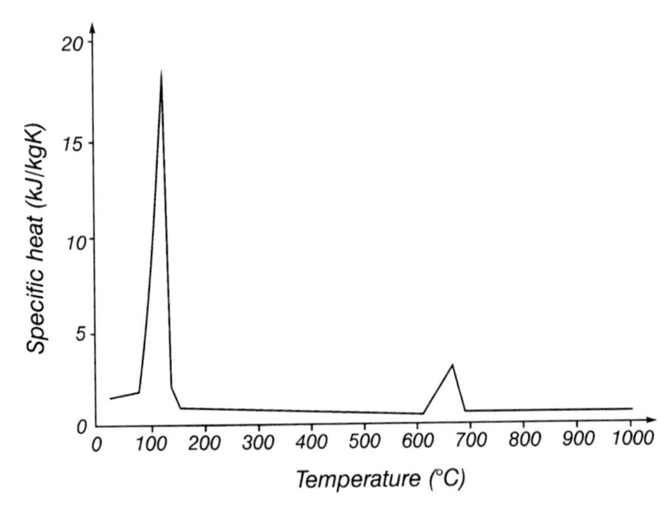

Figure 11.7 Specific heat of gypsum plaster (Reproduced from Sultan (1996) by permission of National Fire Protection Association)

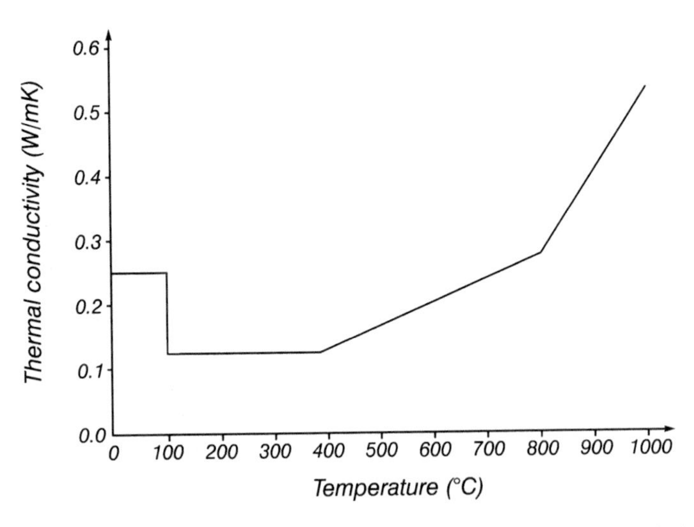

Figure 11.8 Thermal conductivity of gypsum plaster (Reproduced from Sultan (1996) by permission of National Fire Protection Association)

Figure 11.8 shows the thermal conductivity of gypsum plaster as a function of temperature, with a drop at 100°C and a steady rise after temperatures reach 400°C. Thermal conductivity also depends on the density of the gypsum board (Clancy, 1999). The value of thermal conductivity above about 400°C will be affected by the presence of shrinkage cracks in the gypsum board, which will depend on the formulation of the individual board and the type of fire. Cracking may be more severe in a fire with rapidly increasing initial temperatures, such as the 'hydrocarbon' fire described in Chapter 5.

11.5.5 Fire Resistance

Because of the moisture-related reactions described above, all gypsum board products exhibit similar behaviour in fires. When a board is heated from one side, temperatures on the exposed face will increase continuously until about 100°C is reached, at which time there will be a delay while the water of crystallization is driven off. As the heating continues, the 100°C temperature plateau will progress slowly through the board, until the entire board has been dehydrated. After dehydration the gypsum has almost no strength because it has been converted to a powdery form. Any residual strength depends on glass fibre reinforcing to hold the board together.

The fire resistance of assemblies made with gypsum-based panel products depends on several important interrelated properties:

- the insulating capacity of the board, which protects internal structural members and delays temperature rise on unexposed surfaces;
- the ability of the board to remain in place and not disintegrate or fall off after dehydration;
- resistance to shrinkage which usually causes cracking within the board or separation at joints between sheets;
- the ability of the core material to resist ablation from the fire side during extreme fire exposure.

Increased density will generally improve the fire performance of gypsum board, because it is a measure of the greater quantity of gypsum resulting from fewer air voids. Increased density provides more heat absorbing capacity and requires more water of crystallization to be driven off. Richardson and McPhee (1996) refer to tests where a 6% increase in density produced an 8% increase in fire-resistance for otherwise identical construction.

Fire-resisting gypsum boards contain glass fibres which control shrinkage, causing a maze of fine cracks rather than a single large crack which can initiate premature failure of regular board. One of the most critical aspects of fire-resisting gypsum board is the extent to which the glass fibre reinforcing can hold the board together after the gypsum has dehydrated, to prevent the board pulling away from nailed or screwed connections when the board shrinks. Shrinkage can be reduced with various additives such as vermiculite.

Regular gypsum board can fall off a wall or ceiling as soon as the gypsum plaster has dehydrated, at about the same time as charring of the timber studs begins. Boards with glass fibre reinforcing and closely spaced fixings will not fall off until the glass fibres melt, when the entire board reaches a temperature of about 700°C. König and Walleij (2000) report that the critical falling-off temperatures are 600°C for ceiling linings and 800°C for wall linings.

11.5.6 Ablation

Ablation is a term used to describe the slow process whereby dehydrated gypsum powder slowly falls off the heated surface of a fire-exposed gypsum board, resulting in a reduction in board thickness during the fire. In fire-resistant boards, ablation does not occur until after the glass fibres in the board have melted at about 700°C. Ablation is a minor effect, but can be included in finite-element modelling by increasing the thermal conductivity at high temperatures (Thomas, 1997).

11.6 TEMPERATURES WITHIN LIGHT FRAME ASSEMBLIES

11.6.1 Typical Temperatures

Typical temperature profiles within an uninsulated wall during an ISO 834 standard fire-resistance test are shown in Figure 11.9. The temperature on the cavity side of the fire-exposed gypsum board has a long plateau at 100°C as the free water and water of crystallization are driven off and heat is conducted through the board. During this time, the temperature on the cavity side of the unexposed gypsum board has a plateau at a slightly lower temperature, and the temperature on the unexposed face lags much further behind. For a steel stud wall, the gypsum temperatures would be identical, with the temperature of the steel stud being between the temperatures shown for points 2 and 3.

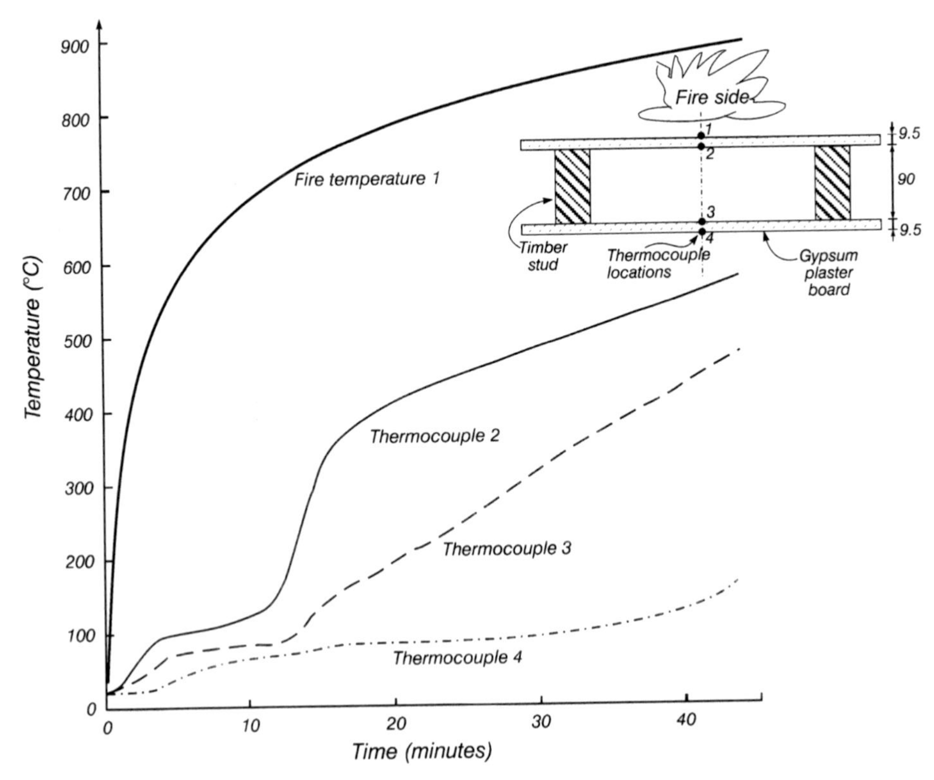

Figure 11.9 Temperature profiles within a cavity wall during a standard fire-resistance test (Thomas, 1997)

11.6.2 Insulation

The insulation criterion for fire-resistance requires that the temperature on the unexposed face remains below a certain critical temperature, so there is no danger of ignition on the unexposed surface and subsequent fire growth. Using the ISO 834 criteria, the assembly is considered to have failed the test when the average temperature rise on the unexposed surface exceeds 140°C, or the maximum temperature rise at any point exceeds 180°C.

The insulating properties of an assembly depend on the geometrical arrangement and the component materials. For assemblies without insulating batts, the highest temperatures on the unexposed face occur remote from the studs or joists. Heat transfer in this region is essentially one dimensional, with heat from the fire passing through the exposed sheet, across the cavity and through the unexposed sheet. The stud material (steel or timber) has no influence on these temperatures, and the distance across the empty cavity (the depth of the studs) is not very significant. Heat transfer across the cavity is by both convection and radiation. The radiative component increases as the temperature increases. Moisture movement may also contribute to heat transfer through the assembly, because moisture continually evaporates in hot regions and condenses on cool surfaces.

For assemblies with insulating batts in the cavities, the overall thickness of the wall becomes more important, and the highest temperatures may occur near the studs, especially if they are steel.

11.6.3 Charring

Charring of the wood begins when its temperature reaches about 300°C. Typical shapes of the charred profile are shown in Figure 11.10 for timber studs in walls with

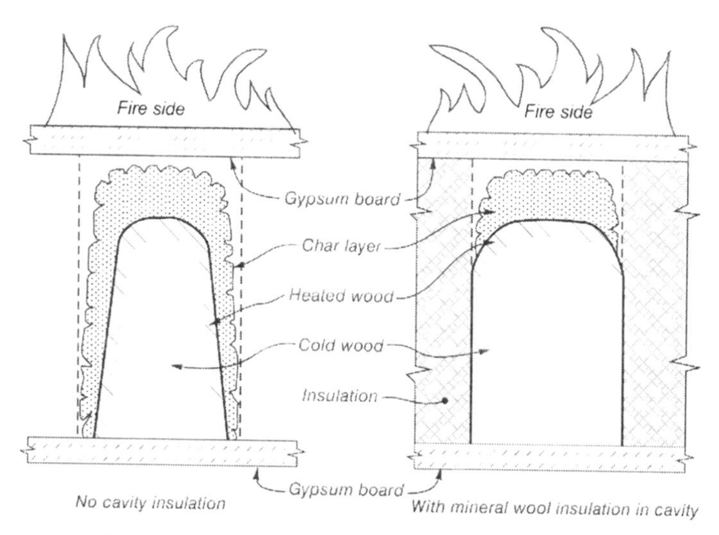

Figure 11.10 Typical charring of timber studs (a) without cavity insulation and (b) with cavity insulation

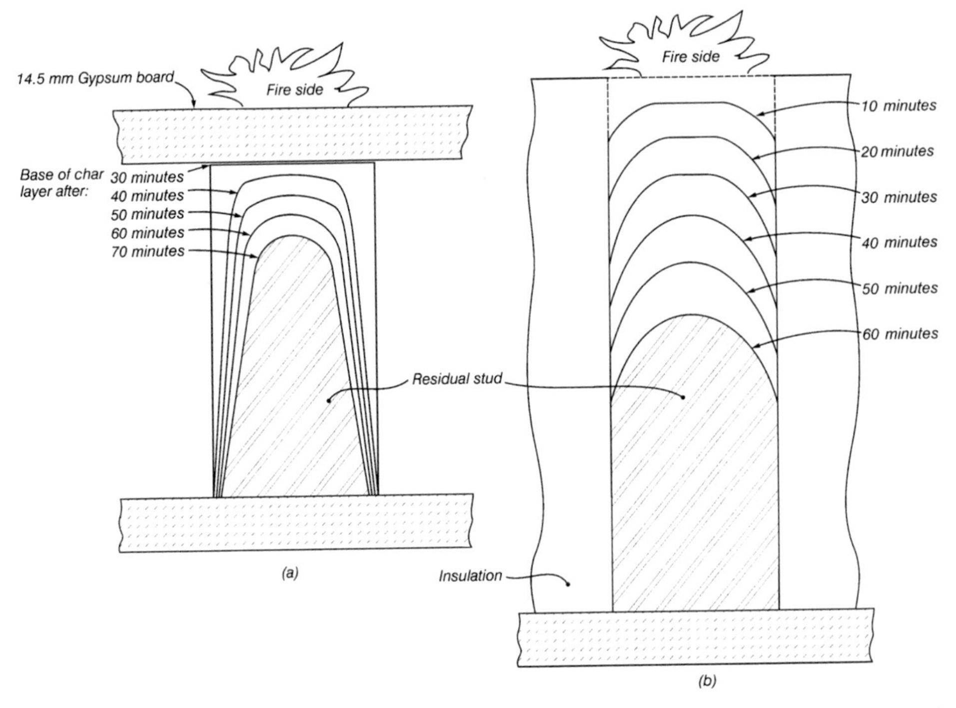

Figure 11.11 Measured char profiles on timber studs: (a) Stud in empty cavity, protected with 14.5 mm gypsum board (Reproduced from Collier (1991(b)) by permission of Building Research Association of New Zealand); (b) Stud in insulated cavity with no protection on the fire-exposed face (Reproduced from König and Walleij (2000) by permission of AB Trätek)

and without cavity insulation. For uninsulated walls, the rate of char on the wide surface facing into the cavity is about half of that on the edge of the stud in contact with the fire-exposed gypsum board. There is no charring on the edge of the stud fixed to the unexposed gypsum board. Figure 11.11 shows char profiles measured at various times for timber studs in walls tested with and without cavity insulation. Figure 11.11(a) is a 90 × 45 mm stud in a wall with gypsum board on both faces but no cavity insulation, and Figure 11.11(b) is a 140 × 45 mm stud in a wall assembly tested with mineral wool cavity insulation but no gypsum board on the fire side. Both sketches are drawn to the same scale.

In load-bearing timber walls it is essential to retain sufficient area of residual stud to carry the applied loads (Figure 11.12). In non load-bearing walls, the studs only need to hold the lining material in place for the duration of the fire, so they may be almost completely burned away by the end of a fire test. It is not possible to reuse a wall after a severe fire if significant charring has occurred, even though the wall may have provided the fire-resistance as expected.

Figure 11.13 shows the rates of charring of timber studs exposed to the standard fire, for a number of different scenarios, adapted from König and Walleij (2000). The basic charring rate β is for large unprotected timber members subjected to one-dimensional charring, as described in Chapter 10. The charring rate β_1 is a faster rate of charring observed for small unprotected timber members (as in Figure 11.11(b))

Figure 11.12 Residual charred studs of a light timber frame wall after a full-scale fire resistance test

where corner effects become significant. The charring rate β_1 increases as the studs become narrower.

The charring rate β_2 shown with the heavy line in Figure 11.13 applies to protected studs (as in Figure 11.11(a)) which begin to char after time t_c. This charring rate is less than β_1 due to the protection of the gypsum board, but it may be greater than β if the stud is narrow. If the protective gypsum board falls off the wall at time t_f the charring rate will increase to β_3 which is greater than the other charring rates. Empirical expressions for all of these charring rates are given by König and Walleij (2000).

11.6.4 Calculation

Calculation of thermal behaviour provides an assessment as to whether an assembly would meet the insulation criterion when exposed to a standard fire-resistance test or

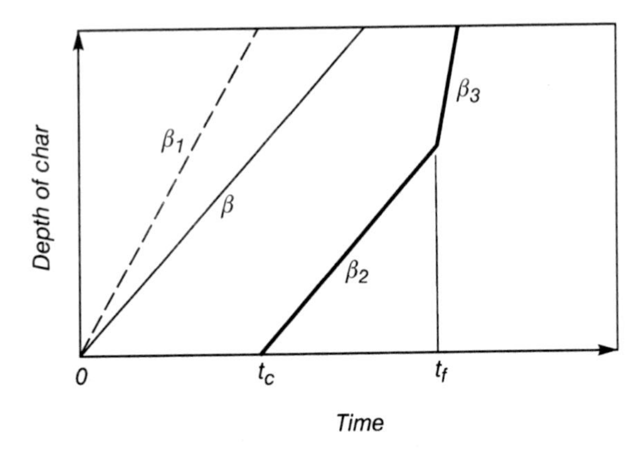

Figure 11.13 Rates of charring of timber studs exposed to the standard fire (Reproduced from König and Walleij (2000) by permission of AB Trätek)

to a simulated real fire. Thermal calculations are not simple because the thermal properties of gypsum and wood are highly temperature dependant, and because assumptions have to be made about heat-transfer coefficients on the exterior surfaces and within the cavity.

A two-dimensional finite-element model is appropriate in most cases, using the thermal properties of gypsum plaster given above, and the thermal properties of wood from Chapter 10. Thermal properties of mineral wool insulation are given by König and Walleij (2000) with thermal conductivity of 0.03 W/mK at 20°C, rising to 0.12 W/mK at 400°C and 0.45 W/mK at 800°C, and the specific heat has a constant value of 1.0 kJ/kgK.

For timber members, the transition from wood to char at 300°C not only effects changes in thermal properties, but also results in shrinkage gap between the timber framing and the lining on the fire side of the assembly. It is necessary to include this gap in any finite-element modelling to get good results. Another significant influence is moisture which evaporates at about 100°C, travels through the wood, the gypsum and the cavity, and condenses on cooler surfaces. Most finite-element modelling does not allow for this explicitly, but the effects are considered to be included implicitly in the thermal properties.

Thomas (1997) used the TASEF finite-element package to get excellent agreement with New Zealand test results for light timber frame walls. König and Walleij (2000) obtained similar results using TEMPCALC. Other recent computer models for predicting thermal behaviour are those by Clancy (1999), Collier (1996) and Takeda and Mehaffey (1998). Sultan (1996) and Cooper (1997) have developed heat-transfer models for light steel framed walls. None of these have yet produced simple formulae for use in hand calculations.

11.7 STRUCTURAL BEHAVIOUR

This section describes structural behaviour of light frame construction in fire conditions, with particular reference to the standard fire-resistance test. The stability criterion, applied to all load-bearing elements, requires that load capacity be maintained throughout the duration of the design fire.

The strength of light frame assemblies is mainly in the timber or steel members themselves and not the lining materials. Lining materials are essential for providing lateral stability to the structural members, but their contribution to overall strength and stiffness is small. For load-bearing timber stud walls, Young (2000) has shown that the lining material on the cold side of the wall increases the flexural stiffness of the wall, hence increasing the resistance to buckling failure during a fire test.

In a standard fire-resistance test, it is essential to ensure that all the typical studs or joists have similar loads, and that the studs or joists at the sides of the furnace do not carry any load, because they are partially protected from the furnace temperatures. A difficulty of comparing fire-resistance tests from different furnaces is that the exact method of applying load is not specified and it is seldom reported.

11.7.1 Timber Stud Walls

Much can be learned about the fire behaviour of timber stud walls from observations in standard fire-resistance tests. Behaviour in real fires will always be different from behaviour in standard tests because of different conditions including fire exposure, support conditions and loading arrangements. In standard furnace tests, load-bearing light timber framed walls almost always deform away from the furnace as shown in Figure 11.14 and eventually fail by buckling in that direction. There may be a small movement towards the furnace in the early stages of the test

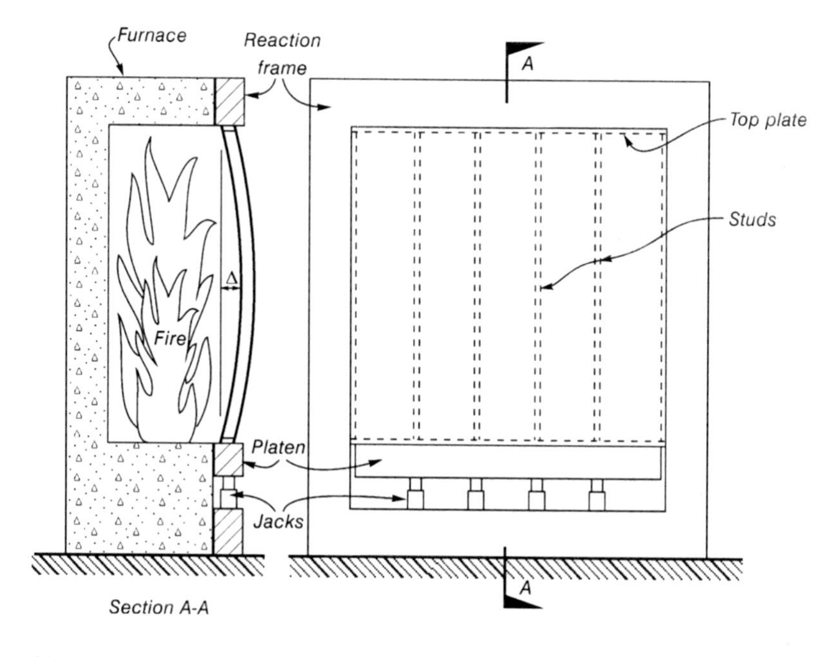

Figure 11.14 Fire-resistance test of light timber frame wall

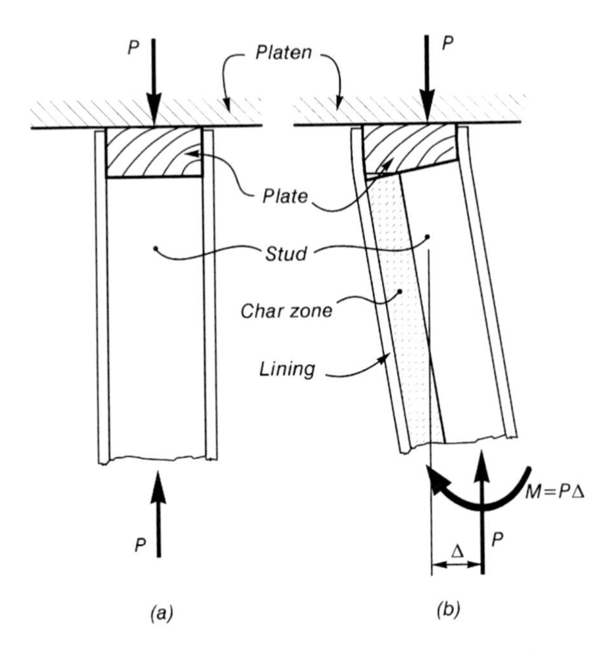

Figure 11.15 Detail of top end of timber stud before and during a fire test

caused by thermal expansion of the gypsum board on the furnace side, but this will be reversed as shrinkage occurs. After the stud starts to char, the centre of resistance moves towards the cool side of the wall, resulting in an eccentricity which causes the wall to deflect away from the fire. Any deflection results in an additional bending moment due to P–Δ effects. The load capacity of the wall depends on the size, temperature and moisture content of the residual cross section of the timber studs.

Figure 11.15 shows the top plate connection before and during a fire test. As the test proceeds, the stud chars and the ends of the stud rotate, causing deformations in the top and bottom plates and causing the line of application of the load to shift away from the fire. As the stud deforms, the bending moment in the stud increases due to P–Δ effects. The P–Δ moment is partly offset by the shift in the line of application of the load shown at the top of Figure 11.15.

Buckling of studs

When timber stud walls experience a structural failure in a fire-resistance test, the failure is caused by buckling of the studs about the strong axis, usually outwards from the furnace. Buckling about the weak axis is prevented mainly by the gypsum board on the unexposed face of the wall. The gypsum board on the exposed face provides very little lateral restraint after it has dehydrated. The cooler gypsum board providing lateral restraint is on the tension edge of the studs, which requires the residual stud to have torsional rigidity, which will reduce as charring occurs. The provision of torsional restraint to studs becomes more important as the depth to width ratio of the stud cross section increases. Torsional restraint of the studs can be enhanced by the use of solid timber blocking between the studs. In New Zealand, all of the approvals listed by Winstone Wallboards (1997) have solid timber blocking at 800 mm spacing between all the studs (see Figure 11.1). Collier (1991(b)) demonstrated the effectiveness of such blocking by testing walls with and without blocking.

Blocking between the studs is not normally used in North America, which may partly explain the poorer fire performance of some North American assemblies compared to those in New Zealand. Thin steel resilient channel fixed to the exposed edge of the studs to improve acoustic performance may provide torsional rigidity to the studs. Blocking between studs and provision of lateral restraint is more difficult for walls where the studs are staggered for acoustic performance.

Because failures occur by buckling under axial loads, the load capacity depends more on the modulus of elasticity of the stud parallel to the grain than on the compression strength of the wood.

Steam softening

Structural failure of timber stud walls in fire is usually a result of reduced stud size and stiffness due to charring. In some cases, failure may occur before charring begins. For example, Young and Clancy (1996) tested load-bearing wall assemblies in which the ends of the studs were fitted with pinned connections allowing free rotation. The studs buckled and the wall failed to carry the applied load after only 35 minutes in two identical tests. When dismantled, the studs were found to have only minor charring, but they were permanently bent as a result of steam softening of the wood. The same wall design with the ends of the studs cut square and butted to the top and bottom plates achieved a fire-resistance of almost 60 minutes. The difference is that the rotational stability of the square-end studs provided enough time for the wood in the studs to pass through the plastic stage without buckling, at which time the wood properties increased due to drying, and eventual failure was a result of charring. Steam softening of wood is a result of plastification of the wood due to a combination of high moisture content and elevated temperature, described in Chapter 10.

Calculation methods

Several calculation methods for assessing the fire resistance of light timber frame construction have been developed in recent years, but these are for research rather than for design purposes because they are much more complicated than simple selection of a proprietary system from a listing of fire ratings. Calculation of structural behaviour is more difficult than calculation of thermal behaviour because of the poor knowledge of mechanical properties of wood at elevated temperatures and changing moisture content. Computational models for walls must include second-order effects in order to predict buckling. It is very difficult to model the effect of sheets of gypsum board falling off a wall during a fire, because of the unpredictable time and computational problems. Clancy (1999) allows the finite-element grid to change if a sheet of board falls off the wall.

Thomas (1997) used the ABAQUS 3-D finite-element package to get good agreement with test results for timber stud walls, exposed to the standard fire, assuming that all the load was carried by the studs without composite action. Clancy (1999) has used the structural model of Young (2000) to do a probabilistic study of timber stud walls exposed to realistic fires, finding that the time to failure of typical walls has a coefficient of variation of about 12%, considering the likely variability, both in mechanical properties and in fuel loads. Young's model includes the effect of composite action between the studs and the gypsum board, allowing for partial slippage in the connectors, but he recommends that full fixity be used in the calculations.

König and Walleij (2000) have used an extensive experimental and analytical research programme to propose a simple design method for walls exposed to the standard fire. The method is conceptually simple, with the following steps:

(1) estimate the time to onset of char of the timber studs,

(2) estimate the rate of char,

(3) calculate the char depth, hence the size of the fire-reduced cross section after a particular time of fire exposure,

(4) estimate the strength and stiffness of the heated wood in the residual cross section,

(5) calculate the load capacity.

The time to onset of char in the stud is estimated from the thickness and density of the gypsum board on the fire side of the wall, or from test results. The rate of charring on the face of the stud in contact with the gypsum board is calculated as described above. The residual stud can be approximated as a rectangular cross section as shown in Figure 11.16, for walls with or without insulation in the cavity.

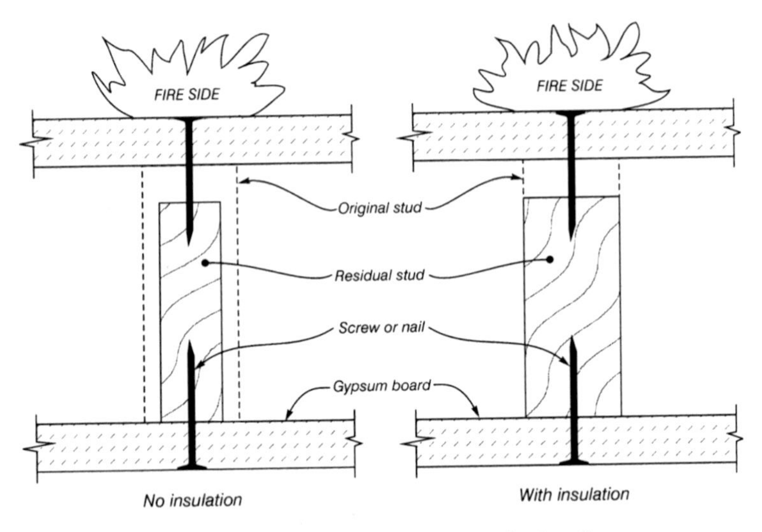

Figure 11.16 Notional residual stud in a light timber wall after fire exposure

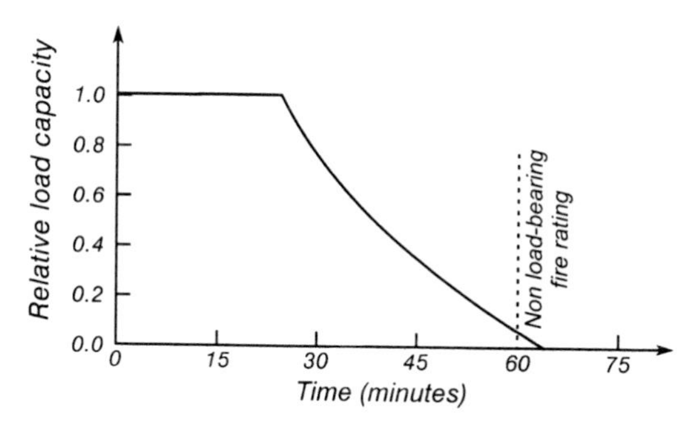

Figure 11.17 Reduction in axial load capacity of a timber stud wall during fire exposure

The mechanical properties of the wood in the residual cross section are calculated from a simple empirical expression. The structural calculation uses the residual cross section with a conventional formula for timber column design, assuming that the stud is restrained against weak axis buckling by the gypsum board on the cooler face of the wall.

This design method requires further work before it can be used for general application, with possible modifications for other species of wood and other types of gypsum board. Figure 11.17 shows the shape of the load capacity curve with time for a typical timber stud wall (with a non load-bearing rating of 60 minutes) where the studs start to char after 25 minutes of fire exposure. Assuming a typical load ratio of about 0.5, it can be seen that this particular wall would have a structural fire resistance of about 40 minutes.

This model, illustrated in Figure 11.17, ignores any loss of strength due to heating of the stud before charring occurs, which could be significant for walls with very high load ratios.

Extrapolation from test results

Because full scale fire-resistance tests are very expensive, it is often necessary to extrapolate from a listed rating to achieve a fire-resistance rating for a wall with different height or different load from that tested. A few furnaces can test walls 4 m high, but most full scale fire-resistance wall furnaces are only 3 m high, which limits the height of a test specimen. Calculation methods are necessary to extrapolate to taller walls. Collier (1991(a)) provides a method based on New Zealand test results, applying the secant formula for column buckling to residual studs such as shown in Figure 11.16.

11.7.2 Steel Stud Walls

Steel stud walls have similar fire performance to timber stud walls in many ways, but there are some important differences, including much larger deflections. In standard fire-resistance tests, timber frame walls bow outwards from the furnace due to loss of

Figure 11.18 Full scale fire-resistance test of a light steel frame wall. The wall has deflected in towards the furnace due to thermal bowing

charred cross section, but light steel frame walls bow in towards the furnace due to differential thermal expansion of the steel studs (Figure 11.18). Bowing inwards helps to improve the load-bearing capacity of steel studs during fire because the compression edge of the stud (rather than the tension edge) is laterally restrained by the plasterboard on the cold side of the wall. Canadian tests of load-bearing steel stud walls (Kodur *et al.*, 1999) showed bowing towards the furnace as expected, but the final sudden failure was away from the furnace after local buckling of the flange on the hot side of the stud. This buckling occurred near the top or bottom of the wall where the compression force in the hot flange was greater than at mid-height, due to the deflected shape of the wall.

Some steel stud walls have horizontal blocking between the studs, which will improve fire-resistance by providing lateral stability. This is important because thin steel channel-section studs have no torsional rigidity. Essential lateral restraint is provided by the gypsum board on the cold side of the wall. Lateral restraint will decrease after this board loses strength due to dehydration and melting of the glass fibres. Gerlich *et al.* (1996) suggest that at least 3mm of the board should remain below 100°C for this reason.

A review of the fire resistance of load-bearing steel stud walls protected with gypsum board has been published by Alfawakhiri and Suttan (1999) and a large experimental programme on such walls is described by Klippstein (1980).

Calculation methods

As with timber structures, calculation of the structural behaviour of steel structures is more difficult than calculation of the thermal behaviour. Computational models for walls must include second-order effects in order to predict overall buckling, and models for steel stud walls must additionally be able to predict local buckling of thin steel sections restrained by degrading lining materials.

Gerlich *et al.* (1996) describe test results and calculation methods, showing how the normal temperature design equations can be modified to calculate the axial load capacity under fire conditions if the temperature of the steel studs is known. They used the TASEF program for calculating the temperatures in the steel, and modified the design equations of AISI (1991) for calculating load capacity at elevated temperatures. This method can be used for standard or real fire exposure.

11.7.3 Timber Joist Floors

Timber joist floors behave similarly to timber stud walls, in that structural failure occurs when charring of the timber joists causes significant loss of load capacity. Whereas wall strength is governed by the modulus of elasticity of the wood, floor strength is governed more by wood strength because buckling is less important. Floor joists are flexural members, not compression members, so they are not subject to buckling about the strong axis. Buckling about the weak axis is prevented by the floor diaphragm on the cooler side of the assembly, fixed to the compression edge of the joist. This statement applies to simply supported floor assemblies exposed to fire from below, which covers almost all design situations. Floors are very susceptible to the fire-exposed gypsum board falling off the underside of the assembly, which cannot be easily modelled because it depends mainly on the strength of the board and the fixing details, as discussed later in this chapter.

Failure stresses within timber floor joists during fire exposure are shown in Figure 11.19 (from König and Walleij, 2000). Figures 11.19(a) and (b) show identical failure stresses in positive and negative bending for a joist at normal temperatures, assuming a bi-linear stress–strain relationship. It can be seen that yielding of the wood has occurred near the compression face, producing a stress distribution similar to that shown in Figure 7.2. If the joist had any defects in the tensile region, it would fail in tension before yielding occurs (Buchanan, 1990).

Figures 11.19(c) and (d) show failure stresses after the loss of 15% of the cross section due to charring, causing the flexural strength to drop by about two thirds. These stresses have been calculated from temperature profiles during charring as shown in Figure 11.11(b) for an insulated cavity, and mechanical properties of wood at elevated temperature as described in Chapter 10. When the fire side is in tension, Figure 11.19(c) shows a parabolic distribution of tensile stresses, with the

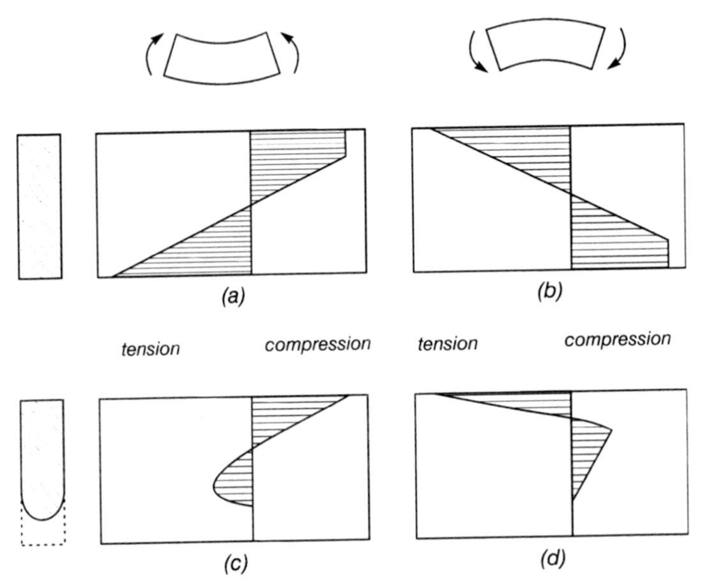

Figure 11.19 Failure stresses in timber floor joists at normal temperatures (a and b) and exposed to fire from below (c and d) (Adapted from König and Walleij (2000) by permission of AB Trätek)

maximum stress well up into the residual joist. When the fire side is in compression, Figure 11.19(d) shows low compressive stresses over most of the residual cross section, with high tensile stresses in the cold wood at the top edge. These calculations confirm those by Thomas (1997), based on the experimental results of König (1995). Similar stress diagrams for studs in walls show much more of the cross section in compression due to the applied compressive loads.

In some countries it is common to cast a thin concrete topping over light timber floors, in order to reduce sound transmission to the room below. Concrete toppings are usually 35 to 50 mm thick, not designed for composite action. Richardson *et al.* (2000) give an opinion that such concrete toppings will not reduce the fire-resistance rating of a timber floor assembly if the cavity has no insulation, or contains mineral wool insulation.

11.7.4 Timber Trusses

Light timber trusses exposed directly to post-flashover fires have negligible fire resistance. Trusses protected by a fire-resisting ceiling can have good fire resistance. Temperatures measured in a ceiling plenum are much less than furnace exposure, typically reaching 327°C after 60 minutes exposure to an ASTM E119 furnace test (Shrestha *et al.*, 1995). The weakest link is usually the truss plate connectors rather than the wood members. Shrestha *et al.* (1994, 1995) have developed models for predicting the temperatures within truss members and the mechanical properties of the wood and the connectors at elevated temperatures.

Deformations in the wood and in the connections affect the stresses in the members. Cramer *et al.* (1993) describe a truss analysis model (SAWTEF) which gives good agreement with full-scale test results for both trusses and truss plate connected joists (Cramer, 1995). This model can allow for non-uniform temperatures within the ceiling plenum and for load sharing between trusses or joists of different stiffnesses.

11.8 DESIGN OF LIGHT FRAME STRUCTURES IN FIRE

This section is a summary of design methods for light frame structures of steel and wood. Most designs will simply be a selection of an assembly from a list of proprietary ratings, comparing the listed rating with the prescribed fire resistance or an equivalent time of fire exposure. Calculations of thermal and structural behaviour in real fires are possible, but difficult, so they are only recommended for research and development purposes. Interpolation between listed ratings is possible, with an expert opinion required for significant changes from the listed assembly.

For realistic fire exposure, a time-equivalent formula from Chapter 5 can be used to calculate the fire severity of a complete burnout, to compare with a listed fire-resistance rating. For timber structures, design to resist a complete burnout of a compartment is not as simple as with non-combustible materials, because charring of the wood may continue even after the fire is out (König, 1998(b)) unless the timber is specially protected to prevent all charring. To prevent collapse if charring occurs, the Fire Service or the owners must intervene to remove damaged linings and extinguish any charring after the fire has burned itself out. This becomes important in some countries, including Norway, where regulations require that multi-storey timber buildings should be designed for a complete burnout of the room of origin.

For load-bearing steel structures, a simple design approach is to make a thermal calculation and ensure that the maximum temperature of the steel does not exceed a limiting temperature of 350°C or 400°C. In this case the normal temperature design methods can be used for fire design. Gerlich *et al.* (1996) justified a temperature of 400°C, showing that a steel strength of 60% of the normal temperature strength gives sufficient safety margin in calculated fire resistance compared with the results of full-scale wall tests.

For steel stud walls, Gerlich *et al.* (1996) have proposed a conservative linear interpolation method for estimating the reduction in fire-resistance rating as the applied load increases. This assumes a linear interaction diagram between the fire resistance of a non-load bearing wall and the load capacity of the wall at normal temperatures as shown in Figure 11.20. If a fire test result is available for a non load-bearing wall, the load-bearing capacity of the wall in a fire of lesser duration can be estimated. For example, a wall with axial load '*A*' (one third of the normal temperature load capacity) would have a load-bearing fire-resistance rating of '*B*' (two thirds of the non load-bearing rating). This approach should not be applied to timber walls.

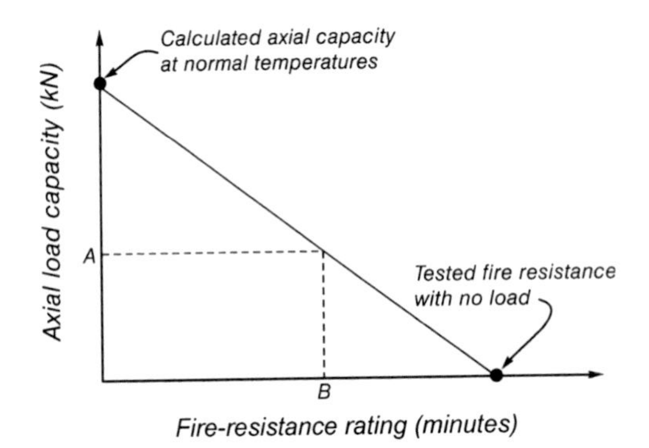

Figure 11.20 Linear interpolation for fire resistance of load-bearing steel-stud walls

11.9 CONSTRUCTION DETAILS

Construction details can have a significant influence on the fire resistance of light frame construction. Several important details are discussed below.

11.9.1 Insulating Batts

Light frame construction often contains insulation in the cavities, to improve thermal, acoustic or fire performance. Insulating batts have a mixed impact on fire performance, depending on the material.

The most common forms of insulation are glass fibre batts or mineral wool (rock fibre) batts. Less common materials include foam plastic, batts made from natural wool, or sprayed insulation containing cellulose fibres. Glass fibre batts are made from thin glass fibres bonded into a mat with an organic binder. Mineral wool batts are made from mineral or ceramic fibres which do not melt at fire temperatures. Mineral wool batts and cellulose fibre insulation can provide a significant increase in fire resistance, but glass fibre batts may lead to reduced fire resistance (Sultan, 1999).

The major negative effect of all cavity insulation is that the gypsum board on the fire-exposed side heats up much faster than for an empty cavity, leading to earlier dehydration and possible falling off of the board. Richardson *et al.* (2000) observed that ceiling linings fell off timber joist floor assemblies 15 minutes earlier when the cavity was insulated. For assemblies with glass fibre batts, the ceiling fall-off was followed by melting of the glass fibres into small droplets within 2 to 3 minutes.

If the fire-exposed board falls off any assembly containing glass fibre batts, the batts will rapidly melt leaving the studs or joists directly exposed to the fire. On the other hand, well-fitting mineral wool batts will remain in place and protect the studs, joists and unexposed lining from the fire. Some positive fixing of the batts is recommended, and the batts must fit well because it has been observed by Sultan and Loughheed (1995) and König (1998(b)) that loose-fitting batts produce a worse result than no batts at all.

11.9.2 Number of Layers

Multiple layers of thin gypsum boards may be cheaper and lighter to fix than one thick board, but multiple layers do not usually provide the same fire resistance as a single layer of the same total thickness, because the outer layers can fall off sequentially, leading to much greater thermal exposure to the inner board. This is a particular problem for regular gypsum board which contains no glass fibres, because it will tend to fall off as soon as large cracks occur or the gypsum becomes dehydrated.

An advantage of multiple boards is that the joints between the sheets can be staggered, reducing the likelihood of early flame penetration into the cavity, especially if sheet joints are not on studs. If more than one layer is used, the inner layer is not usually taped or stopped at the joints.

Some light timber frame walls have additional layers of wood-based sheet material such as plywood or oriented strand board nailed to the studs to improve lateral load resistance. Kodur and Sultan (2000) investigated the fire performance of these 'shear walls' finding that the addition of such materials under the gypsum board increased the fire resistance. This result would be expected from the principles illustrated in Figure 6.1.

11.9.3 Fixing of Sheets

Gypsum board sheets must be fixed to the studs or joists such that the board remains in place for the intended period and gaps or cracks do not appear at joints. Most boards are attached to the studs or joists with screws, although nails are sometimes used on timber framing. The screws at the board edges must be close enough together and far enough from the edge to prevent the board pulling away from the joints due to shrinkage during a fire. The strength of a cut edge is less than that of a machine-finished edge. Butt joints between boards are usually finished with plaster or jointing compound which falls off during a fire test and does not contribute to fire resistance.

The use of slender framing members reduces the edge distance between the nail or screw and the edge of the board. Figure 11.21 shows how little edge distance is available for nailing two sheets of board to a stud 38 mm wide. The edge distance of 10 mm meets current Canadian code requirements, but recent tests show a significant increase in fire resistance if the screws are located at least 35 mm from the edge of the gypsum board (Richardson *et al.*, 2000) as shown in Figure 11.24. For assemblies with double layers of gypsum board good performance can be obtained by screwing the outer sheet to the inner sheet using screws which have a very coarse thread (Type G screws). The more fasteners in a sheet, the better the resistance to falling off of the sheet, hence the better the fire performance, especially for ceiling panels.

Joints between sheets are usually made on main framing members (studs or joists) or on blocking members between them. All approvals in New Zealand are for sheet joints over framing members (Winstone Wallboards, 1997) which improves the fire resistance (Collier, 1991(b)). If the studs are not blocked and the 1.2 m wide sheets of

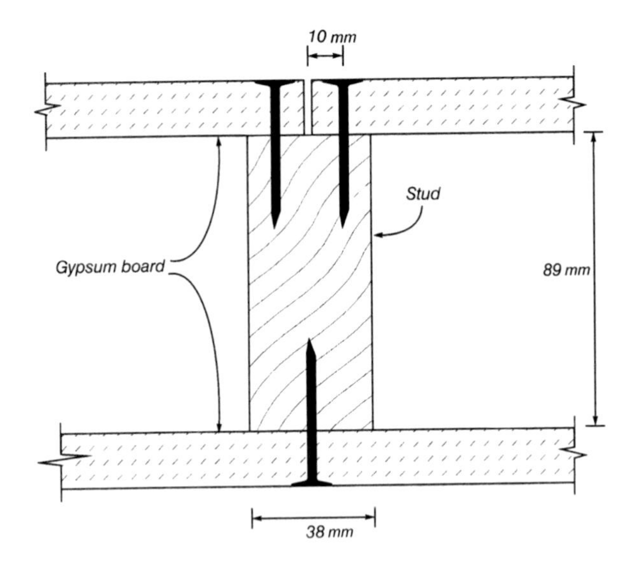

Figure 11.21 Nailing of gypsum board to 38 mm wide timber stud

gypsum board are fixed in a horizontal pattern to the vertical studs, the horizontal sheet joints will have no backing. The same applies if sheets of gypsum board are fixed in a vertical pattern to horizontal resilient channels. The Canadian code (NBCC, 1995) permits joints to have no backing provided that the studs or resilient channels are no more than 400 mm apart, but such joints are weak points for fire resistance. For timber frame construction, König and Walleij (2000) found that the time to onset of char in a stud decreased by 8 minutes if there was a joint between the sheets of gypsum board over the stud. For double layers a similar decrease occurred, but only if the joint was in the outer layer of gypsum board.

To prevent sheets of lining material from falling off timber framing during fires, it is essential that the fasteners are long enough to remain anchored in sound wood after significant charring has occurred. König and Walleij (2000) suggest a minimum length of 10 mm. Figure 11.16 shows how little anchorage length may be left after charring of the stud occurs.

Figure 11.22 shows the type of gap that can occur at the joint between sheets, due to shrinkage in the gypsum board. This can only be prevented by using good-quality fire resisting board containing vermiculite or other additives, fixed with sufficient fasteners preferably kept away from the edges of the sheet.

11.9.4 Resilient Channels

Acoustic performance of gypsum board assemblies can be improved by spacing one or both layers of board off the studs or joists using a pressed steel resilient channel as shown in Figure 11.23. From a fire point of view, this results in improved insulation, but less secure fixing of the gypsum board. Some tests have shown a 12% reduction in fire resistance resulting from the use of resilient channels on the fire side of the wall

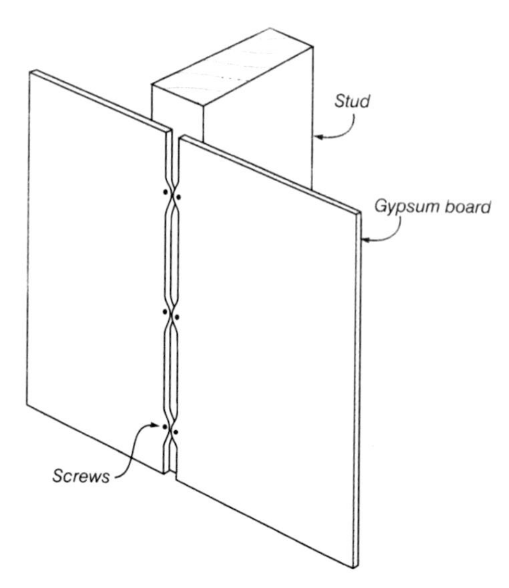

Figure 11.22 Gap between gypsum boards caused by shrinkage of gypsum plaster

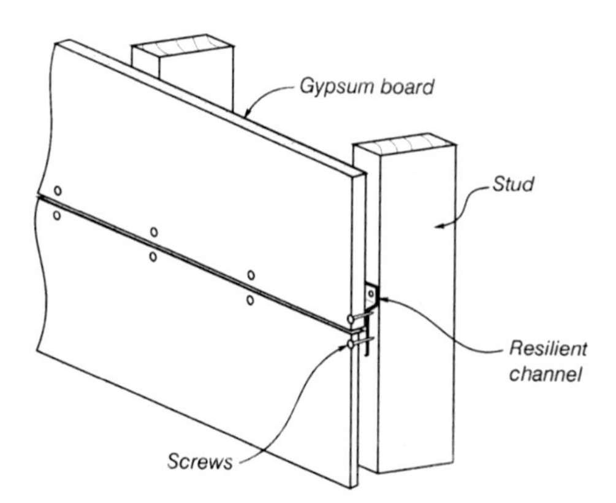

Figure 11.23 Gap between board and stud resulting from the use of resilient channels

(Kodur and Sultan, 2000) but others have found no effect (König and Walleij, 2000). A weakness can occur in walls where one layer of gypsum board is fixed to resilient channels on the fire exposed face. Once fire has penetrated the gap between two sheets of gypsum board, the gap between the board and the stud shown in Figure 11.22 allows hot gases to move throughout the assembly, causing premature charring of the studs and failure of the wall (Richardson and McPhee, 1996). This situation can be worse if the sheets are fixed in a vertical pattern because the joints have no backing.

Resilient channels can be used to increase the distance between fixing screws and the edge of the sheets of gypsum board, as shown in Figure 11.24 resulting in a significant increase in fire resistance (Richardson *et al.*, 2000).

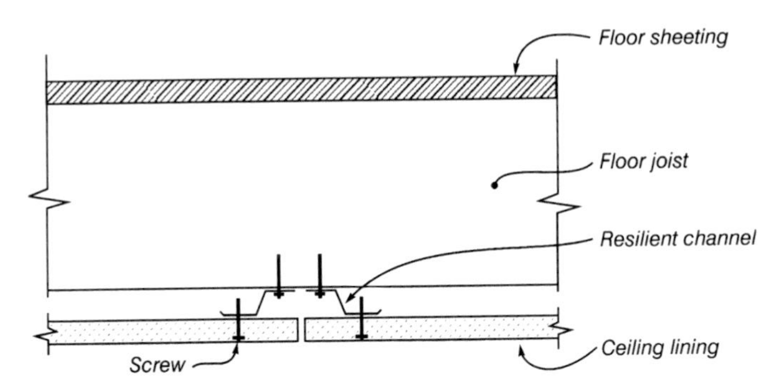

Figure 11.24 Distance between screws and edge of gypsum board increased with the use of resilient channels (Reproduced from Richardson *et al.* (2000) by permission of John Wiley & Sons Ltd)

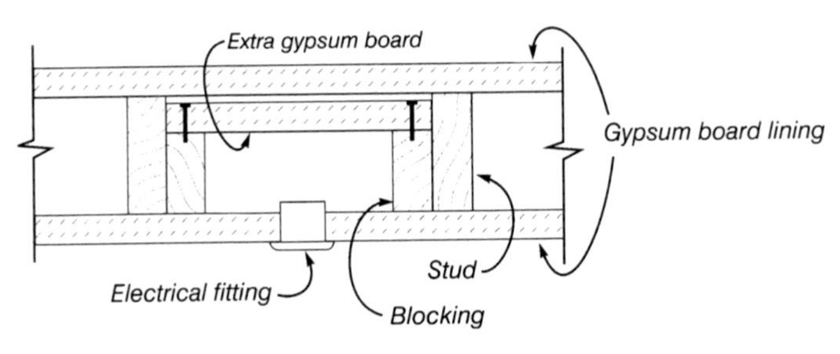

Figure 11.25 Protection of electrical fitting in cavity wall

11.9.5 Penetrations

A major concern about the reduction in fire resistance of light frame assemblies is the effect of penetrations for services or fixtures. Fire resistance may be severely reduced if electrical outlets are placed back-to-back at the same location in a wall. This problem is reduced if the cavity is completely filled with mineral wool insulation. Solid timber blocking or extra layers of gypsum board can be provided behind electrical outlets as shown in Figure 11.25. An alternative protection is to house the electrical fitting in a proprietary pressed-steel box containing intumescent material which will expand when subjected to fire temperatures. Plastic pipe penetrations can be protected with proprietary collars containing intumescent products that will expand to fill any gap produced by melting or burning of the pipe. Many alternative products are available as described in Section 6.7. Fire tests are reported by Parker *et al.* (1975).

11.9.6 Party Walls

Figure 11.26 shows the situation that often occurs in light frame construction, where a fire-resisting wall is required between two occupancies that are not otherwise

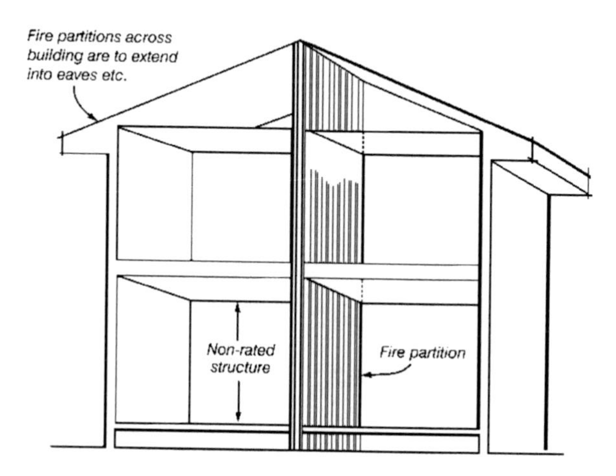

Figure 11.26 Party wall between apartments (Reproduced from (SNZ, 1986) by permission of Standards New Zealand)

required to have fire resistance. The party wall between the two apartments must remain in place and prevent fire spreading from one apartment to the other. If the construction burns down on one side of the wall, the remaining structure must provide lateral support to the party wall. Non load-bearing party walls can be attached to the construction on both sides with aluminium clips, so that a fire on one side will melt the clips and allow the structure to collapse without pulling down the wall (Gypsum Association, 1994).

11.9.7 Fire Stopping, Junctions

Hollow cavity construction should be provided with fire stopping, referred to as draft-stopping in North America, to ensure that any flames or hot gases entering the wall or floor cavity cannot spread into other storeys or other parts of the building. This is particularly important in multi-storey construction where the studs are continuous over more than one storey (balloon framing).

There are many different details for fire stopping. For junctions between floors and a fire-rated wall, Figure 11.27(a) shows solid timber blocking in the floor cavity, to prevent fire from spreading through the separating wall, and Figure 11.27(b) shows a method of connecting timber floors to the fire-resisting wall without reducing the fire resistance of the wall.

A fire-resisting barrier is only as good as its weakest link. Junctions between walls, ceilings and floors must be constructed such that the fire resistance of the barrier is not reduced locally, and so that fire in one cavity cannot spread into an adjacent cavity. Junctions between fire-rated and non fire-rated construction must be constructed so that failure of the non fire-rated element does not allow fire to enter the fire-rated assembly.

Fire stopping in light timber framed buildings can be achieved with solid timber blocking between the studs and within the junctions between floors and walls.

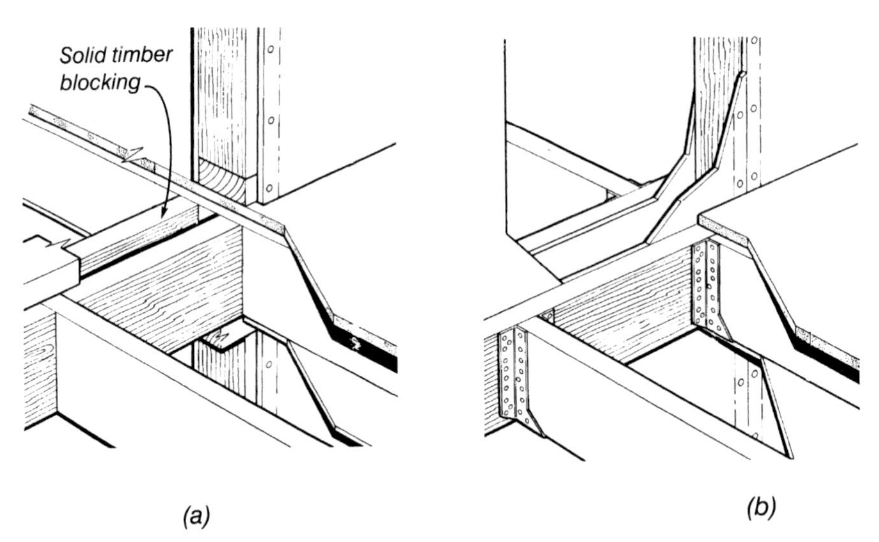

Figure 11.27 Fire stopping details (Reproduced from (SNZ, 1986) by permission of Standards New Zealand)

Typical details are given by manufacturers (e.g. Winstone Wallboards, 1997). Light gauge steel angles behind the gypsum board can be used to enhance the integrity of fire stopping at junctions (Collins, 1998).

In light steel framing, fire stopping can be achieved with blocking between the studs and the use of steel angles to close off paths for fire spread between separate cavities. Long walls are sometimes provided with control joints to allow for longitudinal thermal movements. Tested details of control joints are provided by the Gypsum Association (1994).

Double stud walls are often used to meet acoustic requirements, but there is concern about possible fire spread into the cavity between the two leaves of the wall. In a recent series of fire tests, Sultan (2000) found that small openings from fire-rated floors into wall cavities could be sealed with layers of various materials, including semi-rigid sheets of glass fibre or mineral wool, 0.4 mm steel sheet, or 13 mm oriented-strand-board.

11.9.8 Conflicting Requirements

Difficulties often arise when detailing light frame construction because of conflicting requirements for fire, structural and acoustic performance (NRCC, 1998). In general terms, the structural requirement is usually for all floors and walls to be continuous diaphragms, which also provides good fire safety by eliminating extended cavities where fire could spread. Acoustic requirements are for as much separation as possible, with floors and walls being non-continuous through junctions, and with gaps provided within walls to eliminate transfer of sound through structural elements. Careful consideration is often needed to meet these conflicting requirements without compromising fire safety.

11.10 LIGHTWEIGHT SANDWICH PANELS

Lightweight sandwich panels are becoming a very common building material, especially in buildings such as food-processing facilities where hygiene and thermal insulation are very important. An increasing number of severe fires have recently occurred in sandwich panel buildings. This section is a brief overview of the fire performance of sandwich panels.

11.10.1 Description

Lightweight sandwich panels take many forms, but those referred to here are lightweight panels consisting of outer sheets of thin steel with a core of plastic foam. Cores may be made from a wide variety of foamed plastics, but the most common are polystyrene or polyurethane. Most sandwich panels have no structural connection between the two outer sheets other than adhesion to the lightweight core. Some manufacturers use combustion-modified foams which are more difficult to ignite than normal foams, but these will still burn in post flashover fires. Some panels have cores made from non-combustible mineral wool fibres which perform much better in fire than panels with foamed plastic cores.

Sandwich panels are manufactured in a wide range of sizes and thicknesses, and are often used as structural materials for wall, roof or ceiling construction. The external surfaces of the steel sheets are coated with plastic film or high-performance paint, and some may have a facing of gypsum board which will improve fire resistance.

11.10.2 Structural Behaviour

Under normal temperature conditions, sandwich panels have very good structural properties because of their light weight and high stiffness. The structural stiffness comes from the rigidity of the core material which holds the skins apart and prevents shear deformations. Figure 11.28 shows flexural behaviour of panels under normal temperature conditions.

11.10.3 Fire Behaviour

Lightweight sandwich panels can be a serious problem in fire because of their potential huge contribution to the fire load in the building, hidden fire spread within the panels, and rapid loss of strength when exposed to fire. Fire behaviour depends greatly on the type of foam and its behaviour when exposed to heat. Some plastic foam materials such as polystyrene will melt and shrink away from the heated facing, leading to rapid debonding and poor structural performance (Figure 11.30). Some other foams will remain in place for a longer period of time, although most of the adhesive bonds between the core and the facing will fail at temperatures below 150°C. All plastic foams will burn fiercely when exposed to post-flashover fire temperatures.

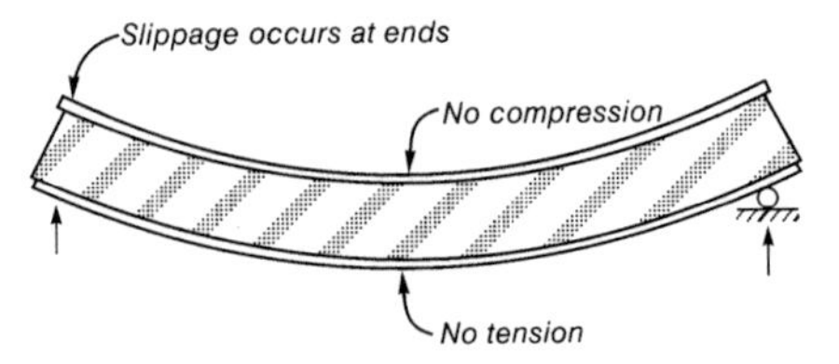

Low-strength adhesive allows facings to slip relative to each other

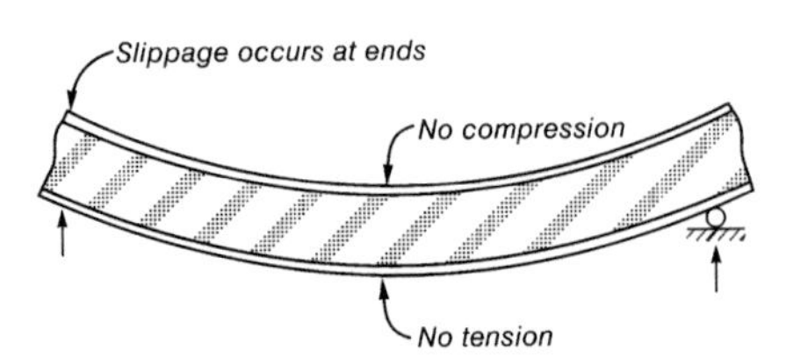

Low-strength modulus of core allows facings to slip relative to each other

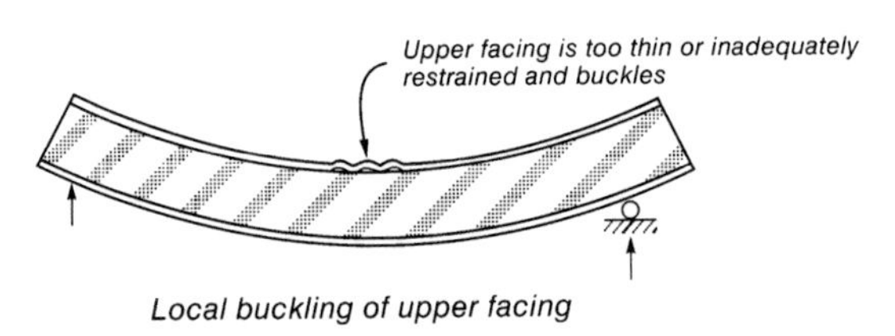

Local buckling of upper facing

Figure 11.28 Flexural behaviour of sandwich panels (Cooke, 1997)

Standard fire-resistance tests or standard reaction to fire tests are not suitable for assessing the real fire performance of lightweight sandwich panels, because the size of specimens cannot accommodate the range and scale of joints between panels in real buildings.

The increased contribution to the fire load of the building from the foam plastic is a serious concern. The heated foam may melt and flow out of the panels, or will be converted to a flammable gas within the heated panel, depending on the temperatures, the type of plastic, and how well the panels are held together. Some foam will melt and produce flaming droplets which can spread fire to other locations. Burning foam plastic produces large volumes of toxic smoke.

11.10.4 Fire Resistance

It may be difficult to see how any fire-resistance ratings can be achieved with such highly combustible material as foamed plastic. However, some manufacturers have obtained non load-bearing fire-resistance ratings of up to 4 hours for sandwich panels containing polystyrene foam, but only meeting the *integrity* criterion. Observations show that the plastic foam melts and escapes in the first few minutes of the fire-resistance test, leaving two skins of thin sheet steel supported by the frame of

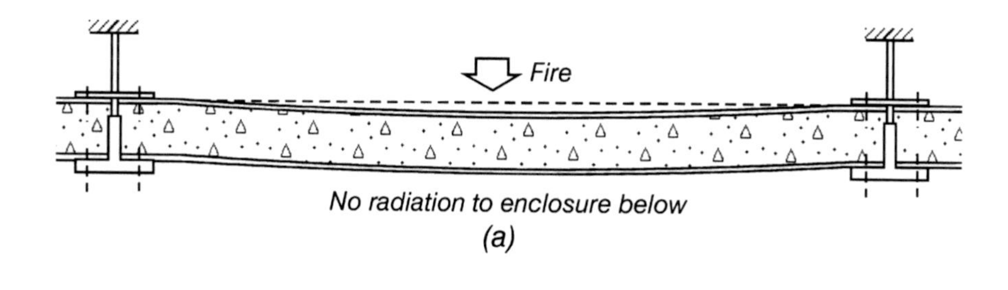

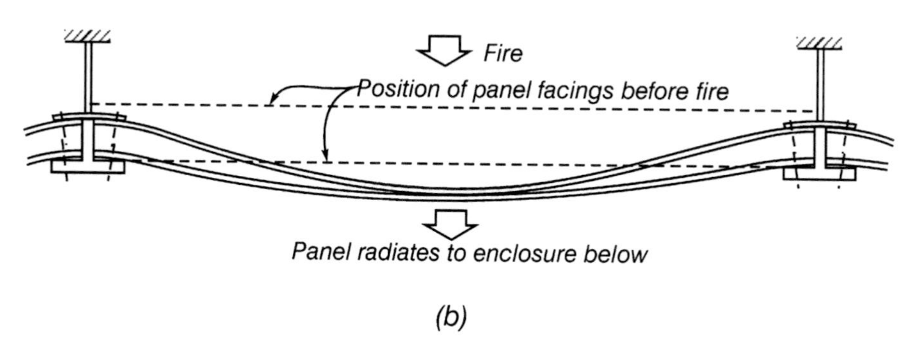

Figure 11.29 Support of sandwich panel ceiling exposed to fire from above (Cooke, 1997)

the furnace. The steel becomes very hot, rapidly exceeding the insulation criterion, but the integrity criterion can be met if the vertical joints between the steel sheets are well connected with an overlapping joint filled with an intumescent strip and fixed with steel (not aluminium) screws to prevent any penetration of flames or hot gases.

This type of fire-resistance rating is only useful where the sandwich panels are supported on all sides by a fully fire-rated structure, and where only an integrity rating is required, such as in a boundary wall situation. Most sandwich panel structures rely on the other sandwich panels or unprotected steel members for structural support, in which case the integrity rating described above is of limited use.

11.10.5 Design

Design of foam plastic sandwich panels for fire-resistance is not possible, other than for an integrity rating as described above. Sandwich panels are often used as structural elements, but these will have negligible structural fire resistance. Design for integrity should ensure that there are fire-resisting mechanical fastenings between the facing panels and the supporting structure, so that panels do not fall apart as soon as the bonding becomes delaminated in a fire. Figure 11.29 shows a sandwich panel ceiling exposed to fire from above. In the initial stages of the fire, the foam plastic will rapidly melt close to the fire-exposed skin, causing delamination and loss of strength. The foam plastic will soon disappear and the panel skins will totally collapse unless they are fixed as shown. Many suitable details and a more

(a)

(b)

Figure 11.30 (a) Fire damage to sandwich panels with foam plastic core, after a factory fire. The building to the right of the photo was completely destroyed by the fire. The panels on the left were partially damaged at the time the fire was extinguished. (b) Steel skins of sandwich panel roofing, draped over the supporting structure after the fire

complete discussion of fire behaviour of sandwich panels are given by Cooke (1997). The use of sandwich panels for structural fire-resisting walls can only be recommended if they are manufactured with mineral wool cores rather than plastic foam cores.

12

Design Recommendations

12.1 OVERVIEW

This chapter gives a brief summary of the recommendations outlined in this book for structural design of fire-exposed buildings constructed from steel, concrete or timber.

12.2 SUMMARY OF MAIN POINTS

The overall design approach is to compare the estimated *fire severity* with the *fire resistance* of the selected structural assembly. The comparison can be made in the *time domain* by comparing fire-resistance times, in the *temperature domain* by comparing the maximum temperature with a critical value, or in the *load domain* by comparing the actual load on the structure during the fire with the minimum load capacity at any point in the fire exposure.

12.2.1 Fire Exposure

There are three levels of fire exposure.

(1) The traditional fire exposure is simply the required fire-resistance time specified in a prescriptive code, based on the *standard fire* test. This can be compared with a generic or proprietary fire-resistance rating.

(2) The next level of fire exposure is the *equivalent fire severity*, which is an estimate of the time of exposure to the standard fire that would produce the same effect in the fire compartment considering fuel load, ventilation and construction materials. Equivalent fire severity should only be used where calculation methods for real fires are not available or where a proprietary rating is to be used to provide fire resistance.

(3) The most accurate and preferred method is to use the predicted time–temperature curve for a *real fire* in the compartment as the basis for calculations. The recommended fire design curve is the Eurocode parametric fire, with modifications as suggested in Chapter 4. Any estimation of real fire temperatures

involves some uncertainties, but using an estimated time–temperature curve eliminates the layer of approximations which is embedded in the equivalent fire-severity method.

12.2.2 Fire Resistance

There are several levels at which fire-resistance can be estimated. The most simple method is to select a *generic* or *proprietary* rating from a listing of fire-resistance ratings. Sometimes it is appropriate to modify or extrapolate from a listed rating for different load or support conditions. The more sophisticated method is to carry out *calculations* of load bearing capacity or critical temperature to compare with conditions expected during the fire.

Selection from these options is different for different materials, because of the differing level of information and different levels of complexity of calculating internal temperatures and calculating structural capacity at elevated temperatures. Design methods for each material are discussed separately.

12.3 SUMMARY FOR MAIN MATERIALS

Table 12.1 lists a hierarchy of design methods for the main materials, depending on which level of fire exposure is selected, or required, by the local building code. It is recommended where possible that a real fire be considered, using a parametric design fire curve.

12.3.1 Structural Steel

For structural steel, the easiest design method is to compare a proprietary listing for protected steelwork with the fire severity prescribed by the code. The next level of accuracy is to use the time-equivalent formula rather than the code-prescribed severity. Most proprietary listings for steel assume that the member is always loaded to its full design capacity, so considerable savings can be made by calculating the residual strength during standard fire exposure and comparing that with the actual design load for fire conditions, especially if the member is over-designed for gravity load.

For most steel members exposed to real fires, the residual strength can be easily calculated if the maximum steel temperature during the fire is known, as described in Chapter 8. For steel members with large thermal gradients or members in a moment-resisting frame, it is necessary to make simplifying assumptions or obtain the use of a specialist fire-design software package.

Calculations of fire resistance of steel members can be made either in the load domain or the temperature domain, with a simple transformation between the two. Calculations in the strength domain, comparing load with load capacity are recommended because they are more familiar to structural engineers.

Table 12.1 Summary of fire design methods for main structural materials

Material	Prescribed fire resistance	Time equivalent formula	Real fire exposure
Structural steel	Compare with proprietary listing	Compare with proprietary listing	
		Calculate residual strength in standard fire	Calculate residual strength in real fire
Reinforced concrete	Compare with generic listing	Compare with generic listing	
		Calculate residual strength in standard fire	Calculate residual strength in real fire
Heavy timber	Calculate strength of residual cross section	Not applicable	Calculate strength of residual cross section
			Use charring rate for parametric fire
Light frame construction	Compare with proprietary listing	Compare with proprietary listing	No simple design methods

12.3.2 Reinforced Concrete

For reinforced concrete, many typical building designs will have sufficient fire resistance to meet prescriptive code requirements or time-equivalent fire severity with no special treatment. The design can be assessed by comparing the required fire-resistance with a listing of generic ratings. If the generic rating does not offer enough fire resistance, the residual strength under exposure to the standard fire can be calculated as described in Chapter 9. In special cases, or for development of new proprietary systems, it will be appropriate to carry out full fire engineering calculations for exposure to a parametric fire.

Considering the structural response, a hierarchy of calculation methods is as follows.

(1) For simply supported slabs or tee-beams exposed to fire from below, concrete in the compression zone remains at normal temperatures, so structural design need only consider the effect of elevated temperatures on the yield strength of the reinforcing steel. Simple hand calculations are possible.

(2) For continuous slabs or beams, some of the fire-exposed surfaces are in compression, so the simple hand calculation methods must consider the effects of elevated temperature on the compression strength of the concrete.

(3) Similar methods can be applied to fire-exposed concrete walls and columns, but these methods are less accurate because of deformations caused by non-uniform heating and the possibility of instability failures.

(4) For moment-resisting frames, or structural members affected by axial restraint or non-uniform heating, it is recommended to use a special-purpose computer program for structural analysis under fire conditions.

12.3.3 Heavy Timber

For heavy timber construction, the residual strength should be calculated using one of the Eurocode methods described in Chapter 10. If the fuel load and geometry are known, the design method for parametric fire exposure should be used. For a prescribed fire resistance, the recommended method is the Eurocode 5 *reduced properties method* with allowance for corner rounding. Time-equivalent formulae have not been validated for heavy timber construction, so they are not recommended.

12.3.4 Light Frame Construction

For light frame construction, the recommended design method consists of selecting a listed proprietary assembly with a fire-resistance rating greater than the fire-resistance prescribed by the code, or an equivalent time of fire exposure calculated for a burnout of the fire compartment. Calculations of thermal and structural behaviour in real fires are possible, but difficult, so they are only recommended for research and development purposes.

12.4 THERMAL ANALYSIS

Most calculation methods require estimation of member temperatures. Table 12.2 summarizes the available tools for calculating internal temperatures in structural assemblies. The recommended design methods are limited by the availability of suitable heat-transfer tools. For example, the simple step-by-step method for calculating temperatures of protected or unprotected steel members can be easily used in all situations where thermal gradients in the member are not important. It becomes important to have access to a finite-element heat-transfer package when calculating temperatures in steel members with non-uniform temperatures and especially for reinforced concrete exposed to real fires, where almost no published temperatures are available and hand calculations may not be accurate.

Table 12.2 Summary of thermal calculation methods for main structural materials

Material	Calculation method	Notes
Structural steel	Charts for standard fire exposure (no thermal gradient in the steel)	Easy to use
	Step-by-step method (no thermal gradient in the steel)	Easy to write a spreadsheet program
	Finite-element program (calculates thermal gradients in the steel)	Requires access to a suitable program
Reinforced concrete	Published temperature contours for standard fire exposure	Widely available, easy to use
	Wickström's formula for standard fire exposure	Easy to use
	Finite-element program for any fire exposure	Requires access to a suitable program
Heavy timber	Thermal analysis not required	
Light frame construction	Finite-element program	Requires access to a suitable program Difficult to estimate thermal properties

Thermal-analysis calculations are not so important for timber construction or light frame structures.

12.5 CONCLUSIONS

Severe fires in large buildings are rare and unpredictable events, but when they occur they can cause great damage and loss of life. Structural design for fire is a small but important part of the overall process of providing fire safety in buildings. Safer buildings can help to reduce the risk of loss of life and property in the event of an unwanted fire.

This book provides simple methods of designing building structures to resist fires, based on an understanding of fire severity, fire resistance, and the behaviour of materials and structures at elevated temperatures.

APPENDIX A
Units and conversion factors

This book uses metric units throughout. These are generally SI (Système International) units. The SI unit for length is the *metre* (m), for time the *second* (s), and for mass the *kilogram* (kg). Weight is expressed using the *Newton* (N) where 1 N is the force that gives a mass of 1 kg an acceleration of 1 m/s^2. On the surface of the earth, 1 kg weighs approximately 9.81 N because the acceleration due to gravity is 9.81 m/s^2.

The SI unit of stress or pressure is the *Pascal* (Pa) which is 1 N/m^2. It is more common to express stress using the megapascal (MPa) which is one 1 MN/m^2 or identically 1 N/mm^2.

The SI unit of heat or energy or work is the *Joule* (J) defined as the work done when the point of application of 1 N is displaced 1 m. Heat or energy is more often expressed in thousands of Joules (kilojoules (kJ)) or millions of Joules (megajoules (MJ)). The basic unit for rate of power or heat release rate is the *Watt* (W); 1 W is 1 J/s , hence 1 kW is 1000 J/s and 1 MW is 1 MJ/s.

The SI prefixes for multiples and submultiples of units are given below:

Factor	Prefix	Symbol
10^{12}	tera	T
10^{9}	giga	G
10^{6}	mega	M
10^{3}	kilo	k
10^{2}	hecto	h
10	deka	da
10^{-1}	deci	d
10^{-2}	centi	c
10^{-3}	milli	m
10^{-6}	micro	μ
10^{-9}	nano	n

Commonly used conversion factors are given in the following table, from Lie (1972). A much more extensive list of units and conversion factors can be found in the SFPE Handbook (SFPE, 1995). Units that are in accordance with Système International d'Unités are marked (SI).

Quantity	Multiply	by	to obtain
Activation energy	J/kg (SI)	4.302×10^{-4}	Btu/lb
Area	m^2 (SI)	10.8	ft^2
	cm^2 (SI)	0.155	in^2
Coefficient of expansion (linear)	m/m K (SI)	0.556	in/in °F
	m/m °C	0.556	in/in °F
Coefficient of expansion (cubic)	m^3/m^3 K (SI)	0.556	in^3/in^3 °F
	m^3/m^3 °C	0.556	in^3/in^3 °F
Coefficient of heat transfer	W/m^2 K (SI)	0.176	Btu/ft^2 h °F
	$kcal/m^2$ h °C	0.205	Btu/ft^2 h °F
	$kcal/m^2$ h °C	1.166	W/m^2 K (SI)
	$kcal/m^2$ h °C	2.78×10^{-5}	cal/cm^2 s °C
	cal/cm^2 s °C	7.364×10^3	Btu/ft^2 h °F
	cal/cm^2 s °C	4.184×10^4	W/m^2 K (SI)
Density	kg/m^3 (SI)	6.24×10^{-2}	lb/ft^3
	g/cm^3 (SI)	62.4	lb/ft^3
	g/cm^3 (SI)	1×10^3	kg/m^3 (SI)
Energy	J (SI)	9.48×10^{-4}	Btu
	kcal	3.966	Btu
	kcal	4.184×10^3	J (SI)
	kcal	1.000	cal
	cal	3.966×10^{-3}	Btu
	cal	4.184	J (SI)
Fire load	kg (SI)	2.205	lb
Fire load density	kg/m^2 (SI)	0.205	lb/ft^2
Flux (heat)	W (SI)	0.948	Btu/s
	kcal/h	3.966	Btu/h
	kcal/h	1.166	W (SI)
	kcal/h	0.278	cal/s
	cal/s	3.966×10^{-3}	Btu/s
	cal/s	14.278	Btu/h
	cal/s	4.184	W (SI)

(Continued)

Quantity	Multiply	by	to obtain
Flow rate	m^3/s (SI)	35.3	ft^3/s
Frequency	Hz (SI)	1	c/s
Force	N (SI) N (SI) kgf	0.225 0.102 2.205	lbf kgf lbf
Heat	see Energy		
Heat of combustion	J/kg (SI) kcal/kg kcal/kg	4.302×10^{-4} 1.8 4.184×10^3	Btu/lb Btu/lb J/kg (SI)
Intensity (heat)	W/m^2 (SI) $cal/cm^2 s$ $cal/cm^2 s$	0.317 1.326×10^4 4.184×10^4	$Btu/ft^2 h$ $Btu/ft^2 h$ W/m^2 (SI)
Latent heat	J/kg (SI) kcal/kg kcal/kg	4.3×10^{-4} 1.8 4.184×10^3	Btu/lb Btu/lb J/kg (SI)
Length	m (SI) cm (SI)	3.281 0.394	ft in
Mass	kg (SI)	2.205	lb
Modulus of elasticity	see Stress		
Opening factor	$m^{1/2}$ (SI)	1.811	$ft^{1/2}$
Power	W (SI) W (SI)	3.41 9.48×10^{-4}	Btu/h Btu/s
Pressure	see Stress		
Proportional limit	see Stress		
Rate of burning	kg/s (SI) kg/h	7.938×10^3 2.205	lb/h lb/h
Rate of heating	see Flux (heat)		

(Continued)

Quantity	Multiply	by	to obtain
Rate of heating per unit area	see Intensity (heat)		
Specific heat	J/kg K (SI)	2.39×10^{-4}	Btu/lb °F
	kcal/kg °C	1	Btu/lb °F
	cal/g °C	4.184×10^{3}	J/kg K (SI)
Specific heat (volumetric)	J/m^3 K (SI)	1.49×10^{-5}	Btu/ft^3 °F
	kcal/m^3 °C	6.234×10^{-2}	Btu/ft^3 °F
Stress	N/m^2 (SI)	2.09×10^{-2}	lbf/ft^2
	N/m^2 (SI)	1.45×10^{-4}	lbf/in^2 (psi)
	MPa	145	lbf/in^2 (psi)
	kgf/m^2	0.205	lbf/ft^2
	kgf/m^2	1.422×10^{-3}	lbf/in^2
	kgf/m^2	9.807	N/m^2 (SI)
	kgf/cm^2	14.22	lbf/in^2
	kgf/cm^2	9.807×10^{-4}	N/m^2 (SI)
Temperature	K (SI)	°C = K − 273.15	°C
	K (SI)	°F = 1.8K − 459.67	°F
	K (SI)	°R = 1.8 K	°R
	°C	°F = 1.8°C + 32	°F
Temperature interval	K (SI)	1.8	°F
	°C	1.8	°F
Thermal capacity	J/K (SI)	5.267×10^{-4}	Btu/°F
	kcal/°C	2.203	Btu/°F
	cal/°C	4.184	J/K (SI)
Thermal conductivity	W/m K (SI)	0.578	Btu/ft h °F
	kcal/m h °C	0.673	Btu/ft h °F
	kcal/m h °C	1.162	W/m K (SI)
	kcal/m h °C	2.78×10^{-3}	cal/cm s °C
	cal/cm s °C	241.8	Btu/ft h °F
	cal/cm s °C	418.4	W/m K (SI)
Thermal diffusivity	m^2/s (SI)	3.875×10^{4}	ft^2/h
	m^2/h	10.765	ft^2/h
	cm^2/s (SI)	0.36	m^2/h
	cm^2/s (SI)	1×10^{-4}	m^2/s (SI)
Velocity	m/s (SI)	3.281	ft/s
	km/h	0.278	m/s (SI)
	cm/min	0.394	in/min

(Continued)

Quantity	Multiply	by	to obtain
Viscosity (dynamic)	N s/m^2 (SI)	2.09×10^{-2}	lbf s/ft^2
	kg/ms	0.672	lb/ft s
Viscosity (kinematic)	m^2/s (SI)	10.8	ft^2/s
Volume	m^3 (SI)	35.32	ft^3
	cm^3 (SI)	6.1×10^{-2}	in^3
Wavelength	m (SI)	10^{10}	Å
	μm (SI)	10^4	Å
Weight	N (SI)	0.225	lbf
	N (SI)	0.102	kgf
	kgf	2.205	lbf

APPENDIX B
Fire Load Energy Densities

This Appendix is extracted from the Design Guide – Structural Fire Safety (CIB, 1986). The table gives average fire load densities using data from Switzerland.

The following values for *fire load densities* (only variable fire load densities) are taken from *Beilage 1:Brandschutztechnische Merkmale verschiedener Nutzungen und Lagerguter* and are defined as density per unit floor area (MJ/m^2).

Note that for the determination of the variable fire load of storage areas, the values given in the following table have to be multiplied by the height of storage in metres. Areas and aisles for transportation have been taken into consideration in an averaging manner.

The values are based on a large investigation carried out during the years 1967–1969 by a staff of 10–20 students under the guidance of the Swiss Fire Prevention Association for Industry and Trade (Brandverhutungsdienst fur Industrie und Gewerbe, Nuschelerstrasse 45, CH-8001 Zurich), with the financial support of the governmental civil defence organisation.

For each type of occupancy, storage and/or building, a minimum of 10–15 samples were analysed; normally, 20 or more samples were available. All values given in the following pages are average values. Unfortunately, it has been impossible to obtain the basic data sheets of this investigation. In order to estimate the corresponding standard deviations and the 80%–90%- and 95%-fractile values, the data from this source were compared with data given in refs 1–5,7–11 of CIB (1986). This comparison results in the following suggestions.

(a) For well-defined occupancies which are rather similar or with very limited differences in furniture and stored goods, e.g. dwellings, hotels, hospitals, offices and schools, the following estimates may suffice:

Coefficient of variation	= 30%–50% of the given average value
90%-fractile value	= (1.35–1.65) × average value
80%-fractile value	= (1.25–1.50) × average value
Isolated peak values	= 2 × average value

(b) For occupancies which are rather dissimilar or with larger differences in furnitures and stored goods, e.g. shopping centres, department stores and industrial occupancies, the following estimates are tentatively suggested:

Coefficient of variation	= 50%–80% of the given average value
90%-fractile value	= (1.65–2.0) × average value
80%-fractile value	= (1.45–1.75) × average value
Isolated peak values	= 2.5 × average value

Type of occupancy	Fabrication (MJ/m²)	Storage (MJ/m²/m)
Academy	300	
Accumulator forwarding	800	
Accumulator mfg	400	800
Acetylene cylinder storage	700	
Acid plant	80	
Adhesive mfg	1000	3400
Administration	800	
Adsorbent plant for combustible vapours	>1700	
Aircraft hangar	200	
Airplane factory	200	
Aluminium mfg	40	
Aluminium processing	200	
Ammunition mfg	Special	
Animal food preparing, mfg	2000	3300
Antique shop	700	
Apparatus forwarding	700	
Apparatus mfg	400	
Apparatus repair	600	
Apparatus testing	200	
Arms mfg	300	
Arms sales	300	
Artificial flower mfg	300	200
Artificial leather mfg	1000	1700
Artificial leather processing	300	
Artificial silk mfg	300	1100
Artificial silk processing	210	
Artificial stone mfg	40	
Asylum	400	
Authority office	800	
Awning mfg	300	1000
Bag mfg (jute, paper, plastic)	500	
Bakery	200	
Bakery, sales	300	
Ball bearing mfg	200	
Bandage mfg	400	
Bank, counters	300	
Bank, offices	800	
Barrel mfg, wood	1000	800
Basement, dwellings	900	
Basketware mfg	300	200
Bed sheeting production	500	1000
Bedding plant	600	
Bedding shop	500	
Beer mfg (brewery)	80	
Beverage mfg, nonalcoholic	80	
Bicycle assembly	200	400
Biscuit factories	200	
Biscuit mfg	200	
Bitumen preparation	800	3400
Blind mfg, venetian	800	300
Blueprinting firm	400	
Boarding school	300	
Boat mfg	600	
Boiler house	200	
Bookbinding	1000	
Bookstore	1000	
Box mfg	1000	600
Brick plant, burning	40	
Brick plant, clay preparation	40	
Brick plant, drying kiln with metal grates	40	
Brick plant, drying kiln with wooden grates	1000	
Brick plant, drying room with metal grates	40	
Brick plant, drying room with wooden grates	400	
Brick plant, pressing	200	
Briquette factories	1600	
Broom mfg	700	400
Brush mfg	700	800
Butter mfg	700	4000
Cabinet making (without woodyard)	600	
Cable mfg	300	600
Cafe	400	
Camera mfg	300	
Candle mfg	1300	22400
Candy mfg	400	1500
Candy packing	800	
Candy shop	400	
Cane products mfg	400	200
Canteen	300	
Car accessory sales	300	
Car assembly plant	300	
Car body repairing	150	
Car paint shop	500	
Car repair shop	300	
Car seat cover shop	700	
Cardboard box mfg	800	2500
Cardboard mfg	300	4200
Cardboard products mfg	800	2500
Carpenter shed	700	
Carpet dyeing	500	
Carpet mfg	600	1700
Carpet store	800	
Cartwright's shop	500	
Cast iron foundry	400	800
Celluloid mfg	800	3400
Cement mfg	1000	
Cement plant	40	
Cement products mfg	80	
Cheese factory	120	
Cheese mfg (in boxes)	170	
Cheese store	100	
Chemical plants (rough average)	300	100
Chemist's shop	1000	
Children's home	400	
China mfg	200	
Chipboard finishing	800	
Chipboard pressing	100	
Chocolate factory, intermediate storage	6000	
Chocolate factory, packing	500	
Chocolate factory, tumbling treatment	1000	
Chocolate factory, all other specialities	500	
Church	200	
Cider mfg (without crate storage)	200	
Cigarette plant	3000	
Cinema	300	
Clay, preparing	50	
Cloakroom, metal wardrobe	80	
Cloakroom, wooden wardrobe	400	
Cloth mfg	400	
Clothing plant	500	
Clothing store	600	
Coal bunker	2500	
Coal cellar	10500	

Type of occupancy	Fabrication (MJ/m^2)	Storage (MJ/m^2/m)
Cocoa processing	800	
Coffee extract mfg	300	
Coffee roasting	400	
Cold storage	2000	
Composing room	400	
Concrete products mfg	100	
Condiment mfg	50	
Congress hall	600	
Contractors	500	
Cooking stove mfg	600	
Coopering	600	
Cordage plant	300	600
Cordage store	500	
Cork products mfg	500	800
Cosmetic mfg	300	500
Cotton mills	1200	
Cotton wool mfg	300	
Cover mfg	500	
Cutlery mfg (household)	200	
Cutting-up shop, leather, artificial leather	300	
Cutting-up shop, textiles	500	
Cutting-up shop, wood	700	
Dairy	200	
Data processing	400	
Decoration studio	1200	2000
Dental surgeon's laboratory	300	
Dentist's office	200	
Department store	400	
Distilling plant, combustible materials	200	
Distilling plant, incombustible materials	50	
Doctor's office	200	
Door mfg, wood	800	1800
Dressing, textiles	200	
Dressing, paper	700	
Dressmaking shop	300	
Dry-cell battery	400	600
Dry cleaning	300	
Dyeing plant	500	
Edible fat forwarding	900	
Edible fat mfg	1000	18900
Electric appliance mfg	400	
Electric appliance repair	500	
Electric motor mfg	300	
Electrical repair shop	600	
Electrical supply storage H < 3 m	1200	
Electro industry	600	
Electronic device mfg	400	
Electronic device repair	500	
Embroidery	300	
Etching plant glass/metal	200	
Exhibition hall, cars including decoration	200	
Exhibition hall, furniture including decoration	500	
Exhibition hall, machines including decoration	80	
Exhibition of paintings including decoration	200	
Explosive industry	4000	

Type of occupancy	Fabrication (MJ/m^2)	Storage (MJ/m^2/m)
Fertiliser mfg	200	200
Filling plant/barrels		
liquid filled and/or barrels incombustible	<200	
liquid filled and/or barrels combustible:		
Risk Class I	>3400	
Risk Class II	>3400	
Risk Class III	>3400	
Risk Class IV	>3400	
Risk Class V (if higher, take into consideration combustibility of barrels)	>1700	
Filling plant/small casks:		
liquid filled and casks incombustible	<200	
liquid filled and/or casks combustible:		
Risk Class I	<500	
Risk Class II	<500	
Risk Class III	<500	
Risk Class IV	<500	
Risk Class V (if higher, take into consideration combustibility of casks)	<500	
Finishing plant, paper	500	
Finishing plant, textile	300	
Fireworks mfg	Spez	2000
Flat	300	
Floor covering mfg	500	6000
Floor covering store	1000	
Flooring plaster mfg	600	
Flour products	800	
Flower sales	80	
Fluorescent tube mfg	300	
Foamed plastics fabrication	3000	2500
Foamed plastics processing	600	800
Food forwarding	1000	
Food store	700	
Forge	80	
Forwarding, appliances partly made of plastic	700	
Forwarding, beverage	300	
Forwarding, cardboard goods	600	
Forwarding, food	1000	
Forwarding, furniture	600	
Forwarding, glassware	700	
Forwarding, plastic products	1000	
Forwarding, printed matters	1700	
Forwarding, textiles	600	
Forwarding, tinware	200	
Forwarding, varnish, polish	1300	
Forwarding, woodware (small)	600	
Foundry (metal)	40	
Fur, sewing	400	
Fur store	200	
Furniture exhibition	500	
Furniture mfg (wood)	600	
Furniture polishing	500	
Furniture store	400	
Furrier	500	

Type of occupancy	Fabrication (MJ/m^2)	Storage (MJ/m^2/m)
Galvanic station	200	
Gambling place	150	
Glass blowing plant	200	
Glass factory	100	
Glass mfg	100	
Glass painting	300	
Glass processing	200	
Glassware mfg	200	
Glassware store	200	
Glazier's workshop	700	
Gold plating (of metals)	800	3400
Goldsmith's workshop	200	
Grainmill, without storage	400	13000
Gravestone carving	50	
Graphic workshop	1000	
Greengrocer's shop	200	
Hairdressing shop	300	
Hardening plant	400	
Hardware mfg	200	
Hardware store	300	
Hat mfg	500	
Hat store	500	
Heating equipment room, wood or coal-firing	300	
Heat sealing of plastics	800	
High-rise office building	800	
Homes	500	
Homes for aged	400	
Hosiery mfg	300	1000
Hospital	300	
Hotel	300	
Household appliances, mfg	300	200
Household appliances, sales	300	
Ice cream plant (including packaging)	100	
Incandescent lamp plant	40	
Injection moulded parts mfg (metal)	80	
Injection moulded parts mfg (plastic)	500	
Institution building	500	
Ironing	500	
Jewellery mfg	200	
Jewellery shop	300	
Joinery	700	
Joiners (machine room)	500	
Joiners (workbench)	700	
Jute, weaving	400	1300
Laboratory, bacteriological	200	
Laboratory, chemical	500	
Laboratory, electric, electronic	200	
Laboratory, metallurgical	200	
Laboratory, physics	200	
Lacquer forwarding	1000	
Lacquer mfg	500	2500
Large metal constructions	80	
Lathe shop	600	
Laundry	200	
Leather goods sales	700	
Leather product mfg	500	

Type of occupancy	Fabrication (MJ/m^2)	Storage (MJ/m^2/m)
Leather, tanning, dressing, etc	400	
Library	2000	2000
Lingerie mfg	400	
Liqueur mfg	400	800
Liquor mfg	500	800
Liquor store	700	
Loading ramp, including goods (rough average)	800	
Lumber room for miscellaneous goods	500	
Machinery mfg	200	
Match plant	300	800
Mattress mfg	500	500
Meat shop	50	
Mechanical workshop	200	
Metal goods mfg	200	
Metal grinding	80	
Metal working (general)	200	
Milk, condensed, evaporated mfg	200	9000
Milk, powdered, mfg	200	10500
Milling work, metal	200	
Mirror mfg	100	
Motion picture studio	300	
Motorcycle assembly	300	
Museum	300	
Musical instrument sales	281	
News stand	1300	
Nitrocellulose mfg	Spez	1100
Nuclear research	2100	
Nursery school	300	
Office, business	800	
Office, engineering	600	
Office furniture	700	
Office, machinery mfg	300	
Office machine sales	300	
Oilcloth mfg	700	1300
Oilcloth processing	700	2100
Optical instrument mfg	200	200
Packing, food	800	
Packing, incombustible goods	400	
Packing material, industry	1600	3000
Packing, printed matters	1700	
Packing, textiles	600	
Packing, all other combustible goods	600	
Paint and varnish, mfg	4200	
Paint and varnish, mixing plant	2000	
Paint and varnish shop	1000	
Painter's workshop	500	
Pain shop (cars, machines, etc)	200	
Paint shop (furniture, etc)	400	
Paper mfg	200	10000
Paper processing	800	1100
Parking building	200	
Parquetry mfg	2000	1200
Perambulator mfg	300	800
Perambulator shop	300	
Perfume sale	400	
Pharmaceutical mfg	300	800

Type of occupancy	Fabrication (MJ/m²)	Storage (MJ/m²/m)
Pharmaceuticals, packing	300	800
Pharmacy (including storage)	800	
Photographic laboratory	100	
Photographic store	300	
Photographic studio	300	
Picture frame mfg	300	
Plaster product mfg	80	
Plastic floor tile mfg	800	
Plastic mfg	2000	5900
Plastic processing	600	
Plastic products fabrication	600	
Plumber's workshop	100	
Plywood mfg	800	2900
Polish mfg	1700	
Post office	400	
Potato, flaked, mfg	200	
Pottery plant	200	
Power station	600	
Precious stone, cutting etc	80	
Precision instrument mfg		
(containing plastic parts)	200	
(without plastic parts)	100	
Precision mechanics plant	200	
Pressing, metal	100	
Pressing, plastics, leather, etc	400	
Preparation briquette production		
Printing, composing room	300	
Printing, ink mfg	700	3000
Printing, machine hall	400	
Printing office	1000	
Radio and TV mfg	400	
Radio and TV sales	500	
Radio studio	300	
Railway car mfg	200	
Railway station	800	
Railway workshop	800	
Record player mfg	300	200
Record repository, documents see also storage	4200	
Refrigerator mfg	1000	300
Relay mfg	400	
Repair shop, general	400	
Restaurant	300	
Retouching department	300	
Rubber goods mfg	600	5000
Rubber goods store	800	
Rubber processing	600	5000
Saddlery mfg	300	
Safe mfg	80	
Salad oil forwarding	900	
Salad oil mfg	1000	18900
Sawmill (without woodyard)	400	
Scale mfg	400	
School	300	
Scrap recovery	800	
Seedstore	600	
Sewing machine mfg	300	
Sewing machine store	300	
Sheet mfg	100	
Shoe factory, forwarding	600	
Shoe factory, mfg	500	
Shoe polish mfg	800	2100

Type of occupancy	Fabrication (MJ/m²)	Storage (MJ/m²/m)
Shoe repair with manufacture	700	
Shoe store	500	
Shutter mfg	1000	
Silk spinning (natural silk)	300	
Silk weaving (natural silk)	300	
Silverwares	400	
Ski mfg	400	1700
Slaughter house	40	
Soap mfg	200	4200
Soda mfg	40	
Soldering	300	
Solvent distillation	200	
Spinning mill, excluding garnetting	300	
Sporting goods store	800	
Spray painting, metal goods	300	
Spray painting, wood products	500	
Stationery store	700	
Steel furniture mfg	300	
Stereotype plate mfg	200	
Stone masonry	40	
Storeroom (workshop storerooms etc)	1200	
Synthetic fibre mfg	400	
Synthetic fibre processing	400	
Synthetic resin mfg	3400	4200
Tar-coated paper mfg	1700	
Tar preparation	800	
Telephone apparatus mfg	400	200
Telephone exchange	80	
Telephone exchange mfg	100	
Test room, electric appliances	200	
Test room, machinery	100	
Test room, textiles	300	
Theatre	300	
Tin can mfg	100	
Tinned goods mfg	40	
Tinware mfg	120	
Tire mfg	700	1800
Tobacco products mfg	200	2100
Tobacco shop	500	
Tool mfg	200	
Toy mfg (combustible)	100	
Toy mfg (incombustible)	200	
Toy store	500	
Tractor mfg	300	
Transformer mfg	300	
Transformer winding	600	
Travel agency	400	
Turnery (wood working)	500	
Turning section	200	
TV studio	300	
Twisting shop	250	
Umbrella mfg	300	400
Umbrellas store	300	
Underground garage, private	>200	
Underground garage, public	<200	
Upholstering plant	500	
Vacation home	500	
Varnishing, appliances	80	
Varnishing, paper	80	

Type of occupancy	Fabrication (MJ/m^2)	Storage (MJ/m^2/m)
Vegetable, dehydrating	1000	400
Vehicle mfg, assembly	400	
Veneering	500	2900
Veneer mfg	800	4200
Vinegar mfg	80	100
Vulcanising plant (without storage)	1000	
Waffle mfg	300	1700
Warping department	250	
Washing agent mfg	300	200
Washing machine mfg	300	40
Watch assembling	300	40
Watch mechanism mfg	40	
Watch repair shop	300	
Watch sales	300	
Water closets	~0	
Wax products forwarding	2100	
Wax products mfg	1300	2100
Weaving mill (without carpets)	300	
Welding shop (metal)	80	
Winding room	400	
Winding, textile fibres	600	
Window glass mfg	700	
Window mfg (wood)	800	
Wine cellar	20	
Wine merchant's shop	200	
Wire drawing	80	
Wire factory	800	
Wood carving	700	
Wood drying plant	800	
Wood grinding	200	
Wood pattern making shop	600	
Wood preserving plant	3000	
Youth hostel	300	

APPENDIX C
Section factors for steel beams

This Appendix provides section factors for standard hot-rolled I-beams. The sections have been selected from published data for:

North American Wide Flange Beams
Australian Universal Beams
UK Universal Beams
Japanese H Sections
IPE Narrow Flange Beams

These tables give the dimensions and weight of each beam, but not the structural section properties which must be obtained from standard section property tables. The section factors have been calculated assuming that all sections are made from rectangular components, with no allowance for tapered flanges and root radii, and assuming that the protective insulation is in contact with the steel.

The tables do not include column sections, box sections, angles and channels. Section factors for these, and other sizes and shapes, can be easily calculated or can be obtained from manufacturers' literature.

The numbers in these tables have been obtained from The Heavy Engineering Research Association of New Zealand (HERA). Neither HERA nor the author guarantee the accuracy of the tabulated data which can be calculated from standard section property tables. The geometrical data were obtained from the following sources. The North American Wide Flange Beam data are from the structural sections catalogue 'British Steel SPCS 4237/99' from British Steel. The Australian Universal Beam data are from the 'BHP Hot-Rolled and Structural Steel Products' catalogue from BHP Steel, 1998. The UK Universal Beam data are from the 'Structural Sections Catalogue' from British Steel, 1996. The values for the Japanese H Sections are in from the structural sections catalogue 'Wide Flange Shapes Cat. No. EXE 210, Dec.1980' from the Nippon Steel Corporation.

The basic geometry for a hot-rolled I-beam is shown in Figure C.1

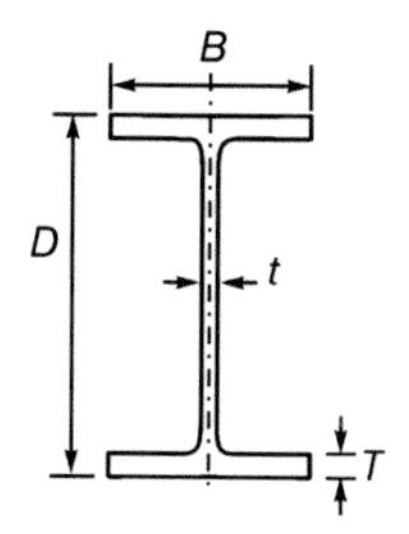

Figure C.1 Geometry of hot-rolled section

North American Wide Flange Beams

Size	Metric size	Section depth	Section width	Thickness Flange	Thickness Web	Section factor Contour 3 Sides		Contour 4 Sides		Hollow 3 Sides		Hollow 4 Sides	
		D	B	T	t	F/V	V/F	F/V	V/F	F/V	V/F	F/V	V/F
in × lb/ft	mm × kg/m	mm	mm	mm	mm	m^{-1}	mm	m^{-1}	mm	m^{-1}	mm	m^{-1}	mm
W24 × 84	W610 × 125	612	229	19.6	11.9	118	8.4	133	7.5	91	11.0	106	9.5
W24 × 76	W610 × 113	608	228	17.3	11.2	130	7.7	146	6.9	100	10.0	116	8.6
W24 × 68	W610 × 101	603	228	14.9	10.5	144	6.9	162	6.2	111	9.0	128	7.8
W24 × 62	W610 × 92	603	179	15.0	10.9	146	6.8	162	6.2	118	8.5	133	7.5
W24 × 55	W610 × 82	599	178	12.8	10.0	164	6.1	181	5.5	132	7.6	149	6.7
W20 × 73	W530 × 109	540	211	18.8	11.6	122	8.2	137	7.3	93	10.8	108	9.3
W20 × 68	W530 × 101	537	210	17.4	10.9	130	7.7	146	6.8	99	10.1	116	8.7
W20 × 62	W530 × 92	533	209	15.6	10.2	142	7.0	160	6.3	108	9.2	126	7.9
W20 × 57	W530 × 85	535	167	16.5	10.3	143	7.0	159	6.3	114	8.7	130	7.7
W20 × 50	W530 × 74	529	166	13.6	9.7	161	6.2	179	5.6	129	7.8	146	6.8
W20 × 44	W530 × 66	525	165	11.4	8.9	182	5.5	202	4.9	145	6.9	165	6.1
W18 × 65	W460 × 97	466	193	19.0	11.4	121	8.3	137	7.3	92	10.9	107	9.3
W18 × 60	W460 × 89	463	192	17.7	10.5	130	7.7	147	6.8	98	10.2	115	8.7
W18 × 55	W460 × 82	460	191	16.0	9.9	141	7.1	159	6.3	106	9.4	125	8.0
W18 × 50	W460 × 74	457	190	14.5	9.0	155	6.4	175	5.7	117	8.6	137	7.3
W18 × 46	W460 × 68	459	154	15.4	9.1	156	6.4	174	5.8	123	8.1	141	7.1
W18 × 40	W460 × 60	455	153	13.3	8.0	178	5.6	198	5.0	140	7.1	160	6.2
W18 × 35	W460 × 52	450	152	10.8	7.6	202	4.9	225	4.4	159	6.3	182	5.5
W16 × 50	W410 × 75	413	180	16.0	9.7	141	7.1	160	6.2	106	9.5	124	8.0
W16 × 45	W410 × 67	410	179	14.4	8.8	156	6.4	177	5.7	116	8.6	137	7.3
W16 × 40	W410 × 60	407	178	12.8	7.7	176	5.7	199	5.0	131	7.6	154	6.5
W16 × 36	W410 × 53	403	177	10.9	7.5	194	5.1	220	4.5	144	6.9	170	5.9
W16 × 31	W410 × 46	403	140	11.1	7.0	207	4.8	231	4.3	161	6.2	185	5.4
W16 × 26	W410 × 39	399	140	8.8	6.4	241	4.1	269	3.7	188	5.3	216	4.6
W14 × 38	W360 × 57	358	172	13.1	7.9	169	5.9	192	5.2	123	8.1	147	6.8
W14 × 34	W360 × 51	355	171	11.6	7.2	188	5.3	214	4.7	137	7.3	163	6.1
W14 × 30	W360 × 45	352	171	9.8	6.9	210	4.8	240	4.2	153	6.6	182	5.5
W14 × 26	W360 × 39	353	128	10.7	6.5	216	4.6	242	4.1	168	6.0	193	5.2
W14 × 22	W360 × 33	349	127	8.5	5.8	256	3.9	286	3.5	198	5.1	228	4.4
W12 × 22	W310 × 33	313	102	10.8	6.6	220	4.5	244	4.1	174	5.7	199	5.0
W12 × 19	W310 × 28	309	102	8.9	6.0	252	4.0	280	3.6	199	5.0	227	4.4
W12 × 16	W310 × 24	305	101	6.7	5.6	297	3.4	330	3.0	234	4.3	267	3.7
W10 × 30	W250 × 45	266	148	13.0	7.6	168	6.0	194	5.2	119	8.4	145	6.9
W10 × 26	W250 × 39	262	147	11.2	6.6	193	5.2	223	4.5	136	7.3	166	6.0
W10 × 22	W250 × 33	258	146	9.1	6.1	226	4.4	261	3.8	159	6.3	194	5.2
W10 × 19	W250 × 28	260	102	10.0	6.4	224	4.5	252	4.0	171	5.8	200	5.0
W10 × 17	W250 × 25	257	102	8.4	6.1	250	4.0	281	3.6	191	5.2	222	4.5
W10 × 15	W250 × 22	254	102	6.9	5.8	282	3.5	318	3.1	214	4.7	250	4.0
W8 × 21	W200 × 31	210	134	10.2	6.4	202	4.9	236	4.2	139	7.2	172	5.8
W8 × 18	W200 × 27	207	133	8.4	5.8	237	4.2	276	3.6	161	6.2	201	5.0

Australian Universal Beams						Section factor							
						Contour				Hollow			
						3 Sides		4 Sides		3 Sides		4 Sides	
Size	Mass	Section depth	Section width	Thickness Flange	Thickness Web								
		D	B	T	t	F/V	V/F	F/V	V/F	F/V	V/F	F/V	V/F
	kg/m	mm	mm	mm	mm	m^{-1}	mm	m^{-1}	mm	m^{-1}	mm	m^{-1}	mm
610 UB	125	612	229	19.6	11.9	118	8.5	132	7.6	91	11.0	105	9.5
610 UB	113	607	228	17.3	11.2	129	7.7	145	6.9	99	10.1	115	8.7
610 UB	101	602	228	14.8	10.6	144	7.0	161	6.2	110	9.1	128	7.8
530 UB	92.4	533	209	15.6	10.2	142	7.1	159	6.3	108	9.3	126	8.0
530 UB	82.0	528	209	13.2	9.6	158	6.3	178	5.6	120	8.3	140	7.1
460 UB	82.1	460	191	16.0	9.9	140	7.1	158	6.3	106	9.5	124	8.1
460 UB	74.6	457	190	14.5	9.1	154	6.5	174	5.7	116	8.6	136	7.4
460 UB	67.1	454	190	12.7	8.5	170	5.9	192	5.2	128	7.8	150	6.7
410 UB	59.7	406	178	12.8	7.8	174	5.7	197	5.1	130	7.7	153	6.5
410 UB	53.7	403	178	10.9	7.6	192	5.2	218	4.6	143	7.0	169	5.9
360 UB	56.7	359	172	13.0	8.0	168	5.9	192	5.2	123	8.1	147	6.8
360 UB	50.7	356	171	11.5	7.3	187	5.3	214	4.7	136	7.3	163	6.1
360 UB	44.7	352	171	9.7	6.9	210	4.8	240	4.2	153	6.5	183	5.5
310 UB	46.2	307	166	11.8	6.7	185	5.4	213	4.7	132	7.6	160	6.3
310 UB	40.4	304	165	10.2	6.1	209	4.8	241	4.1	148	6.7	180	5.6
310 UB	32.0	298	149	8.0	5.5	253	4.0	289	3.5	183	5.5	219	4.6
250 UB	37.3	256	146	10.9	6.4	197	5.1	228	4.4	139	7.2	169	5.9
250 UB	31.4	252	146	8.6	6.1	232	4.3	268	3.7	162	6.2	199	5.0
250 UB	25.7	248	124	8.0	5.0	262	3.8	300	3.3	190	5.3	228	4.4
200 UB	29.8	207	134	9.6	6.3	210	4.8	245	4.1	143	7.0	179	5.6
200 UB	25.4	203	133	7.8	5.8	246	4.1	287	3.5	167	6.0	208	4.8
200 UB	22.3	202	133	7.0	5.0	276	3.6	323	3.1	187	5.3	233	4.3
200 UB	18.2	198	99	7.0	4.5	295	3.4	338	3.0	213	4.7	256	3.9
180 UB	22.2	179	90	10.0	6.0	218	4.6	250	4.0	159	6.3	191	5.2
180 UB	18.1	175	90	8.0	5.0	265	3.8	304	3.3	191	5.2	230	4.3
180 UB	16.1	173	90	7.0	4.5	298	3.4	342	2.9	214	4.7	258	3.9
150 UB	18.0	155	75	9.5	6.0	227	4.4	260	3.8	167	6.0	200	5.0
150 UB	14.0	150	75	7.0	5.0	289	3.5	331	3.0	211	4.7	253	4.0

UK Universal Beams						Section factor							
						Contour				Hollow			
						3 Sides		4 Sides		3 Sides		4 Sides	
Size	Mass	Section depth	Section width	Thickness Flange	Thickness Web								
		D	B	T	t	F/V	V/F	F/V	V/F	F/V	V/F	F/V	V/F
	kg/m	mm	mm	mm	mm	m^{-1}	mm	m^{-1}	mm	m^{-1}	mm	m^{-1}	mm
762 × 267 UB	197	770	268	25.4	15.6	92	10.9	103	9.7	72	13.9	83	12.1
762 × 267 UB	173	762	267	21.6	14.3	104	9.6	116	8.6	81	12.3	94	10.7
762 × 267 UB	147	754	265	17.5	12.8	122	8.2	136	7.4	95	10.5	109	9.2
686 × 254 UB	170	693	256	23.7	14.5	98	10.2	110	9.1	76	13.2	87	11.4
686 × 254 UB	152	688	255	21.0	13.2	109	9.2	122	8.2	84	11.9	97	10.3
686 × 254 UB	140	684	254	19.0	12.4	118	8.5	132	7.6	91	11.0	105	9.5
686 × 254 UB	125	678	253	16.2	11.7	132	7.6	147	6.8	101	9.9	117	8.5
610 × 305 UB	238	636	311	31.4	18.4	72	14.0	82	12.2	52	19.1	63	16.0
610 × 305 UB	179	620	307	23.6	14.1	94	10.7	107	9.3	68	14.7	81	12.3
610 × 305 UB	149	612	305	19.7	11.8	111	9.0	127	7.8	81	12.4	97	10.4
533 × 210 UB	122	545	212	15.6	10.1	146	6.9	164	6.1	111	9.0	129	7.7
533 × 210 UB	109	540	211	13.2	9.6	161	6.2	181	5.5	123	8.1	143	7.0
533 × 210 UB	101	537	210	21.3	12.7	108	9.2	122	8.2	83	12.1	96	10.4
533 × 210 UB	92	533	209	18.8	11.6	120	8.3	135	7.4	92	10.9	107	9.4
533 × 210 UB	82	528	209	17.4	10.8	129	7.8	145	6.9	98	10.2	114	8.8
457 × 191 UB	98	467	193	19.6	11.4	119	8.4	135	7.4	90	11.1	106	9.5
457 × 191 UB	89	463	192	17.7	10.5	130	7.7	147	6.8	98	10.2	115	8.7
457 × 191 UB	82	460	191	16.0	9.9	142	7.1	160	6.2	107	9.4	125	8.0
457 × 191 UB	74	457	190	14.5	9.0	155	6.4	175	5.7	117	8.6	137	7.3
457 × 191 UB	67	453	190	12.7	8.5	171	5.9	193	5.2	128	7.8	150	6.6
406 × 178 UB	74	413	180	16.0	9.5	142	7.0	161	6.2	106	9.4	125	8.0
406 × 178 UB	67	409	179	14.3	8.8	156	6.4	177	5.6	117	8.6	138	7.3
406 × 178 UB	60	406	178	12.8	7.9	174	5.7	197	5.1	130	7.7	153	6.5
406 × 178 UB	54	403	178	10.9	7.7	192	5.2	217	4.6	142	7.0	168	5.9
356 × 171 UB	67	363	173	15.7	9.1	144	7.0	164	6.1	105	9.5	126	8.0
356 × 171 UB	57	358	172	13.0	8.1	168	6.0	191	5.2	122	8.2	146	6.8
356 × 171 UB	51	355	172	11.5	7.4	186	5.4	213	4.7	136	7.4	162	6.2
356 × 171 UB	45	351	171	9.7	7.0	210	4.8	240	4.2	153	6.6	182	5.5
305 × 165 UB	54	310	167	13.7	7.9	161	6.2	185	5.4	114	8.7	139	7.2
305 × 165 UB	46	307	166	11.8	6.7	187	5.4	215	4.6	133	7.5	161	6.2
305 × 165 UB	40	303	165	10.2	6.0	212	4.7	245	4.1	150	6.6	183	5.5
305 × 102 UB	33	313	102	10.8	6.6	220	4.5	244	4.1	174	5.7	199	5.0
305 × 102 UB	28	309	102	8.8	6.0	254	3.9	282	3.5	200	5.0	229	4.4
305 × 102 UB	25	305	102	7.0	5.8	286	3.5	318	3.1	225	4.4	257	3.9
254 × 146 UB	43	260	147	12.7	7.2	173	5.8	200	5.0	122	8.2	149	6.7
254 × 146 UB	37	256	146	10.9	6.3	199	5.0	230	4.4	139	7.2	171	5.9
254 × 146 UB	31	251	146	8.6	6.0	234	4.3	271	3.7	163	6.1	200	5.0
254 × 102 UB	28	260	102	10.0	6.3	226	4.4	254	3.9	173	5.8	201	5.0
254 × 102 UB	25	257	102	8.4	6.0	253	4.0	284	3.5	193	5.2	224	4.5
254 × 102 UB	22	254	102	6.8	5.7	286	3.5	323	3.1	218	4.6	254	3.9
203 × 133 UB	30	207	134	9.6	6.4	210	4.8	245	4.1	143	7.0	178	5.6
203 × 133 UB	25	203	133	7.8	5.7	248	4.0	290	3.4	169	5.9	210	4.8

Japanese H Sections						Section factor							
						Contour				Hollow			
						3 Sides		4 Sides		3 Sides		4 Sides	
Size	Mass	Section depth	Section width	Thickness Flange	Thickness Web								
		D	B	T	t	F/V	V/F	F/V	V/F	F/V	V/F	F/V	V/F
	kg/m	mm	mm	mm	mm	m^{-1}	mm	m^{-1}	mm	m^{-1}	mm	m^{-1}	mm
800 × 300 H	241	808	302	30.0	16.0	81	12.4	91	11.0	62	16.0	72	13.9
800 × 300 H	210	800	300	26.0	14.0	92	10.8	104	9.6	71	14.1	82	12.2
800 × 300 H	191	792	300	22.0	14.0	101	9.9	113	8.8	77	12.9	90	11.1
700 × 300 H	215	708	302	28.0	15.0	84	11.9	95	10.5	63	15.9	74	13.5
700 × 300 H	185	700	300	24.0	13.0	97	10.4	109	9.1	72	13.9	85	11.8
700 × 300 H	166	692	300	20.0	13.0	107	9.4	121	8.3	80	12.6	94	10.7
600 × 300 H	175	594	302	23.0	14.0	93	10.8	106	9.4	67	14.9	81	12.4
600 × 300 H	151	588	300	20.0	12.0	107	9.4	122	8.2	77	13.0	92	10.8
600 × 300 H	137	582	300	17.0	12.0	117	8.6	134	7.5	84	11.9	101	9.9
600 × 200 H	134	612	202	23.0	13.0	106	9.5	118	8.5	84	12.0	95	10.5
600 × 200 H	120	606	201	20.0	12.0	117	8.5	131	7.7	93	10.8	106	9.4
600 × 200 H	106	600	200	17.0	11.0	132	7.6	147	6.8	104	9.6	119	8.4
600 × 200 H	95	596	199	15.0	10.0	147	6.8	163	6.1	115	8.7	132	7.6
500 × 300 H	128	488	300	18.0	11.0	113	8.8	132	7.6	78	12.8	96	10.4
500 × 300 H	114	482	300	15.0	11.0	127	7.9	147	6.8	87	11.5	107	9.3
500 × 200 H	103	506	201	19.0	11.0	121	8.2	137	7.3	92	10.8	108	9.3
500 × 200 H	90	500	200	16.0	10.0	138	7.2	156	6.4	105	9.5	123	8.2
500 × 200 H	80	496	199	14.0	9.0	155	6.4	175	5.7	118	8.5	137	7.3
450 × 300 H	124	440	300	18.0	11.0	112	9.0	131	7.6	75	13.3	94	10.6
450 × 300 H	106	434	299	15.0	10.0	129	7.7	151	6.6	86	11.6	109	9.2
450 × 200 H	76	450	200	14.0	9.0	153	6.5	174	5.8	114	8.8	134	7.4
450 × 200 H	66	446	199	12.0	8.0	175	5.7	198	5.0	129	7.7	153	6.5
400 × 300 H	107	390	300	16.0	10.0	122	8.2	144	6.9	79	12.6	101	9.9
400 × 300 H	94	386	299	14.0	9.0	137	7.3	162	6.2	89	11.2	114	8.8
400 × 200 H	66	400	200	13.0	8.0	165	6.1	188	5.3	119	8.4	143	7.0
400 × 200 H	57	396	199	11.0	7.0	191	5.2	218	4.6	137	7.3	165	6.1
350 × 250 H	80	340	250	14.0	9.0	139	7.2	164	6.1	92	10.9	116	8.6
350 × 250 H	69	336	249	12.0	8.0	159	6.3	187	5.3	104	9.6	133	7.5
350 × 175 H	50	350	175	11.0	7.0	192	5.2	220	4.6	139	7.2	166	6.0
350 × 175 H	41	346	174	9.0	6.0	228	4.4	261	3.8	164	6.1	197	5.1
300 × 200 H	65	298	201	14.0	9.0	143	7.0	168	6.0	97	10.3	121	8.3
300 × 200 H	57	294	200	12.0	8.0	162	6.2	190	5.3	109	9.2	137	7.3
300 × 150 H	37	300	150	9.0	6.5	222	4.5	254	3.9	160	6.2	192	5.2
300 × 150 H	32	298	149	8.0	5.5	253	4.0	289	3.5	183	5.5	219	4.6
250 × 175 H	44	244	175	11.0	7.0	178	5.6	209	4.8	118	8.5	149	6.7
250 × 125 H	30	250	125	9.0	6.0	229	4.4	262	3.8	166	6.0	199	5.0
250 × 125 H	26	248	124	8.0	5.0	263	3.8	300	3.3	190	5.3	228	4.4
200 × 150 H	31	194	150	9.0	6.0	212	4.7	250	4.0	138	7.3	176	5.7
200 × 100 H	21	200	100	8.0	5.5	254	3.9	291	3.4	184	5.4	221	4.5
200 × 100 H	18	198	99	7.0	4.5	295	3.4	338	3.0	214	4.7	256	3.9

IPE Narrow Flange Beams

Size	Mass	Section depth	Section width	Thickness: Flange	Thickness: Web	Section factor: Contour 3 Sides		Section factor: Contour 4 Sides		Section factor: Hollow 3 Sides		Section factor: Hollow 4 Sides	
		D	B	T	t	F/V	V/F	F/V	V/F	F/V	V/F	F/V	V/F
	kg/m	mm	mm	mm	mm	m^{-1}	mm	m^{-1}	mm	m^{-1}	mm	m^{-1}	mm
600R IPE	144	608	218	23.0	14.0	100	10.0	112	8.9	78	12.8	90	11.1
600O IPE	155	610	224	24.0	15.0	95	10.6	106	9.4	73	13.6	85	11.8
600A IPE	108	597	220	17.5	9.8	134	7.5	150	6.7	103	9.7	119	8.4
600 IPE	122	600	220	19.0	12.0	118	8.5	132	7.6	91	11.0	105	9.5
500R IPE	111	508	198	20.0	12.6	112	9.0	126	8.0	85	11.7	99	10.1
500O IPE	107	506	202	19.0	12.0	116	8.6	131	7.6	89	11.3	103	9.7
500A IPE	79.4	497	200	14.5	8.4	156	6.4	176	5.7	118	8.5	138	7.2
500 IPE	90.7	500	200	16.0	10.2	136	7.3	153	6.5	103	9.7	121	8.3
400R IPE	81.5	407	178	17.0	10.6	128	7.8	145	6.9	95	10.5	113	8.9
400O IPE	75.7	404	182	15.5	9.7	138	7.2	157	6.4	103	9.7	122	8.2
400A IPE	57.4	397	180	12.0	7.0	181	5.5	205	4.9	133	7.5	158	6.3
400 IPE	66.3	400	180	13.5	8.6	157	6.4	178	5.6	116	8.6	137	7.3
360R IPE	70.3	366	168	16.0	9.9	136	7.4	154	6.5	100	10.0	119	8.4
360O IPE	66.0	364	172	14.7	9.2	146	6.9	166	6.0	107	9.3	127	7.8
360A IPE	50.2	357	170	11.5	6.6	189	5.3	216	4.6	138	7.2	165	6.1
360 IPE	57.1	360	170	12.7	8.0	167	6.0	190	5.3	122	8.2	146	6.9
300R IPE	51.7	306	147	13.7	8.5	157	6.4	180	5.6	115	8.7	137	7.3
300O IPE	49.3	304	152	12.7	8.0	167	6.0	191	5.2	121	8.3	145	6.9
300A IPE	36.5	297	150	9.2	6.1	222	4.5	254	3.9	160	6.3	192	5.2
300 IPE	42.2	300	150	10.7	7.1	193	5.2	220	4.5	139	7.2	167	6.0
270R IPE	44.0	276	133	13.1	7.7	167	6.0	191	5.2	122	8.2	146	6.8
270O IPE	42.3	274	136	12.2	7.5	175	5.7	200	5.0	127	7.9	152	6.6
270A IPE	30.7	267	135	8.7	5.5	237	4.2	272	3.7	171	5.8	206	4.9
270 IPE	36.1	270	135	10.2	6.6	203	4.9	232	4.3	147	6.8	176	5.7
240R IPE	37.3	245	118	12.3	7.5	175	5.7	199	5.0	128	7.8	153	6.5
240O IPE	34.3	242	122	10.8	7.0	191	5.2	219	4.6	139	7.2	167	6.0
240A IPE	26.2	237	120	8.3	5.2	247	4.0	283	3.5	178	5.6	214	4.7
240 IPE	30.7	240	120	9.8	6.2	212	4.7	242	4.1	153	6.5	184	5.4
220R IPE	31.6	225	108	11.8	6.7	189	5.3	216	4.6	139	7.2	166	6.0
220O IPE	29.4	222	112	10.2	6.6	205	4.9	235	4.3	149	6.7	179	5.6
220A IPE	22.2	217	110	7.7	5.0	266	3.8	305	3.3	192	5.2	231	4.3
220 IPE	26.2	220	110	9.2	5.9	227	4.4	260	3.8	165	6.1	198	5.1

APPENDIX D
Generic fire-resistance ratings for reinforced concrete

Generic fire-resistance ratings for reinforced concrete beams, slabs and columns are given in the following pages. These tables show minimum dimensions and minimum cover for fire-resistance ratings from 0.5 hour up to 4 hours. Fire exposure is the standard fire curve (ISO 834, ASTM E119 or similar).

This information is a summary of tabular data, extracted from a large number of international documents. The tables are extracted from Wade (1991(a)). The original sources are listed below.

MP9: 1989. Fire Properties of Building Materials and Elements of Structure. Miscellaneous Publication No. 9. Standards New Zealand, Wellington.

CP 110: 1972. Code of Practice for the Structural Use of Concrete. CP 110, Part 1. British Standards Institution, UK.

BS 8110: 1985. Structural Use of Concrete. BS 8110. British Standards Institution, UK.

BRE: 1988. Guidelines for the Construction of Fire-Resisting Elements. BRE Report. Fire Research Station, Building Research Establishment, Watford, UK.

Forrest, J.C.M and Law, M., 1984. Guidance for the Application of Tabular Data for Fire Resistance of Concrete Elements. The Institution of Structural Engineers, London.

FIP/CEB: 1975. Guides to Good Practice. FIP/CEB Recommendations for the Design of Reinforced and Prestressed Concrete Structural Members for Fire Resistance. FIP/1/1. Federation Internationale de la Precontrainte. Wexham Springs, Slough, UK.

CEB: 1987. Model Code for Fire Design of Concrete Structures. Bulletin d'Information No 174. Comite Euro-International du Beton, Lausanne, Switzerland.

NBCC Supplement: 1985. Supplement to the National Building Code of Canada. National Research Council of Canada, Ottawa.

UBC: 1988. Uniform Building Code. International Conference of Building Officials, Whittier, California, USA.

AS 3600: 1988. Concrete Structures. AS 3600-1988. Standards Association of Australia.

Table D.1 Reinforced concrete beams

Fire-resistance rating:	0.5 hour		1.0 hour		1.5 hours		2.0 hours		3.0 hours		4.0 hours		
	Width (mm)	Cover (mm)	Width (mm)	Cover (mm)	Width (mm)	Cover (mm)	Width (mm)	Cover (mm)	Width (mm)	Cover (mm)	Width (mm)	Cover (mm)	Notes
Document, standard or code													
MP9:1989 (NZ)	–	–	100	25	140	35	180	45	240	55	280	65	Siliceous aggregate
CP110:1972 (UK)	80	15	110	25	140	35	180	45	240	55	280	65	Siliceous aggregate
BS8110:1985 (UK)	80	20	120	30	150	40	200	50	240	70	280	80	Dense aggregate
BRE:1998 (UK)	80	20	120	30	150	40	200	50	240	70	280	80	Dense aggregate, effective cover
FORREST AND LAW:1984 (UK)	200	15	200	20	200	30	200	50	250	70	300	80	Dense aggregate, minimum width = 200 mm
FIP/CEB:1975 (EUR)	80 120 160 200	20 10 10 10	120 160 200 300	35 30 25 20	150 200 280 400	50 40 35 30	200 240 300 500	60 50 45 40	240 300 400 600	75 65 60 55	280 350 500 700	85 75 70 65	Dense aggregate, critical temperature = 500°C
CEB:1987 (EUR)	80	25	120 160 200	40 35 30	150 200 280 400	55 45 40 35	200 240 300 500	65 55 50 45	240 300 400 600	80 70 65 60	280 350 500 700	90 80 75 70	Cover is to steel axis
NBC SUPPLEMENT: 1985 (CAN)	100	20	100	20	100	25	100	25	140	39	165	50	Types S, N, L, concrete, Table 2.9A
UBC:1988 (USA)	–	–	305	38	–	–	305	38	305	38	305	50	Siliceous aggregate not monolithic with slab, Table 43-A
AS 3600:1988 (AUS)	80 700	20 15	120 160 230	30 25 20	150 200 300 700	45 35 30 25	200 240 375 700	55 45 40 30	240 300 700	70 60 45	280 350 700	80 70 55	

All data relate to three-sided exposure. Beams are simply supported.

Table D.2 Prestressed concrete beams

Fire-resistance rating:	0.5 hour		1.0 hour		1.5 hours		2.0 hours		3.0 hours		4.0 hours		
	Width (mm)	Cover (mm)	Width (mm)	Cover (mm)	Width (mm)	Cover (mm)	Width (mm)	Cover (mm)	Width (mm)	Cover (mm)	Width (mm)	Cover (mm)	Notes
Document, standard or code													
MP9:1989 (NZ)	–	25	–	32	–	50	–	65	–	80	–	100	Unprotected beam
CP110:1972 (UK)	80	25	110	40	140	50	180	65	240	85	280	100	Siliceous aggregate
BS8110:1085 (UK)	100	25	120	40	150	55	200	70	240	80	280	90	Dense aggregate
BRE:1998 (UK)	100	25	120	40	150	55	200	70	240	80	280	90	Dense aggregate, effective cover
FORREST AND LAW:1984 (UK)	200	20	200	30	200	45	200	70	250	80	300	90	Dense aggregate minimum width = 200
FIP/CEB:1975 (EUR)	80	30	120	45	150	60	200	70	240	85	280	95	Dense aggregate critical temperature = 400°C
	120	20	160	40	200	50	240	60	300	75	350	85	
	160	20	200	35	280	45	300	55	400	70	500	80	
	200	10	300	30	400	40	500	50	600	65	700	75	
CEB:1987 (EUR)	80	25	120	40	150	55	200	65	240	80	280	90	Cover is to steel axis
			160	35	200	45	240	55	300	70	350	80	
			200	30	280	40	300	50	400	65	500	75	
					400	35	300	45	600	60	700	70	
NBC SUPPLEMENT: 1985 (CAN)	–	25	–	50	–	64	–	–	–	–	–	–	260–970 cm^2
	–	25	–	39	–	45	–	64	–	–	–	–	970–1940 cm^2
	–	25	–	39	–	39	–	50	–	77	–	102	$>1940\ cm^2$
													Table 2.10A, type S,N
UBC:1988 (USA)	–	–	203	44	–	–	203	64	203	114	–	–	Siliceous aggregate
			305	38	–	–	305	50	305	64	305	76	
AS 3600:1988 (AUS)	80	25	120	35	150	55	200	65	240	80	280	90	
	700	20	160	30	200	45	240	55	300	70	350	80	
			230	25	300	40	375	50	700	55	700	65	
					700	35	700	40					

Al data relate to three-sided exposure. Beams are simply supported.

Table D.3 Reinforced concrete columns

Fire-resistance rating:	0.5 hour		1.0 hour		1.5 hours		2.0 hours		3.0 hours		4.0 hours		
Document, standard or code	Width (mm)	Cover (mm)	Width (mm)	Cover (mm)	Width (mm)	Cover (mm)	Width (mm)	Cover (mm)	Width (mm)	Cover (mm)	Width (mm)	Cover (mm)	Notes
MP9:1989 (NZ)	–	–	150	20	200	30	250	38	300	50	400	50	Siliceous aggregate
CP110:1972 (UK)	150	–	200	–	250	–	300	–	400	–	450	–	Siliceous aggregate
BS8110:1085 (UK)	150	20	200	25	250	30	300	35	400	35	450	35	Dense aggregate
BRE:1998 (UK)	150	20	200	25	250	30	300	35	400	35	450	35	Dense aggregate
FORREST AND LAW:1984 (UK)	150	20	200	25	250	30	300	35	400	35	450	35	Dense aggregate
FIP/CEB:1975 (EUR)	150	10	200	20	240	30	300	35	400	35	450	35	Dense aggregate, Table la
CEB:1987 (EUR)	150	25	200	35	240	50	300	50	400	50	450	55	Cover is to steel axis, 3 m long, 20 MPa concrete
NBC SUPPLEMENT: 1985(CAN)	150	13	200	25	250	38	300	50	400	63	500	75	Overdesign factor = 1 types S, N, $kh<3.7$
UBC:1988 (USA)	–	–	305	38	–	–	305	38	305	38	305	50	Siliceous aggregate, Table 43-A
AS 3600:1988 (AUS)	150	10	200 240	20 15	240 300	35 25	300 400	45 35	400 500	60 50	450 600	70 60	

All data relate to a fully exposed (4 sides) column

Table D.4 Reinforced concrete slabs

Fire-resistance rating:	0.5 hour		1.0 hour		1.5 hours		2.0 hours		3.0 hours		4.0 hours		
	Width (mm)	Cover (mm)	Width (mm)	Cover (mm)	Width (mm)	Cover (mm)	Width (mm)	Cover (mm)	Width (mm)	Cover (mm)	Width (mm)	Cover (mm)	Notes
Document, standard or code													
MP9:1989 (NZ)	60	15	80	20	100	20	120	20	150	25	175	25	Simply supported
CP110:1972 (UK)	100	15	100	15	125	20	125	20	150	25	150	25	Siliceous or calcareous solid slabs, average cover
BS8110:1085 (UK)	75	15	95	20	110	25	125	35	150	45	170	55	Dense aggregate, plain soffit, simply supported
BRE:1998 (UK)	75	15	95	20	110	25	125	35	150	45	170	55	Dense aggregate, plain soffit, simply supported, effective cover
FORREST AND LAW:1984 (UK)	75	15	95	20	110	25	125	35	150	45	170	55	Dense aggregate, plain soffit, simply supported
FIP/CEB:1975 (EUR)	60	10	80	20	100	30	120	40	150	55	175	65	Dense aggregate, one-way span, simply supported, critical temperature = 550°C
CEB:1987 (EUR)	60	10	80	25	100	35	120	45	150	60	175	70	Dense aggregate, one-way span, simply supported, critical temperature = 550°C, cover to steel axis
NBC SUPPLEMENT: 1985 (CAN)	60	20	90	20	112	20	130	25	158	32	180	39	Type S concrete
UBC:1988 (USA)	–	–	89	19	–	–	127	25	157	25	–	32	Siliceous aggregate
AS 3600:1988 (AUS)	60	15	80	20	100	25	120	30	150	45	170	55	One-way simply supported

Table D.5 Prestressed concrete slabs

Fire-resistance rating:	0.5 hour		1.0 hour		1.5 hours		2.0 hours		3.0 hours		4.0 hours		
	Width (mm)	Cover (mm)	Width (mm)	Cover (mm)	Width (mm)	Cover (mm)	Width (mm)	Cover (mm)	Width (mm)	Cover (mm)	Width (mm)	Cover (mm)	Notes
Document, standard or code													
MP9:1989 (NZ)	60	13	80	25	100	32	120	38	150	50	175	64	Normal-weight, simply supported
CP110:1972 (UK)	90	15	100	25	125	30	125	40	150	50	150	65	Siliceous or calcareous solid slabs, average cover
BS8110:1985 (UK)	75	20	95	25	110	30	125	40	150	55	170	65	Dense aggregate, plain soffit, simply supported
BRE:1998 (UK)	75	20	95	25	110	30	125	40	150	65	170	65	Dense aggregate, plain soffit, simply supported, effective cover
FORREST AND LAW:1984 (UK)	75	20	95	25	110	30	125	40	150	55	170	65	Dense aggregate, plain soffit, simply supported
FIP/CEB:1975 (EUR)	60	20	80	30	100	40	120	50	150	65	175	75	Dense aggregate, one-way span, simply supported, critical temperature = 400°C
CEB:1987 (EUR)	60	20	80	35	100	45	120	55	150	70	175	80	Dense aggregate, one-way span, simply supported, critical temperature = 400°C, cover to steel axis
NBC SUPPLEMENT: 1985 (CAN)	60	20	90	25	112	32	130	39	158	50	180	64	Type S concrete
UBC:1988 (USA)	–	–	89	30	–	–	127	48	157	61	–	–	Siliceous aggregate, pretensioned
AS 3600:1988 (AUS)	60	20	80	25	100	35	120	40	150	55	170	65	One-way simply supported

Table D.6 Reinforced concrete walls

Fire-resistance rating:	0.5 hour		1.0 hour		1.5 hours		2.0 hours		3.0 hours		4.0 hours		
	Width (mm)	Cover (mm)	Width (mm)	Cover (mm)	Width (mm)	Cover (mm)	Width (mm)	Cover (mm)	Width (mm)	Cover (mm)	Width (mm)	Cover (mm)	Notes
Document, standard or code													
MP9:1989(NZ)	-	-	75	-	100	-	120	-	150	-	175	-	Siliceous
CP110:1972 (UK)	75	-	75	-	100	-	100	-	150	-	180	-	>1% reinforcement
BS8110:1985 (UK)	100 75	25 15	120 75	25 15	140 100	25 25	160 100	25 25	200 150	25 25	240 150	25 25	0.4 to 1% reinforcement >1% reinforcement Dense aggregate
BRE:1998 (UK)	100 75	25 15	120 75	25 15	140 100	25 25	160 100	25 25	200 150	25 25	240 180	25 25	0.4 to 1% reinforcement >1% reinforcement Dense aggregate
FORREST AND LAW:1984 (UK)	100 75	25 15	120 75	25 15	140 100	25 25	160 100	25 25	200 150	24 25	240 185	25 25	0.4 to 1% reinforcement >1% reinforcement Dense aggregate
FIP/CEB:1975 (EUR)	100 60	10 –	120 80	10 –	140 100	15 –	160 120	25 –	200 150	25 –	240 175	25 –	Load-bearing Non load-bearing Dense aggregate
CEB:1987 (EUR)	120 60	10 –	120 80	15 –	140 100	25 –	160 120	35 –	200 150	55 –	240 175	75 –	Load-bearing, concrete stress $<0.15\times$characteristic strength, cover to steel axis, load-bearing dense aggregate
NBC SUPPLEMENT: 1985(CAN)	60	–	90	–	112	–	130	–	158	–	180	–	Type S concrete, load-bearing and non load-bearing
UBC:1988 (USA)	–	–	89N	–	–	–	127	–	157	–	178	–	Siliceous aggregate, Table 43-B
AS 3600:1988 (AUS)	60	20	80	20	100	35	120	40	150	45	170	50	Limits on slenderness depending on axial force

References

Aarnio, M. (1979). 'Glulam timber construction and the fire resistance of the joints' [in Finnish]. Helsinki School of Technology, Division of Building Engineering, Diploma Work. Otnas.

Aarnio, M., and Kallioniemi, P. (1983). 'Fire safety in joints of load bearing timber construction' [in Finnish]. *VTT Research Report No. 233*. Technical Research Centre of Finland.

Abrams, M.S., and Gustaferro, A.H. (1968). 'Fire endurance of concrete slabs as influenced by thickness, aggregate type and moisture'. *Research Bulletin 223*. Research and Development Laboratories of the Portland Cement Association. Skokie, Illinois, USA.

Abrams, M.S., and Gustaferro, A.H. (1971). 'Fire tests of poke-through assemblies'. *Fire Journal*, May, 1971.

ACI (1981). 'Guide for determining the fire endurance of concrete elements'. *ACI 216R-81*. American Concrete Institute, Detroit, USA.

AFPA (1999). 'Calculating the fire resistance of exposed wood members'. *Technical Report No. 10*. American Forest and Paper Association Inc., Washington DC, USA.

Ahmed, G.N., and Hurst, J.P. (1995). 'Modelling the thermal behaviour of concrete slabs subjected to the ASTM E119 Standard Fire condition'. *Journal of Fire Protection Engineering*, **7**, 4, 125–132.

AISI (1991). 'Load and resistance factor design specification for cold-formed steel structural members'. American Iron and Steel Institute, Washington DC, USA.

Aldea, C-M., Franssen, J-M., and Dotreppe, J-C. (1997). 'Fire test on normal and high strength reinforced concrete columns'. *Proceedings of the International Workshop on Fire Performance of High Strength Concrete*, pp 109–124. NIST Special Publication 919. National Institute of Standards and Technology, USA.

Alfawakhiri, F., and Sultan, M.A. (1999). 'Fire resistance of load-bearing steel-stud walls protected with gypsum board: a review'. *Fire Technology*, **35**, 4, 308–335.

Al-Jabri, K.S. *et al.* (1998). 'Behaviour of steel and composite beam – column connections in fire'. *Journal of Constructional Steel Research*, **46**, 1–3, Paper 180.

Allen, L.W. (1970). 'Fire endurance of selected non-loadbearing concrete masonry walls'. *DBR Fire Study No. 25*. Division of Building Research, National Research Council of Canada, Ottawa, Canada.

Almand, K. (1989). 'The role of passive systems in fire safety design in buildings'. *Fire Safety and Engineering – International Symposium Papers*, pp 103–129. The Warren Centre, University of Sydney, Australia.

Anderberg, Y. (1976). 'Fire exposed hyperstatic concrete structures – an experimental and theoretical study'. Division of Structural Mechanics and Concrete Construction, Lund Institute of Technology, Lund, Sweden.

Anderberg, Y. (1986). 'Measured and predicted behaviour of steel beams and columns in fire'. Lund Institute of Technology, Lund, Sweden.

Anderberg, Y. (1988). 'Modelling steel behaviour'. *Fire Safety Journal*, **13**, 1, 17–26.

Anderberg, Y. (1989). 'Fire engineering design based on PC'. *Nordic Mini-Seminar on Fire Resistance of Concrete Structures*, Trondheim. Fire Safety Design, Lund, Sweden.

Anderberg, Y. (1993). 'Computer simulations and a design method for fire exposed concrete columns'. *Report 92-50*. Fire Safety Design, Lund, Sweden.

Anderberg, Y., and Forsen, N.E. (1982). 'Fire resistance of concrete structures'. *Nordic Concrete Research Publication No. 1*, Oslo. Reprinted as *Report LUTVDG /(TVBB-3009)*, Lund University, Sweden.

Andersen, N.E., and Laurisden, D.H. (1999). 'TT-roof slabs'. 'Hollow core concrete slabs'. *Technical Report X52650, Parts 1 and 2 respectively*. Danish Institute of Fire Technology, Denmark.

Armer, G.S.T., and O'Dell, T. (Editors) (1996). 'Fire, static and dynamic tests of building structures'. *Proceedings of the Second Cardington Conference*. E & F.N. Spon, London, UK.

ASCE (1995). 'Standard for load and resistance factor design (LRFD) for engineered wood construction'. *AF&PA/ASCE 16-95*. American Forest and Paper Association, American Society for Civil Engineers, New York, USA.

ASFPCM (1988). 'Fire protection for structural steel in buildings'. Association of Specialist Fire Protection Contractors and Manufacturers, UK.

ASTM (1988(a)). 'Standard test methods for fire tests of building construction and materials'. *E119-88*. American Society for Testing and Materials.

ASTM (1988(b)). 'Fire tests of through-penetration fire stops'. *E814-88*. American Society for Testing and Materials.

ASTM (1995). *ASTM C-36*. American Society for Testing and Materials.

Babrauskas, V. (1979). 'COMPF2: a program for calculating post-flashover fire temperatures'. *NBS Technical Note 991*. National Bureau of Standards.

Babrauskas, V. (1981). 'A closed form approximation for post flashover compartment fires'. *Fire Safety Journal*, **4**, 63–73.

Babrauskas, V. (1995). 'Burning rates'. Chapter 3-1. *SFPE Handbook of Fire Protection Engineering, Second Edition*. Society of Fire Protection Engineers, USA.

Babrauskas, V., and Grayson, S.J. (1992). *Heat Release in Fires*. Elsevier Applied Science.

Babrauskas, V., and Williamson, R.B. (1978(a)). 'Post-flashover compartment fires – basis of a theoretical model'. *Fire and Materials*, **2**, 2, 39–53.

Babrauskas, V., and Williamson, R.B. (1978(b)). 'Temperature measurement in fire test furnaces'. *Fire Technology*, **14**, 3, 226–238.

Babrauskas, V., and Williamson, R.B. (1978(c)). 'The historical basis of fire resistance testing – Part I'. *Fire Technology*, **14**, 3, 184–205.

Babrauskas, V., and Williamson, R.B. (1978(d)). 'The historical basis of fire resistance testing – Part II'. *Fire Technology*, **14**, 4, 304–316.

Bailey, C.G., Burgess, I.W., and Plank, R.J. (1996(a)). 'Analyses of the effects of cooling and fire spread on steel-framed buildings'. *Fire Safety Journal*, **26**, 273–293.

Bailey, C.G., Burgess, I.W., and Plank, R.J. (1996(b)). 'The lateral torsional buckling of unrestrained steel beams in fire'. *Journal of Constructional Steel Research*, **36**, 2, 101–119.

Barnfield, J.R., and Porter, A.M. (1984). 'Historic buildings and fire: fire performance of cast-iron structural elements'. *The Structural Engineer*, **62A**, 12, 373–380.

Bazant, Z.P., and Kaplan, M.F. (1996). *Concrete at High Temperatures – Material Properties and Mathematical Models*. Concrete Design and Construction Series. Longman Group Ltd, UK.

BCA (1996). *Building Code of Australia*. Australian Building Codes Board.

Beck, V.R., and Yung, D. (1994). 'The development of a risk–cost assessment model for the evaluation of fire safety in buildings'. *Proceedings of the Fourth International Conference on Fire Safety Science*, Ottawa, Canada, pp 817–828.

Bennetts, I.D., Poh, K.W., and Thomas, I.R. (1999). *Economical Carparks – A Guide to Fire Safety*. BHP Steel, Australia.

Berto, A.F., and Tomina, J.C. (1988). *Lições di incêndio da sede adminsitrativa da Cesp*. Technologia de Edificações – Instituto de Pesguisas Technologicas do Estado de São Paolo, Brasil.

BIA (1992). *New Zealand Building Code and Approved Documents*. New Zealand Building Industry Authority, Wellington, New Zealand.

Bodig, J., and Jayne, B.A. (1982). *Mechanics of Wood and Wood Composites*. Van Nostrand Reinhold, New York, USA.

Bolonius Olesen, F., and Hansen, T. (1992). *Full-scale Tests on Loaded Glulam Beams Exposed to Natural Fires*. Aalborg University. Aalborg, Denmark.

Bolonius Olesen, F., and König, J. (1992). 'Tests on glued laminated beams in bending exposed to natural fires'. *Proceedings, CIB-W18 Meeting*, Aarhus, Denmark (*Trätek Report No. I9210061*).

Bond, G.V.L. (1975). *Fire and Steel Construction – Water Filled Hollow Columns*. Constrado, UK.

Botting, R. (1998). 'The impact of post earthquake fire on the built urban environment'. *Fire Engineering Research Report 98/1*. University of Canterbury, New Zealand.

BRE (1988). *Guidelines for the Construction of Fire-Resisting Elements*. BRE Report, Fire Research Station, Building Research Establishment, Watford, UK.

Bresler, B., and Iding, R.H.J. (1982). 'Effect of fire exposure on steel frame buildings'. *Final Report to American Iron and Steel Institute*. Wiss, Janney Elstner Associates, Emeryville, California, USA.

British Steel (1999). *The Behaviour of Multi-Storey Steel Framed Buildings in Fire*. British Steel, Swinden Technology Centre, UK.

BSI (1972). *Code of Practice for the Structural Use of Concrete, CP 110, Part 1*. British Standards Institution, UK.

BSI (1978). *Code of Practice for the Structural Use of Timber, BS 5268, Section 4.1*. Methods of Calculating the Fire Resistance of Timber Members. British Standards Institution, UK.

BSI (1985). *Structural Use of Concrete, BS 8110*. British Standards Institution, UK.

BSI (1987). *Fire Tests on Building Materials and Structures, BS 476 (Parts 1 to 23)*. British Standards Institution, UK.

BSI (1990(a)). *Structural Use of Steelwork in Buildings, BS 5950. Part 8, Code of Practice for Fire Resistant Design*. British Standards Institution, UK.

BSI (1990(b)). *Structural Use of Timber, BS 5268. Part 4, Section 4.2. Recommendations for calculating the fire resistance of timber stud walls and joisted floor constructions*. British Standards Institution, UK.

BSI (1997). *Fire Safety Engineering in Buildings. Draft for Development DD240:1997*. British Standards Institution, UK.

Buchanan, A.H. (1990). 'The bending strength of lumber'. *Journal of Structural Engineering, USA*, **116**, 5, 1213–1229.

Buchanan, A.H. (1994(a)). 'Fire engineering for a performance based code'. *Fire Safety Journal*, **23**, 1, 1–16.

Buchanan, A.H. (1994(b)). 'Structural design for fire in New Zealand'. *Proceedings of the Australasian Structural Engineering Conference*, Sydney, Australia, pp 643–649.

Buchanan, A.H. (1997). 'Modelling post flashover fires with FASTLite'. *Journal of Fire Protection Engineering*, **9**, 3, 1–11.

Buchanan, A.H. (1998). 'Mechanical properties of wood exposed to fires'. *Proceedings of the Fourth World Conference on Timber Engineering*, Montreux, Switzerland. Vol. 2, pp 238–245.

Buchanan, A.H. (1999) 'Implementation of performance-based fire codes'. *Fire Safety Journal*, **32**, 377–383.

Buchanan, A.H. (Editor) (2001). *Fire Engineering Design Guide*. Second Edition. Centre for Advanced Engineering, University of Canterbury, New Zealand.

Buchanan, A.H., and Barber, D.J. (1994). 'Fire resistance of epoxied steel rods in glulam'. *Proceedings of the 1994 Pacific Timber Engineering Conference*, Gold Coast, Australia. Vol. 1, 590–598.

Buchanan, A.H., and King, A.B. (1991). 'Fire performance of gusset connections in glue-laminated timber'. *Fire and Materials*, **15**, 137–143.

Bukowski, R.W. (1997). 'Fire hazard analysis', Section 11, Chapter 7, *Fire Protection Handbook, 18th Edition*. National Fire Protection Association, Quincy, MA, USA.

Butcher, E.G., Chitty, T.B., and Ashton, L.A. (1966). 'The temperature attained by steel in building fires'. *Fire Research Technical Paper No. 15*. HMSO, London, UK.

Caldwell, C.A., Buchanan, A.H., and Fleischmann, C.M. (1999). 'Documentation for performance-based fire engineering design in New Zealand'. *Journal of Fire Protection Engineering*, **10**, 2, 24–31.

Carling, O. (1989). 'Fire resistance of joint details in load bearing timber construction – a literature survey' [translated from Swedish]. *BRANZ Study Report SR 18*. Building Research Association of New Zealand.

Carlson, C.C., Selvaggio, S.L., and Gustaferro, A.H. (1965). 'A review of studies of the effects of restraint on the fire resistance of prestressed concrete'. *PCA Research Department Bulletin 206*., Portland Cement Association, USA.

CEB (1987). 'Model code for fire design of concrete structures'. *Bulletin d'Information No 174*. Comité Euro-International du Beton, Lausanne, Switzerland.

CIB (1986). 'Design guide – structural fire safety, CIB-W14'. *Fire Safety Journal*, **10**, 2, 75–138.

Clancy, P. (1999). 'Time and probability of failure of timber framed walls in fire'. *PhD thesis*. Victoria University of Technology, Victoria, Australia.

Clifton, G.C. (1996). 'Fire models for large firecells'. *HERA Report R4-83*. New Zealand Heavy Engineering Research Association, Auckland, New Zealand.

Coakley, C. *et al.* (1982). *The Day the MGM Grand Hotel Burned*. Lyle Stuart Inc, Secaucus, N.J., USA.

Collier, P.C.R. (1991(a)). 'Design of light timber framed walls and floors for fire resistance'. *BRANZ Technical Recommendation No. 9*. Building Research Association of New Zealand.

Collier, P.C.R. (1991(b)). 'Design of load-bearing light timber frame walls for fire resistance: Part 1'. *BRANZ Study Report SR36*. Building Research Association of New Zealand.

Collier, P.C.R. (1992). 'Charring rates of timber'. *BRANZ Study Report No. 42*. Building Research Association of New Zealand.

Collier, P.C.R. (1996). 'A model for predicting the fire-resisting performance of small scale cavity walls in realistic fires'. *Fire Technology*, **32**, 2, 120–136.

Collins, G. (1998). 'Fire resistance of inter-tenancy wall connections in multi-residential timber-frame buildings'. *Proceedings of the Fourth World Conference on Timber Engineering*, Montreux, Switzerland, Vol. 2, pp 114–121.

Collins, M.J. (1983). 'Density conversions for radiata pine'. *FRI Bulletin No. 49*. Forest Research Institute, Rotorua, New Zealand.

Comeau, E. (1999). 'Roof collapse kills three'. *NFPA Journal*, **94**, 4, 77–80.

Cooke, G.M.E. (1988). 'Thermal bowing in fire and how it affects building design'. *BRE Information Paper IP 21/88*. Building Research Establishment, UK.

Cooke, G.M.E. (1996). 'A review of compartment fire tests to explore the behaviour of structural steel. Fire, static and dynamic tests of building structures'. *Proceedings, Second Cardington Conference*. E & F.N. Spon, London, pp 17–32.

Cooke, G.M.E. (1997). 'The behaviour of sandwich panels exposed to fire'. *Building Engineer*, **72**, 6, 14–29.

Cooper, L.Y. (1997). 'The thermal response of gypsum panel steel stud wall systems exposed to fire environments'. *NISTR 6027*. National Institute of Standards and Technology, USA.

Cooper, L.Y., and Steckler, K.D. (1996). 'Methodology for developing and implementing alternative temperature–time curves for testing the fire resistance of barriers for nuclear power plant applications'. *NISTIR 5842*. National Institute of Standards and Technology, USA.

Cosgrove, B.W. (1996). 'Fire design of single storey industrial buildings'. *Fire Engineering Research Report No 96/3*. University of Canterbury, New Zealand.

Cramer, S.M. (1995). 'Fire endurance modelling of wood floor/ceiling assemblies'. *Proceedings of the Fire and Materials Conference*, Washington DC, USA, pp 105–114.

Cramer, S.M., Shrestha, D., and Mtenga, P.V. (1993). 'Computation of member forces in metal plate connected wood trusses'. *Structural Engineering Review*, **5**, 3, 209–217.

CSIRO (1993). *FIRECALC*. Division of Building Construction and Engineering. CSIRO, North Ryde, NSW, Australia.

Custer, R.L.P., and Meacham, B.J. (1997). *Introduction to Performance-based Fire Safety*. Society of Fire Protection Engineers and National Fire Protection Association.

Deal, S. (1993). 'Technical reference guide for FPEtool Version 3.2'. *NISTR 5486*. National Institute of Standards and Technology, USA.

Del Senno, M., Cont, S., and Piazza, M. (1998). 'Charring rate slowing by means of fibreglass-reinforcing coatings'. *Proceedings of the Fourth World Conference on Timber Engineering*, Montreux, Switzerland. Vol. 2, pp 230–237.

DIN (1996). 'Baulicher Brandschutz im Industrialbau, Teil 1'. *DIN 18230-1*. (Structural fire protection in industrial buildings – Part 1: Analytically required fire resistance time). Deutsches Institut für Normung e.V. Berlin.

Dotreppe, J-C. *et al.* (1996). 'Experimental research of the determination of the main parameters affecting the behaviour of reinforced concrete columns under fire conditions'. *Magazine of Concrete Research*, **49**, 179, 117–127.

Drysdale, D. (1998). *An Introduction to Fire Dynamics. Second Edition*. John Wiley & Sons, Chichester, UK.

Dunn, V. (1988). *Collapse of Burning Buildings; a Guide to Fireground Safety*. Fire Engineering, PenWell Books, New York, USA.

EC1 (1994). *Eurocode 1: Basis of Design and Design Actions on Structures. Part 2-2: Actions on Structures Exposed to Fire. ENV 1991-2-2*. European Committee for Standardization, Brussels, Belgium.

EC2 (1993). *Eurocode 2: Design of Concrete Structures. ENV 1992-1-2: General Rules – Structural Fire Design*. European Committee for Standardization, Brussels, Belgium.

EC3 (1995). *Eurocode 3: Design of Steel Structures. ENV 1993-1-2: General Rules – Structural Fire Design*. European Committee for Standardization, Brussels, Belgium.

EC4 (1994). *Eurocode 4: Design of Composite Steel and Concrete Structures. ENV 1993-1-2: General Rules – Structural Fire Design*. European Committee for Standardization, Brussels, Belgium.

EC5 (1994). *Eurocode 5: Design of Timber Structures. ENV 1995-1-2: General Rules – Structural Fire Design*. European Committee for Standardization, Brussels, Belgium.

EC6 (1995). *Eurocode 6: Design of Masonry Structures. ENV 1996-1-2: General Rules – Structural Fire Design*. European Committee for Standardization, Brussels, Belgium.

ECCS (1983). 'Calculation of the fire resistance of composite concrete slabs with profiled steel sheet exposed to the standard fire'. *Publication No. 32*. European Commission for Constructional Steelwork, Brussels, Belgium.

ECCS (1985). *Design Manual on the European Recommendations for the Fire Safety of Steel Structures*. European Commission for Constructional Steelwork, Brussels, Belgium.

ECCS (1988). 'Calculation of the fire resistance of centrally-loaded composite steel–concrete columns exposed to the standard fire'. *Technical Note No. 55*. European Commission for Constructional Steelwork, Brussels, Belgium.

ECCS (1995). 'Fire resistance of steel structures'. *ECCS Technical Note No. 89. Technical Committee 3*. European Commission for Constructional Steelwork, Brussels, Belgium.

Ellingwood, B.R., and Corotis, R.B. (1991). 'Load combinations for buildings exposed to fires'. *Engineering Journal, American Institute of Steel Construction*, **28**, 1, 37–44.

El-Rimawi, J.A., Burgess, I.W., and Plank, R.J. (1996). 'The treatment of strain reversal in structural members during the cooling phase of a fire'. *Journal of Constructional Steel Research*, **37**, 2, 115–135.

England, J.P., Young, S.A., Hui, M.C., and Kurban, N. (2000). *Guide for the design of fire resistant barriers and structures*. Building Control Commission, Melbourne, Australia.

FCRC (1996). *Fire Engineering Guidelines*. Fire Code Reform Centre, Sydney, Australia.

Feasey, R. (1999). 'Post-flashover design fires'. *Fire Engineering Research Report 99/6*. University of Canterbury, New Zealand.

Feasey, R., and Buchanan, A.H. (2000). 'Post-flashover Fires for Structural Design'. Submitted to *Fire Safety Journal*.

FIP/CEB (1975). *Guides to Good Practice. FIP/CEB Recommendations for the Design of Reinforced and Prestressed Concrete Structural Members for Fire Resistance*. FIP/1/1. Fédération Internationale de la Précontrainte. Wexham Springs, Slough, UK.

Fleischmann, C.M. (1995). 'Analytical methods for determining fire resistance of concrete members'. Chapter 4-10. *SFPE Handbook of Fire Protection Engineering, Second Edition*. Society of Fire Protection Engineers, USA.

Fontana, M., and Borgogno, W. (1995). *Brandverhalten von Slim-Floor-Verbunddecken. Stahlbau*, **64**, 168–174. Ernst + Sohn, Berlin, Germany.

Fontana, M., and Frangi, A. (1999). 'Fire behaviour of timber–concrete composite slabs'. *Proceedings of the Sixth International Symposium on Fire Safety Science*, Poitiers, France.

Forsen, N.E. (1982). 'A theoretical study of the fire resistance of concrete structures'. *FCB-SINTEF Report STF65 A82062*, Trondheim, Norway.

Franssen, J-M. (1990). 'The unloading of building materials submitted to fire'. *Fire Safety Journal*, **16**, 213–227.

Franssen, J-M. (1999(a)). 'A comparison between the parametric fire of Eurocode 1 and experimental tests'. *Proceedings of the Sixth International Symposium on Fire Safety Science*, Poitiers, France.

Franssen, J-M. (1999(b)). 'Steel structures in fire: from the material level to the calculation methods'. *Proceedings of the Second National Conference on Steel and Composite Construction*, Coimbra, Portugal.

Franssen, J-M. (2000). 'Design of concrete columns based on EC2 tabulated data: a critical review'. *Proceedings of the First International Workshop on Structures in Fire*, Copenhagen, Denmark.

Franssen, J-M., Kodur, V.K.R., and Mason, J. (2000). 'User's manual for SAFIR: a computer program for analysis of structures submitted to fire'. *Internal Report SPEC/2000_03, 2000*, University of Liège, Ponts et Charpentes, Belgium.

Franssen, J-M. *et al.* (1999). 'Competitive steel buildings through natural fire safety concept'. *Draft Final Report, Part 2*. Profil Arbed Centre de Recherches, Luxembourg.

Franssen, J-M., and Bruls, A. (1997). 'Design and tests of prestressed concrete beams'. *Proceedings of the Fifth International Symposium of Fire Safety Science*, Melbourne, Australia, pp 1081–1092.

Franssen, J-M., Cajot, L-G., and Schleich, J-B. (1998). 'Natural fires in large compartments: effects caused on the structure by localized fires in large compartments'. *Proceedings of the Eurofire'98 Conference*, Belgium.

Franssen, J-M., Schleich, J-B., and Cajot, L-G. (1995). 'A simple model for the fire resistance of axially loaded members according to Eurocode 3'. *Journal of Constructional Steel Research*, **35**, 49–69.

Franssen, J-M. *et al.* (1996). 'A simple model for the fire resistance of axially loaded members – comparison with experimental results'. *Journal of Constructional Steel Research*, **37**, 3, 175–204.

Fredlund, B. (1979). 'Structural design of fire exposed rectangular laminated wood beams with respect to lateral buckling'. *Report No. 79-5*. Department of Structural Mechanics, Lund University of Technology, Sweden.

Fredlund, B. (1993). 'Modelling of heat and mass transfer in wood structures during fire'. *Fire Safety Journal*, **20**, 1, 36–69.

Fuller, J.J., Leichti, R.J., and White, R.H. (1992). 'Temperature distribution in a nailed gypsum-stud joint exposed to fire'. *Fire and Materials*, **16**, 95–99.

Gamble, W.L. (1989). 'Predicting protected steel member fire endurance using spreadsheet programs'. *Fire Technology*, **25**, 3, 256–273.

Gerhards, C.C. (1982). 'Effect of the moisture content and temperature on the mechanical properties of wood: an analysis of immediate effects'. *Wood and Fibre*, **14**, 1, 4–36.

Gerlich, J.T., Collier, P.C.R., and Buchanan, A.H. (1996). 'Design of light steel framed walls for fire resistance'. *Fire and Materials*, **20**, 79–96.

Gilvary, K.R., and Dexter, R.J. (1997). 'Evaluation of alternative methods for fire rating structural elements'. *NIST-GCR-97-718*. National Institute of Standards and Technology, USA.

Glos, P., and Henrici, D. (1991). 'Bending strength and MOE of structural timber at temperatures up to 150°C' [in German]. *Holz als Roh- und Werkstoff*, **49**, 417–422. Translated by USDA Forest Products Laboratory, Madison, WI, USA, 1992.

Götz, K.H. *et al.* (1989). *Timber Design and Construction Sourcebook*. McGraw Hill Book Company, New York, USA.

Gustaferro, A., and Martin, L.D. (1988). *Design for Fire Resistance of Precast Prestressed Concrete, Second Edition*. Prestressed Concrete Institute, Illinois, USA.

Gypsum Association (1994). *Fire Resistance Design Manual, 14th Edition*. Gypsum Association, USA.

Hadvig, S. (1981). *Charring of Wood in Building Fires, Report*. Laboratory of Heating and Air Conditioning, Technical University of Denmark.

Harmathy, T.Z. (1965). 'Ten rules of fire endurance rating'. *Fire Technology*, **1**, 2, 93–102.

Harmathy, T.Z. (1993). *Fire Safety Design and Concrete*. Concrete Design and Construction Series. Longman Scientific and Technical, UK.

Hasemi, Y. *et al.* (1995). 'Fire safety of building components exposed to a localized fire'. *Proceedings of the First Asia-Flam Conference, Hong Kong*. Interscience Publications Ltd, London, UK.

Hayashi, Y. *et al.* (1999). 'Full-scale fire test of a wooden three-storey apartment building exposed to simulated city fire' [in Japanese]. *Building Research Data* No. 93. Building Research Institute, Ministry of Construction, Japan.

HERA (1996). 'HERA fire protection manuals, Sections 7 and 8, passive/active fire protection of steel'. *Report R4-89*. New Zealand Heavy Engineering Research Association, Auckland, New Zealand.

Hertz, K. (1981). 'Simple temperature calculations of fire-exposed concrete constructions'. *Report No 159*, Institute of Building Design, Technical University of Denmark.

Hertz, K. (1982). 'The anchorage capacity of deformed reinforcing bars at normal temperature and high temperature'. *Magazine of Concrete Research*, **34**, 121.

HMSO (1968). *Report of the Inquiry into the Collapse of Flats at Ronan Point, Canning Town*. HMSO, UK.

HMSO (1961). 'Protection of structural steel against fire'. *Fire Research Station Note No. 2*. Joint Fire Research Organisation, Her Majesty's Stationery Office.

Hosser, D., Dorn, T., and Richter, E. (1994). 'Evaluation of simplified calculation methods for structural fire design'. *Fire Safety Journal*, **22**, 249–304.

Hviid, N.J. (1979). 'Fire resistance of timber structures – research on load bearing and heated steel connections and the rate of fire penetration in timber' [in Finnish]. *Nordisk Timber Symposium*. State Building Research Institute.

Iding, R., Bresler, B., and Nizamuddin, Z. (1977(a)). 'FIRES-T3: a computer program for the fire response of structures – thermal'. *Fire Research Group Report No. UCB FRG 77-15*. University of California, Berkeley, USA.

Iding, R., Bresler, B., and Nizamuddin, Z. (1977(b)). 'FIRES-RC II: a computer program for the fire response of structures – reinforced concrete frames'. *Fire Research Group Report No. UCB FRG 77-8*. University of California, Berkeley, USA.

IFCI (2000). *International Building Code*. International Fire Code Institute, USA.

IISI (1993). *International Fire Engineering Design for Steel Structures: State of the Art*. International Iron and Steel Institute, Brussels, Belgium.

Ingberg, S.H. (1928). 'Tests of the severity of building fires'. *National Fire Protection Quarterly*, **22**, 1, 43–61.

Inwood, M. (1999). 'Review of NZS 3101 for high strength and lightweight concrete exposed to fire'. *Fire Engineering Research Report 99/10*. University of Canterbury, New Zealand.

ISE (1978). *Design and Detailing of Concrete Structures for Fire Resistance*. The Institution of Structural Engineers, London, UK.

ISO (1975). 'Fire resistance tests – elements of building construction'. *ISO 834-1975*. International Organization for Standardization.

ISO (1993). Fire Tests—Reaction to Fire—Rate of Heat Release from Building Products. ISO 5660. International Organisation for Standardisation, Geneva.

ISO (1998). 'Fire safety engineering'. *Draft Technical Report of ISO Technical Committee 92, SC4*. International Organization for Standardization.

James, M., and Buchanan A.H. (2000). 'Fire resistance of seismic gaps'. *Proceedings of the 12th World Conference on Earthquake Engineering*, Auckland, New Zealand.

Janssens, M.L. (1994). 'Thermo-physical properties for wood pyrolysis models'. *Proceedings of the Pacific Timber Engineering Conference*, Gold Coast, Australia, pp 607–618.

Janssens, M.L. (1997). 'A method for calculating the fire resistance of exposed timber decks'. *Proceedings of the Fifth International Symposium on Fire Safety Science*, Melbourne, Australia, pp 1189–1200.

Janssens, M.L., and White, R.H. (1994). 'Temperature profiles in wood members exposed to fire'. *Fire and Materials*, **18**, 263–265.

Jönsson, R., and Pettersson, O. (1985). 'Timber structures and fire – a review of the existing state of knowledge and research requirements'. *Document D3:1985*. Swedish Council for Building Research.

Karlsson, B., and Quintiere, J.G. (2000). *Enclosure fire dynamics*. CRC Press, Boca Raton, FL, USA.

Kawagoe, K. (1958). 'Fire behaviour in rooms'. *Report No. 27*, Building Research Institute, Tokyo, Japan.

Kay, T.R., Kirby, B.R., and Preston, R.R. (1996). 'Calculation of the heating rate of an unprotected steel member in a standard fire-resistance test'. *Fire Safety Journal*, **26**, 327–350.

Khoury, G.A., Grainger, B.N., and Sullivan, P.J.E. (1985). 'Strain of concrete during first heating to 600°C'. *Magazine of Concrete Research*, **37**, 195–215.

Khoury, G.A. and Sullivan, P.J.E. (1988). 'Research at Imperial College on the effect of elevated temperatures on concrete'. *Fire Safety Journal*, **13**, 69–72.

Kim, A.K., Taber, B.C., and Loughheed, G.D. (1998). 'Sprinkler protection of exterior glazing'. *Fire Technology*, **34**, 2, 116–138.

Kirby, B.R. (1999). *The behaviour of multi-storey steel framed buildings*. Swinden Technology Centre, British Steel plc, UK.

Kirby, B.R., and Preston, R.R. (1988). 'High temperature properties of hot-rolled structural steels for use in fire engineering design studies'. *Fire Safety Journal*, **13**, 27–37.

Kirby, B.R., Lapwood, D.G., and Thomson, G. (1993). *The Reinstatement of Fire Damaged Steel and Iron Framed Structures*. British Steel Technical Swinden Laboratories, UK.

Kirby, B.R. *et al.* (1994). *Natural Fires in Large-Scale Compartments – A British Steel Technical Fire Research Station Collaborative Project*. British Steel Technical Swinden Laboratories, UK.

Kirby, B.R. *et al.* (1999). 'Natural fires in large scale compartments'. *International Journal on Engineering Performance Based-Fire Codes*, **1**, 2, 43–58.

Klippstein, K.H. (1980). 'Behaviour of cold-formed steel studs in fire tests'. *Proceedings of the Fifth International Specialty Conference on Cold-Formed Steel Structures*, St Louis, Missouri, USA, pp 275–300.

Klote, J.H., and Milke, J.A. (1992). 'Design of smoke management systems'. *ASHRAE Special Publication*. American Society of Heating, Refrigerating and Air Conditioning Engineers, USA.

Knudson, R.M., and Schneiwind, A.P. (1975). 'Performance of wood structural members exposed to fire'. *Forest Products Journal*, **25**, 2, 23–32.

Kodur, V.K.R. (1997). 'Studies on the fire resistance of high strength concrete at the National Research Council of Canada'. *Proceedings of the International Workshop on Fire Performance of High Strength Concrete*, pp 75–86. NIST Special Publication 919. National Institute of Standards and Technology, USA.

Kodur, V.K.R. (1999). 'Performance-based fire resistance design of concrete filled steel columns'. *Journal of Constructional Steel Research*, **51**, 21–36.

Kodur, V.K.R., Sultan, M.A., and Alfawakhiri, F. (1999). 'Fire resistance tests on load-bearing steel stud walls'. *Proceedings of the Third International Conference on Fire Research and Engineering*, Chicago, Society of Fire Protection Engineers, USA.

Kodur, V.K.R., and Sultan, M.A. (2000). 'Performance of wood stud shear walls exposed to fire'. *Fire and Materials*, **24**, 1, 9–16.

Kollman, F.F.P., and Schulz, F. (1944). 'Versuche uber den Einfluss der Temperatur auf die Festigkeitwerte von Flugzeugholzbaustoffen' [In German]. *Teilbericht 1 und 2, Eberswalde, 23 bzw. 25* S.

König, J. (1995). 'Fire resistance of timber joists and load-bearing wall frames'. *Report No. I 9412071*. Trätek, Swedish Institute for Wood Technology Research, Stockholm, Sweden.

König, J. (1998(a)). 'Revision of ENV 1995-1-2: charring and degradation of strength and stiffness'. *Proceedings of the CIB-W18 Meeting*, Savonlinna, Finland.

König, J. (1998(b)). 'Structural stability of timber structures in fire-performance and requirements'. *Proceedings of the COST Action E5 Workshop on Timber Frame Building Systems*, Building Research Establishment, UK.

König, J., and Walleij, L. (1999). 'One-dimensional charring of timber exposed to standard and parametric fires in initially unprotected and postprotected situations'. *Report No. I 9908029*. Trätek, Swedish Institute for Wood Technology Research, Stockholm, Sweden.

König, J., and Walleij, L. (2000). 'Timber frame assemblies exposed to standard and parametric fires. Part 2: a design model for standard fire exposure'. *Report No. I 0001001*. Trätek, Swedish Institute for Wood Technology Research, Stockholm, Sweden.

Kordina, K., and Meyer-Ottens, C. (1983). *Holz Brandschutz Handbuch* [in German]. Deutsche Gesellschaft für Holzforschüng, Ernst & Son, Berlin, Germany.

Kordina, K., and Meyer-Ottens, C. (1995). *Holz Brandschutz Handbuch, 2nd Edition* [in German]. Deutsche Gesellschaft für Holzforschung, Ernst & Son, Berlin, Germany.

Kulak, G.L., Adams, P.F., and Gilmor, M.I. (1995). *Limit States Design in Structural Steel*. Canadian Institute of Steel Construction, Canada.

Lau, P.W.C., and Barrett, J.D. (1997). 'Modelling tension strength behaviour of structural lumber exposed to elevated temperatures'. *Proceedings of the Fourth International Symposium on Fire Safety Science*, Melbourne, Australia, pp 1177–1188.

Lau, P.W.C., White, R., and Van Zeeland, I. (1999). 'Modelling the charring behaviour of structural lumber'. *Fire and Materials*, **23**, 209–216.

Law, M. (1971). 'A relationship between fire grading and building design and contents'. *Fire Research Note No. 877*. Fire Research Station, UK.

Law, M. (1973). 'Prediction of fire resistance'. Paper in Symposium No. 5, *Fire-Resistance Requirements of Buildings, a New Approach*. Department of the Environment and Fire Offices Committee Joint Fire Research Organization. HMSO, London, UK.

Law, M. (1983). 'A basis for the design of fire protection of building structures'. *The Structural Engineer*, **61A**, 1.

Law, M. (1997). 'A review of formulae for T-equivalent'. *Proceedings of the Fourth International Symposium on Fire Safety Science*, Melbourne, Australia, pp 985–996.

Law, M., and O'Brien, T. (1989). 'Fire safety of bare external structural steel'. *Report No. SCI-P-OC9*. Steel Construction Institute, Ascot, UK.

Lawson, R.M. (1985). 'Fire resistance of ribbed concrete floors'. *CIRIA Report 107*. Construction Industry Research and Information Association, London, UK.

Lawson, R.M. (1990). 'Behaviour of steel beam-to-column connections in fire'. *The Structural Engineer*, **68**, 14/17, 263–271.

Lawson, R.M. *et al.* (1991). *Investigation into the Broadgate Phase 8 Fire*. The Steel Construction Institute, Ascot, UK.

Leicester, R.H, Seath, C., and Pham, L. (1979). 'The fire resistance of metal connectors'. *Proceedings of the Nineteenth Forest Products Research Conference*, Melbourne, Australia.

Lennon, T., Bullock, M.J., and Enjily, V. (2000). 'The fire resistance of medium-rise timber frame buildings'. *Proceedings, World Conference on Timber Engineering*, Whistler, Canada.

LeVan, S., and Winandy, J.E. (1990). 'Effects of fire retardant treatments on wood strength: a review'. *Wood and Fibre Science*, **22**, 1, 113–131.

Lewis, K.R. (2000). 'Fire design of steel members'. *Fire Engineering Research Report 00/7*. University of Canterbury, New Zealand.

Lie, T.T. (1972). *Fire and Buildings*. Applied Science Publishers Ltd, London, UK.

Lie, T.T. (1977). 'A method for assessing the fire resistance of laminated timber beams and columns'. *Canadian Journal of Civil Engineering*, **4**, 161–169.

Lie, T.T. (Editor) (1992). 'Structural fire protection'. *ASCE Manuals and Reports of Engineering Practice, No 78*. American Society of Civil Engineers, New York.

Lie, T.T. (1995). 'Fire temperature–time relations'. Chapter 4-8. *SFPE Handbook of Fire Protection Engineering, Second Edition*. Society of Fire Protection Engineers, USA.

Lie, T.T., and Irwin, R.J. (1993). 'Method to calculate the fire resistance of reinforced concrete columns with rectangular cross section'. *ACI Structural Journal*, **90**, 1, 52–60.

Lie, T.T., and Kodur, V.K.R. (1996). 'Fire resistance of steel columns filled with bar-reinforced concrete'. *Journal of Structural Engineering*, **122**, 1, 30–36.

Lim, L. (2000). 'Stability of precast concrete tilt panels in fire'. *Fire Engineering Research Report 00/8*. University of Canterbury, New Zealand.

Liu, T.C.H. (1999). 'Moment–rotation–temperature characteristics of steel/composite connections'. *Journal of Structural Engineering*, **125**, 10, 1188–1197.

Madsen, B., and Buchanan, A.H. (1986). 'Size effects in timber explained by a modified weakest link theory'. *Canadian Journal of Civil Engineering*, **13**, 2, 218–232.

Magnusson, S.E. (1972). 'Probabilistic analysis of fire safety'. *ASCE–IABSE International Conference on Planning and Design of Tall Buildings*. Conference preprints Vol DS, p 424. Lehigh University, Bethlehem, Pennsylvania, USA.

Magnusson, S.E., and Thelandersson, S. (1970). *Temperature–Time Curves of Complete Process of Fire Development; Theoretical Study of Wood Fuel Fires in Enclosed Spaces*. Acta Polytechnica Scandinavica. Civil Engineering and Building Construction Series 65.

Majamaa, J. (1991). 'Calculation models of wooden beams exposed to fire'. *VTT Research Note No. 1282*. Technical Research Centre of Finland.

Malhotra, H.L. (1982). *Design of Fire-Resisting Structures*. Surrey University Press, UK.

Malhotra, H.L. (1984). 'Spalling of concrete in fires'. *CIRIA Technical Note No. 118*. Construction Industry Research and Information Association, London, UK.

Malhotra, H.L., and Rogowski, B.F.W. (1970). 'Fire resistance of laminated timber columns'. *Fire and Structural Use of Timber in Buildings, Symposium No. 3*. HMSO, London, UK, pp 17–44.

Marchant, E.W. (1972). *A Complete Guide to Fire and Buildings*. Medical and Technical Publishing Co. Ltd., Lancaster, UK.

Martin, D.M., and Moore, D.B. (1997). 'Introduction and background to the research programme and major fire tests at BRE Cardington'. *Proceedings of the National Steel Construction Conference*, London, UK.

Martin, Z.A., and Tingley, D.A. (2000). 'Fire resistance of FRP reinforced glulam beams'. *Proceedings of the World Conference on Timber Engineering*, Whistler, B.C., Canada.

McCaffrey, B.J., Quintiere, J.G., and Harkleroad, M.F. (1981). 'Estimating room fire temperatures and the likelihood of flashover using fire test data correlations'. *Fire Technology*, **17**, 2, 98–119.

Meyer-Ottens, C. (1983). 'Junctions in wood structures – total construction'. *Proceedings of the International Seminar on Three Decades of Structural Fire Safety*. Building Research Establishment, UK.

Mikkola, E. (1990). 'Charring of wood'. *VTT Research Report 689*. Technical Research Centre of Finland.

Milke, J.A. (1995). 'Analytical methods for determining fire resistance of steel members'. Chapter 4-9, *SFPE Handbook of Fire Protection Engineering, Second Edition*. Society of Fire Protection Engineers, USA.

Milke, J.A., and Hill, S. (1996). *Initial Development of Draft Performance-based Fire Protection Standard on Construction*. American Society of Civil Engineers.

Moore, D.B., and Lennon, T. (1997). 'Fire engineering design of steel structures'. *Progress in Structural Engineering and Materials*, **1**, 1, 4–9.

NBCC (1995). *National Building Code of Canada*. National Research Council of Canada, Ottawa, Canada.

Nelson, H.E. (1986). 'FIREFORM: a computerised collection of convenient fire safety computations'. *NBSIR 86-3308*. National Bureau of Standards.

Neville, A.M. (1997). *Properties of Concrete, Fourth and Final Edition*. John Wiley and Sons, New York, USA.

Newman, G.M. (1990). *The Behaviour of Steel Portal Frames in Boundary Conditions, Second Edition*. The Steel Construction Institute, UK.

Newman, G.M., and Lawson, R.M. (1991). 'Fire resistance of composite beams'. *Technical Report SCI-P-109*. Steel Construction Institute, UK.

NFPA (1997). *Fire Protection Handbook, 18th Edition*. National Fire Protection Association, Quincy, MA, USA.

NFPA (1997). 'Guideline on fire ratings of archaic materials and assemblies. Appendix L'. *NFPA 909 Standard for the Protection of Cultural Resources*. National Fire Protection Association, Quincy, MA, USA.

Nielsen, P.C., and Olesen, F.B. (1982). 'Tensile strength of finger joints at elevated temperatures'. *Report No.8205*. Institute of Building Technology and Structural Engineering, Aalborg, Denmark.

NKB (1994). 'Performance requirements for fire safety and technical guide for verification by calculation'. *Report 1994:07 E*. NKB Fire Safety Committee. Nordic Committee on Building Regulations, Helsinki, Finland.

Norén, J.B. (1988). 'Failure of structural timber when exposed to fire'. *Proceedings of the International Conference on Timber Engineering*, Seattle, USA, pp 5–14.

Norén, J.B. (1996). 'Load-bearing capacity of nailed joints exposed to fire'. *Fire and Materials*, **20**, 133–143.

NRCC (1998). 'Fire stops in walls can provide both fire resistance and sound insulation'. *Construction Innovation, Spring 1998*. National Research Council of Canada.

Nyman, C. (1980). 'The effect of temperature and moisture on the strength of wood and glue joists' [in Finnish]. *VTT Forest Products Report No.6*. Technical Research Centre of Finland.

O'Connor, M.A., and Martin, D.M. (1998). 'Behaviour of a multi-storey steel framed building subjected to fire attack'. *Journal of Constructional Steel Research*, **46**, 1–3.

O'Hara, M. (1994). 'Understanding through penetration protection systems'. *NFPA Journal*, Jan/Feb 1994.

O'Meagher, A.J., and Bennetts, I.D. (1991). 'Modelling of concrete walls in fire'. *Fire Safety Journal*, **17**, 313–335.

O'Meagher, A.J. *et al.* (1992). 'Design of single storey industrial buildings for fire resistance'. *Journal of Australian Institute of Steel Construction*, **26**, 2, 2–17.

Ohlemiller, T.J. (1995). 'Smoldering combustion'. Chapter 2-11, *SFPE Handbook of Fire Protection Engineering, Second Edition*. Society of Fire Protection Engineers, USA.

Ohsawa, J., and Yoneda, Y. (1978). 'Shear test of woods as a model of defibration' [in Japanese]. *Journal of the Japanese Wood Research Society*, **24**, 4, 230–236.

Okuyama, T. (1974). 'Effect of strain rate on the mechanical properties of wood. IV. On the influence of the rate of deflection and the temperature to bending strength of wood' [in Japanese]. *Journal of the Japanese Wood Research Society*, **20**, 5, 210–216.

Okuyama, T., Suzuki, S., and Terazawa, S. (1977). 'Effect of temperature on orthotropic properties of wood. I. On the transverse anisotropy in bending' [in Japanese]. *Journal of the Japanese Wood Research Society*, **23**, 12, 609–616.

Oleszkiewicz, I. (1991). 'Vertical separation of windows using spandrel walls and horizontal projections'. *Fire Technology*, **27**, 4, 334–340.

Östman, B.A. (1985). 'Wood tensile strength at temperatures and moisture contents simulating fire conditions'. *Wood Science and Technology*, **19**, 103–116.

Östman, B.A., König, J., and Norén, J. (1994). 'Contribution to fire resistance of timber frame assemblies by means of fire protective boards'. *Proceedings of the Fire and Materials Conference*, Washington, DC, USA (Trätek Report No. I 9412074).

Papaioannou, K. (1986). 'The conflagration of two large department stores in the centre of Athens'. *Fire and Materials*, **10**, 171–177.

Park, R., and Paulay, T. (1975). *Reinforced Concrete Structures*. John Wiley and Sons, NY, USA.

Parker, W.J. *et al.* (1975). 'Fire endurance of gypsum board walls and chases containing plastic and metallic drain, waste and vent plumbing systems'. *NBS Building Science Series 72*. National Bureau of Standards, Washington DC, USA

Parker, W.J., and Tran, H.C. (1992). 'Wood materials'. Chapter 11, *Heat Release in Fires* (Babrauskas, V. and Grayson, S.J., editors). Elsevier Applied Science.

Peacock, R.D. *et al.* (1993). 'CFAST, the consolidated model of fire growth and smoke transport'. *Technical Note 1299*. National Institute of Standards and Technology.

Pettersson, O. (1973). 'The connection between a real fire exposure and the heating conditions according to standard fire-resistance tests – with special application to steel structures'. *Document CECM 3-73/73*. European Commission for Constructional Steelwork.

Pettersson, O., Magnusson, S.E., and Thor, J. (1976). 'Fire engineering design of steel structures'. *Publication No. 50*, Swedish Institute of Steel Construction.

Phan, L.T. (1996). 'Fire performance of high-strength concrete: a report of the state-of-the-art'. *NISTIR 5934*. National Institute of Standards and Technology.

Pitts, D.R., and Sissom, L.E. (1977). *Schaum's Outline Series: Theory and Problems of Heat Transfer*. McGraw Hill, New York, USA.

Poh, K.W. (1996). 'Modelling elevated temperature properties of structural steel'. *Report BHPR/SM/R/055*. Broken Hill Proprietary Company Ltd, Australia.

Poh, K.W., and Bennetts, I.D. (1995). 'Analysis of structural members under elevated temperature conditions'. *Journal of Structural Engineering*, **121**, 4, 664–675.

Portier, R.W., Peacock, R.D., and Reneke, P.A. (1996). 'FASTLite: engineering tools for estimating fire growth and transport'. *Special Publication 899*. National Institute of Standards and Technology.

Preusser, R. (1968). 'Plastic and elastic behaviour of wood affected by heat in open systems' [in German]. *Holztechnologie*, **9**, 4, 229–231.

Purkiss, J.A. (1996). *Fire Safety Engineering Design of Structures*. Butterworth Heinemann, Oxford, UK.

Quintiere, J.G. (1995). 'Compartment fire modelling'. Chapter 3-5, *SFPE Handbook of Fire Protection Engineering, Second Edition*. Society of Fire Protection Engineers, USA.

Quintiere, J.G. (1998). *Principles of fire behaviour*. Delmar Publishers, Albany, NY, USA.

Richardson, L.R., and McPhee, R.A. (1996). 'Fire resistance and sound-transmission-class ratings for wood frame walls'. *Fire and Materials*, **20**, 123–131.

Richardson, L.R., McPhee, R.A., and Batisita, M. (2000). 'Sound-transmission class and fire-resistance ratings for wood frame floors'. *Fire and Materials*, **24**, 1, 17–28.

Rose, P.S. *et al.* (1998). 'The influence of floor slabs on the structural performance of the Cardington Frame in fire'. *Journal of Constructional Steel Research*, **46** 1–3, 181.

Rotter, J.M. *et al.* (1999). 'Structural performance of redundant structures under local-fires'. *Proceedings of the Interflam '99 Conference*, Edinburgh, UK, pp 1069–1080.

SAA (1990(a)). 'Steel structures'. *AS 4100-1990*. Standards Association of Australia.

SAA (1990(b)). 'Timber structures, Part 4: fire resistance of structural timber members'. *AS 1720.4-1990*. Standards Association of Australia.

SAA (1990(c)). 'Fire-resistance tests of elements of structure'. *AS 1530.4-1990*. Standards Association of Australia.
SAA (1994). 'Concrete structures'. *AS 3600-1994*. Standards Association of Australia.
SAA (1997). 'Timber structures code, Part 1 – design methods'. *AS 1720.1-1997*. Standards Association of Australia.
Sano, E. (1961). 'Effects of temperature on the mechanical properties of wood. I. Compression parallel-to-grain. II. Tension parallel-to-grain. III. Torsion test' [in Japanese]. *Journal of the Japanese Wood Research Society*, **7**, 4, 147–150; **7**, 5, 189–193.
Scawthorn, C.L. (1992). 'Fire following earthquake'. Chapter 4, *Fire Safety in Tall Buildings*. Council on Tall Buildings and Urban Habitat, McGraw Hill Inc., New York, USA.
Schaffer, E.L. (1967). 'Charring rate of selected woods – transverse to grain'. *US Forest Service Research Paper FPL69*. Forest Products Laboratory, Madison, WI, USA.
Schaffer, E.L. (1973). 'Effect of pyrolytic temperature on the longitudinal strength of dry Douglas Fir'. *Journal of Testing and Evaluation*, **1**, 4, 319–329.
Schaffer, E.L. (1977). 'State of structural timber fire endurance'. *Wood and Fibre*, **9**, 2, 145–190.
Schaffer, E.L. (1984). 'Structural fire design: wood'. *Research Paper FPL 450*. US Forest Products Laboratory, Madison, WI, USA.
Schaffer, E.L. (1992). 'Fire-resistive structural design'. *Proceedings of the International Seminar on Wood Engineering*. US Forest Products Laboratory, Madison, WI, USA.
Schaffer, E.L. *et al.* (1986). 'Strength validation and fire endurance of glued laminated timber beams'. *Research Paper FPL 467*. US Forest Products Laboratory, Madison, WI, USA.
Schleich, J.B. (1996). 'A natural fire safety concept for buildings – 1. Fire, static and dynamic tests of building structures'. *Proceedings of the Second Cardington Conference*. E. & F.N. Spon, London, UK, pp 79–104.
Schleich, J.B. (1999). 'Competitive steel buildings through natural fire safety concept'. *Draft Final Report, Part 1*. Profil Arbed Centre de Recherches, Luxembourg.
Schleich, J.B. *et al.* (1999). 'Development of design rules for steel structures subjected to natural fires in closed carparks'. *European Commission, Technical Steel Research, EUR 18867, Final Report*.
Schneider, U. (1986). *Properties of Materials at High Temperatures – Concrete, Second Edition*. RILEM Report, Germany.
Schneider, U. (1988). 'Concrete at high temperatures – a general review'. *Fire Safety Journal*, **13**, 55–68.
Schneider, U., Kersken-Bradley, M., and Max, U. (1990). *Neuberechnung der Wärmeabzugsfaktoren für die DINV 18320 Teil-Baulicher Brandschutz Industribau* [in German]. Arbeitsgemeinschaft Brandsicherheit Munchen/Kassel.
Schneider, U., Morita, T., and Franssen, J-M. (1994). 'A concrete model considering the load history applied to centrally loaded columns under fire attack'. *Proceedings of the Fourth International Symposium on Fire Safety Science*, pp 1101–1112.
Seigel, L.G. (1970). 'Designing for fire safety with exposed structural steel'. *Fire Technology*, **6**, 4, 269–278.
SFPE (1995). *SFPE Handbook of Fire Protection Engineering*. Society of Fire Protection Engineers, USA.
SFPE (1996). *International Conference on Performance-based Codes and Fire Safety Design Methods*, Ottawa, Canada. Society of Fire Protection Engineers, USA.
SFPE (1998). *Second International Conference on Performance-based Codes and Fire Safety Design Methods*, Maui, Hawaii, USA. Society of Fire Protection Engineers.
SFPE (2000). *SFPE Engineering Guide to Performance-based Fire Protection Analysis and Design of Buildings*. Society of Fire Protection Engineers, USA.

Shestopal, V. (1998). *FIREWIND Computer Software for the Fire Engineering Professional.* Fire Modelling and Computing, NSW 2074, Australia.

Shrestha, D., Cramer, S.M., and White, R.H. (1994). 'Time–temperature profile across a lumber section exposed to pyrolytic temperatures'. *Fire and Materials*, **20**, 211–220.

Shrestha, D., Cramer, S.M., and White, R.H. (1995). 'Simplified models for the properties of dimension lumber and metal plate connections at elevated temperatures'. *Forest Products Journal*, **45**, 7/8, 35–42.

Smith, D., and Shaw, K. (1999). 'The single burning item (SBI) test, the Euroclasses and transitional arrangements'. *Proceedings of the Interflam '99 Conference*, **1**, 1–10.

SNZ (1986). 'The design of fire walls and partitions to ensure stability and integrity'. *Standards Magazine*, **32**, 8, 140–19. Standards New Zealand, Wellington, New Zealand.

SNZ (1991). 'MP9: fire properties of building materials and elements of structure'. *Miscellaneous Publication No. 9*. Standards New Zealand, Wellington, New Zealand.

SNZ (1992). 'Code of practice for general structural design and design loadings for buildings'. *NZS 4203: 1992*. Standards New Zealand, Wellington, New Zealand.

SNZ (1993). 'Code of practice for timber design'. *NZS 3603: 1993*. Standards New Zealand, Wellington, New Zealand.

SNZ (1995). 'Code of practice for the design of concrete structures'. *NZS 3101: 1995*. Standards New Zealand, Wellington, New Zealand.

SNZ (1996). 'Steel structures standard'. *NZS 3404: 1996*. Standards New Zealand, Wellington, New Zealand.

Srpcic, S. (1995). 'The influence of the material hardening on the behaviour of a steel plane frame in fire'. *Math. Mech.*, **75**, S179–S180.

Steinbrugge, K.V. (1982). *Earthquakes, Volcanos and Tsunamis – An Anatomy of Hazards*. Skandia America Group.

Sterner, E., and Wickström, U. (1990). 'TASEF – Temperature analysis of structures exposed to fire'. *Fire Technology SP Report 1990: 05*. Swedish National Testing Institute.

Stevens, W.C., and Turner, N. (1970). *Wood Bending Handbook*. HMSO, London, UK.

Sullivan, P.J.E., Terro, M.J., and Morris, W.A. (1994). 'Critical review of fire-dedicated thermal and structural computer programs'. *Journal of Applied Fire Science*, **3**, 2, 113–135.

Sultan, M.A. (1996). 'A model for predicting heat transfer through non-insulated unloaded steel-stud gypsum board wall assemblies exposed to fire'. *Fire Technology*, **32**, 3, 239–259.

Sultan, M.A. (1999). 'Factors affecting fire-resistance performance of lightweight frame floor assemblies'. *Proceedings of the Interflam '99 Conference*, Edinburgh, UK, pp 897–910.

Sultan, M.A. (2000). 'Fire spread via wall/floor joints in multi-family dwellings'. *Fire and Materials*, **24**, 1, 1–8.

Sultan, M.A., and Lougheed, G.D. (1995). 'Fire resistance of gypsum plasterboard wall assemblies'. *Construction Canada*, **37**, 2.

SWRI (1996). *Directory of Listed Products*. Department of Fire Technology, Southwest Research Institute, San Antonio, Texas, USA.

Takeda, H., and Mehaffey, J.R. (1998). 'WALL2D: A model for predicting heat transfer through wood-stud walls exposed to fire'. *Fire and Materials*, **22**, 133–140.

Thomas, G.C. (1997). 'Fire resistance of light timber frame walls'. *PhD thesis: Fire Engineering Research Report 97-7*. University of Canterbury, New Zealand.

Thomas, G.C. *et al.* (1995). 'Light timber framed walls exposed to compartment fires'. *Journal of Fire Protection Engineering*, **7**, 1, 25–35.

Thomas, G.C., Buchanan, A.H., and Fleischmann, C.M. (1997). 'Structural fire design: the role of time equivalence'. *Proceedings of the Fifth International Symposium on Fire Safety Science*, Melbourne, Australia, pp 607–618.

Thomas, I.R., and Bennetts, I.D. (1999). 'Fires in enclosures with single ventilation openings – comparison of long and wide enclosures'. *Proceedings of the Sixth International Symposium on Fire Safety Science*, Poitiers, France.

Thomas, P.H., and Heselden, A.J.M. (1972). 'Fully developed fires in single compartments'. *CIB Report No 20. Fire Research Note 923*, Fire Research Station, UK.

Thomas, P.H. *et al.* (1963). 'Investigations into the flow of hot gases in roof venting'. *Fire Research Technical Paper No. 7*. London, UK.

Tide, R.H.R. (1998). 'Integrity of structural steel after exposure to fire'. *Engineering Journal*, First Quarter, 1998, 26–38.

Tomasson, B. (1998). *High Performance Concrete – Design Guidelines, Report*. Department of Fire Safety Engineering, Lund University, Sweden.

TRADA (1976). 'The fire at Morrison Printing Inks Machinery Ltd., Section 2d-1'. *Wood Products Design Manual*, New Zealand Timber Research and Development Association.

Twilt, L., and Witteveen, J. (1974). 'The fire resistance of wood-clad steel columns'. *Fire Prevention Science and Technology, No. 11*.

UBC (1997). *Uniform Building Code*. International Conference of Building Officials, Whittier, California, USA.

UL (1996). *Fire Resistance Directory*. Underwriters Laboratories Inc, USA.

ULC (1989). 'Standard methods of fire endurance tests of building construction and materials'. *CAN/ULC-S101-M89*. Underwriters Laboratories of Canada, Ontario, Canada.

Wade, C.A. (1991(a)). 'Fire engineering design of reinforced and prestressed concrete elements'. *BRANZ Study Report No. 33*. Building Research Association of New Zealand.

Wade, C.A. (1991(b)). 'Method for fire engineering design of structural concrete beams and floor systems'. *BRANZ Technical Recommendation No. 8*. Building Research Association of New Zealand.

Wade, C.A. (1994). 'Performance of concrete floors exposed to real fires'. *Journal of Fire Protection Engineering*, **6**, 3, 113–124.

Wade, C.A., and Barnett, J. (1997). 'A room-corner fire model including fire growth on linings and enclosure smoke filling'. *Journal of Fire Protection Engineering*, **8**, 4, 183–193.

Walker, B. 1982. Earthquake. Time-Life Books, Amsterdam.

Walton, W.D., and Thomas, P.H. (1995). 'Estimating temperatures in compartment fires'. Chapter 3-6. *SFPE Handbook of Fire Protection Engineering*. Society of Fire Protection Engineers, USA.

Wang, Y.C. (1996). 'Tensile membrane action in slabs and its application to the Cardington Fire tests: Fire, static and dynamic tests of building structures'. *Proceedings of the Second Cardington Conference*. E. & F.N. Spon, London, UK, pp 55–67.

Wang, Y.C., Lennon, T., and Moore, D.B. (1995). 'The behaviour of steel frames subjected to fire'. *Journal of Constructional Steel Research*, **35**, 291–322.

Watts, J.M. (1997). 'Probabilistic fire models'. *Fire Protection Handbook, 18th Edition*. National Fire Protection Association, Quincy, MA, USA.

White, R.H. (1984). 'Use of coatings to improve fire resistance of wood'. *ASTM Special Technical Publication 826*. Philadelphia, USA.

White, R.H. (1988). 'Charring rates of different wood species'. *PhD thesis*, University of Wisconsin, Madison, USA.

White, R.H. (1995). 'Analytical methods for determining fire resistance of timber members'. Chapter 4-11. *SFPE Handbook of Fire Protection Engineering, Second Edition*. Society of Fire Protection Engineers, USA.

White, R.H., and Cramer, S.M. (1994). 'Improving the fire endurance of wood truss systems'. *Proceedings of the Pacific Timber Engineering Conference*, Gold Coast, Australia, pp 582–589.

White, R.H., and Nordheim, E.V. (1992). 'Charring rate of wood for ASTM E-119 exposure'. *Fire Technology*, **28**, 1, 5–30.

White, R.H., and Schaffer, E.L. (1980). 'Transient moisture gradient in fire-exposed wood slab'. *Wood and Fibre*, **13**, 1, 17–38.

White, R.H., Schaffer, E.L., and Woeste, F.E. (1984). 'Replicate fire-endurance tests of an unprotected wood joist floor assembly'. *Wood and Fibre Science*, **16**, 3, 374–390.

Wickström, U. (1986). 'A very simple method for estimating temperatures'. In *Fire Exposed Structures, in New Technology to Reduce Fire Losses and Costs* (eds S.J. Grayson and D.A. Smith)'. Elsevier Applied Science, London, UK, pp 186–194.

Winandy, J.E. (1995). 'Effects of fire-retardant treatments after 18 months of exposure at 150°F (66°C)'. *Research Note FPL-RN-0264*. US Forest Products Laboratory, Madison, WI, USA.

Winstone Wallboards (1997). *Gib® Fire Rated Systems*. Winstone Wallboards Ltd, Auckland, New Zealand.

Woeste, F.E., and Schaffer, E.L. (1979). 'Second moment reliability analysis of fire-exposed wood joist floor assemblies'. *Fire and Materials*, **3**, 3, 126.

Wood Handbook (1987). *Wood Handbook: Wood as an Engineering Material*. US Department of Agriculture, Forest Products Laboratory, Madison, WI, USA.

Young, S.A. (2000). 'Structural modelling of plasterboard-clad light timber framed walls in fire'. *PhD Thesis*, Victoria University of Technology, Victoria, Australia.

Young, S.A., and Clancy, P. (1996). 'Compression load deformation of timber walls in fire'. *Proceedings of the Wood and Fire Safety Conference*, Slovak Republic, pp 127–136.

Young, S.A., and Clancy, P. (1998). 'Degradation of the mechanical properties in compression of Radiata Pine in fire'. *Proceedings of the Fourth World Conference on Timber Engineering*, Montreux, Switzerland, Vol. 2, pp 246–253.

Index